Multi-Omics Analysis of the Human Microbiome

Indra Mani • Vijai Singh
Editors

Multi-Omics Analysis of the Human Microbiome

From Technology to Clinical Applications

Springer

Editors
Indra Mani
Department of Microbiology
Gargi College, University of Delhi
New Delhi, India

Vijai Singh
Department of Biosciences, School of Science
Indrashil University
Rajpur, Mehsana, Gujarat, India

ISBN 978-981-97-1843-6 ISBN 978-981-97-1844-3 (eBook)
https://doi.org/10.1007/978-981-97-1844-3

This Springer imprint is published by the registered company Springer Nature Singapore Pte Ltd.
The registered company address is: 152 Beach Road, #21-01/04 Gateway East, Singapore 189721, Singapore

Foreword

I am thrilled to pen down the foreword for the book titled *Multi-omics Analysis of the Human Microbiome: From Technology to Clinical Applications*, a timely and significant contribution to the rapidly evolving landscape of biological science. This comprehensive volume encompasses a myriad of topics, delving into the general characteristics of multi-omics and their diverse applications.

Within the pages of this book, you will find an exploration of the introduction, technology, tools, and databases instrumental in analyzing multi-omics data. The scope includes metagenomics, metatranscriptomics, metaproteomics, metabolomics, fluxomics, phenomics, and various facets of the human microbiome. It extends its reach to applications in microbial diversity, mycobiome analysis, and studies on the gut, skin, oral, lung, and viral microbiomes.

Moreover, the book elucidates the pivotal role of multi-omics in clinical applications and the prospecting of genes with biotechnological and pharmaceutical significance. The integration of artificial intelligence and machine learning for multi-omics analysis is also a noteworthy aspect covered in this comprehensive resource, providing a holistic understanding of molecular microbiology and its diverse components.

The inclusion of fundamental and advanced aspects of multi-omics, such as high-throughput technology, in silico tools, databases, and the utilization of artificial intelligence, speaks to the depth of knowledge offered by the contributors—eminent scientists with extensive experience in their respective research domains.

I extend my sincere appreciation to Dr. Indra Mani and Dr. Vijai Singh for their unwavering dedication in bringing forth this exceptional book in collaboration with Springer Nature. I firmly believe that this resource will not only serve as a valuable guide for beginners but will also prove indispensable to students, researchers, scientists, clinicians, practitioners, policymakers, and stakeholders. It offers a comprehensive exploration of the potential of multi-omics, spanning from the fundamentals of molecular microbiology to its wide-ranging applications.

Hariom Yadav

USF Center for Microbiome Research,
Microbiomes Institute, Neurosurgery and Brain Repair
and Internal Medicine-Digestive Diseases and Nutrition,
University of South Florida Morsani College of Medicine
Tampa, FL, USA

Preface

The word "omics" is informally related to the field of biology, such as genomics, transcriptomics, proteomics, metabolomics, or fluxomics, where the suffix -omics refers to the field of study of the genome, transcripts, protein, metabolites, or fluxome, respectively. Microbes are ubiquitous, and bacteria are the most abundant on earth. A microbiome is the ecological community of commensal, symbiotic, and pathogenic microorganisms that literally share our body space or the microbiome is the sum of the microbes and their genomic elements in a particular environment. It is an inspiring time in the arena of Microbiology.

With the rapid adoption of high-throughput omics approaches to analyze biological samples such as genomics, transcriptomics, proteomics, and metabolomics, each analysis can generate terabyte to petabyte-sized data files on a daily basis. These data file sizes, together with differences in nomenclature among these data types, make the integration of these multi-dimensional omics data into biologically meaningful context challenging. With the growth of the era of OMICS, we have seen an explosion of modern multi-omics technologies. These include metagenomics, metatranscriptomics, metaproteomics, metabolomics, fluxomics, and metaphenomics in diverse aspects of basic and applied research.

The structure of this book covers fundamental to applied aspects. Chapter 1 offers an overview on multi-omics research in microbiome analysis. Chapters 2 and 3 discuss emerging technology for studying multi-omics, and Chap. 4 explores bioinformatics databases and tools for analysis of multi-omics. Chapter 5 confers metagenomics in census of microbial diversity. Chapter 6 offers metatranscriptomics, metaproteomics, and metabolomics approaches for microbiome characterization. Chapter 7 explores metagenomics insight into mycobiome analysis. Chapter 8 describes fluxomics and metabolic flux analysis. Chapters 9–13 cover various aspects of multi-omics in the study of gut microbiome, skin microbiome, oral microbiome, lung microbiome, and viral microbiome. Chapter 14 discusses multi-omics and its clinical applications. Chapter 15 describes the use of multi-omics in the prospecting of genes of biotechnological importance, Chapter 16 explores artificial intelligence and machine learning for the analysis of multi-omics.

Therefore, this book provides a comprehensive source of multi-omics approaches in microbiome analysis. This book is a valuable source not only for beginners but also for students, researchers, clinicians, stakeholders, and policy makers interested in the potential of the multi-omics approach at different levels in their fields.

On behalf of all the authors, we hope that this book will become a source of inspiration and new ideas for our readers. Wishing you a pleasant reading!

New Delhi, India — Indra Mani
Rajpur, Mehsana, Gujarat, India — Vijai Singh

Acknowledgments

Indra Mani

I would like to express my sincere gratitude and deep appreciation to Prof. Sangeeta Bhatia, Principal, Gargi College, the University of Delhi, India, for extending her outstanding support and motivation to complete this book. I am thankful to my friend, colleague, and co-editor Dr. Vijai Singh, for his great source of inspiration and motivation to complete this project. I am delighted to thank all the authors for their excellent contributions to this book.

I would like to thank Mrs Swati Sharma (Associate Editor—Biomedicine) and Mrs Aishwarya Thyagarajan (Production Editor) from Springer for their excellent management of this project. I would also like to thank Prof. Kailash N. Pandey, Prof. Janardan Yadav, Dr. Satya Prakash, Dr. Vipin Chauhan, Dr. Anurag Tiwari, Dr. Jay Prakash Verma, Dr. Ajay Kumar Singh, Dr. Santosh Kumar Mishra, Dr. Ashok Saini, Dr. Vimal C. Pandey, and those who have directly or indirectly helped in shaping this project.

I wish to express my heartfelt gratitude to my parents and all members of my family who always filled my mind with affirmations and crowded out the negative thoughts. Most of all, I thank my wife Reena Gond and sons Ankur Kumar Gond and Ankit Kumar Gond, who have helped me during this project. I would like to warmly thank the Faculty and Staff of the Department of Microbiology, Gargi College, University of Delhi, for providing a great working environment.

And finally, I express my sincere gratitude to all my friends and well-wishers for providing me moral support during the process of writing this book. Last but not least, my sincere thanks to GOD for his supreme POWER, to live with joy and victory in the shape of this book.

Vijai Singh

I am thrilled to express my sincere gratitude and deep appreciation to Dr. J.S. Yadav, Director (Research), Indrashil University, India, for extending his outstanding support and motivation to complete this book. I would like to give many thanks to co-editor Dr. Indra Mani of this book who gave me outstanding personal and professional support as well as inspiration to finish this book.

I am delighted to thank all the authors for their excellent contributions to this book. I would like to thank Mrs Swati Sharma (Associate Editor—Biomedicine) and Mrs Priyanga Kabali (Production Editor) from Springer for their excellent management of this project.

I would like to thank Prof. Rakesh Rawal, Prof. Bharat Maitreya. Prof. Pawan K. Dhar, Dr. Poonam Bhargava, Dr Madhvi Joshi, Dr. Bhabatosh Das, Dr. Pablo Carbonell, Dr. Rupesh Maurya, Dr. Satya Prakash, Dr. Vimal C. Pandey, Dr. Suresh Ramakrishna, Dr. Dinh-Toi Chu, Dr. Mueksh Kumar Awasthi, and those whose names do not feature here but have directly or indirectly contributed in shaping this project.

I greatly appreciate the support of my students Dr. Nisarg Gohil. Mr. Khushal Khambhati, and Dr. Gargi Bhattacharjee, whose discussion and comments helped to shape this book.

I wish to express my gratitude to my beloved wife Pritee Singh for her endless support, patience, and inspiration. Lots of affection for my kids Aaradhya and Ayush who missed me during this project. I would like to warmly thank the faculty and staff of Indrashil University for providing a great working environment.

I am aware that, even despite our best efforts, the first version always comes with some error that may have crept into the compilation. I would be delighted to receive feedback from readers to further improve the future book. Last but not least, my sincere thanks to GOD for his supreme POWER for endowing me to live with joy and victory in the shape of this book.

Contents

About the Editors

Indra Mani is an Assistant Professor in the Department of Microbiology, Gargi College, University of Delhi (DU), New Delhi, India. He received an M.Sc. in Microbiology from Bundelkhand University (BU), Jhansi, UP, India, and a Ph.D. in Biochemistry (2010) from Banaras Hindu University (BHU), Varanasi, UP, India. He completed postdoctoral training in the Department of Physiology, School of Medicine, Tulane University, New Orleans, Louisiana, USA. He has also worked as a Senior Research Officer (SRO) in the Department of Medicine, All India Institute of Medical Sciences (AIIMS), New Delhi, India. He has more than 12 years of research and teaching experience in Microbiology, Molecular Genetics, Cell Signaling, and Bioinformatics. He has published 56 research articles in peer-reviewed journals, 48 book chapters, and 6 books. He has presented several papers in different national and international conferences. He has extensive experience in microbial characterization, molecular genetics, cell culture, fluorescence/confocal microscopy, and in silico analysis. He serves as a reviewer of peer-reviewed journals and a member of national and international societies.

Vijai Singh is a Professor and Head of the Department of Biosciences, School of Science at Indrashil University, Rajpur, Mehsana, Gujarat, India. He has worked as an Associate Professor in the Department of Biosciences, Indrashil University, Rajpur, Mehsana, Gujarat, India. He was an Assistant Professor in the Department of Biological Sciences and Biotechnology at the Institute of Advanced Research, Gandhinagar, India, and an Assistant Professor in the Department of Biotechnology at the Invertis University, Bareilly, India. Prior to that, he was a Postdoctoral Fellow in the Synthetic Biology Group at the Institute of Systems and Synthetic Biology, Paris, France, and School of Energy & Chemical Engineering at the Ulsan National Institute of Science and Technology, Ulsan, South Korea. He received his Ph.D. in Biotechnology (2009) from the National Bureau of Fish Genetic Resources, Uttar Pradesh Technical University, Lucknow, India, with a research focus on the development of molecular and immunoassays for diagnosis of *Aeromonas hydrophila*. His research interests are focused on building novel biosynthetic pathways for production of medically and industrially important biomolecules. Additionally, his laboratory is working on CRISPR-Cas9 tools for genome editing. He has more than 12 years of research and teaching experience in synthetic biology, metabolic engineering, bioinformatics, microbiology, and industrial microbiology. He has published 105 articles, 75 chapters, 25 books, and 3 patents. He serves as an associate editor, editorial board member, and reviewer of several peer-reviewed journals. He is also a member of the Board of Study and Academic Council of Indrashil University and is the Member Secretary of the Institutional Biosafety Committee (IBSC) at the same University.

Overview on Multi-omics Research in Microbiome Analysis

1

Anshu Mathuria, Namra Ali, Indra Mani, and Vijai Singh

1.1 Introduction

A broad coalition of microorganisms includes bacteria, fungi, archaea, protozoa, viruses, and their combined genetic material, which collectively form microbiome (Shreiner et al. 2015; Turnbaugh et al. 2007; Gupta et al. 2022). On the other hand, a microbiome is a metagenome of microbiota. The term "microbiota" is sometimes used interchangeably with microbiome (Marchesi and Ravel 2015). However, the community of microorganisms is intricate, comprising viruses, archaea, bacteria, fungi, and protozoa, which reside in particular environments (Gilbert et al. 2018). The microbiome is not a passive bystander; it actively contributes to various aspects of the host's well-being, such as regulating energy balance, influencing metabolism, maintaining the health of the gut epithelium, modulating the immune system, and even contributing to neurodevelopment. It's important to recognize that the microbiome is not a static entity. Instead, it is very active and adaptable to many situations throughout the host's life. These factors can include dietary choices, environmental conditions, medical interventions, and the presence of various disease states. As a result, the microbiome's makeup and purpose can undergo significant changes over time, highlighting its complex connection to the host (Shreiner et al. 2015; Cho and Blaser 2012).

All mammals' digestive tracts are home to a rich and varied collection of microorganisms called the gastrointestinal microbiome, which includes bacteria, archaea, fungi, protozoa, and viruses. Extensive research, conducted in both humans and

A. Mathuria · N. Ali · I. Mani (✉)
Department of Microbiology, Gargi College, University of Delhi, New Delhi, India
e-mail: indra.mani@gargi.du.ac.in

V. Singh
Department of Biosciences, School of Science, Indrashil University, Rajpur, Mehsana, Gujarat, India

I. Mani, V. Singh (eds.), *Multi-Omics Analysis of the Human Microbiome*,
https://doi.org/10.1007/978-981-97-1844-3_1

other mammals, has consistently highlighted the microbiome's pivotal role in numerous physiological processes crucial for the overall health of the host. These vital processes encompass maintaining energy balance, regulating metabolism, ensuring the health of the gut lining (epithelium), orchestrating immune responses, and even influencing neurobehavioural development. Remarkably, the microbial genome carries an extensive repertoire of metabolic capabilities that surpass those of the host organism by itself. This essentially transforms the gut microbiome into proactive participant in the host's physiological functions, contributing significantly to the overall well-being of the host (Barko et al. 2018). Despite the fact that the small intestine, stomach, and large intestine make up the human digestive system, the majority of research focuses exclusively on the large intestine's microbial community, which is evaluated by faeces samples. This is because the microbial biomass of the large intestine is disproportionately higher than that of any other organs or site in the human body. In fact, every millilitre of the large intestine includes approximately 10^{11} microbial cells, whereas the small intestine has only 10^{8} cells (Walter and Ley 2011). When we delve into the microbiome of an individual, research indicates that approximately 150–400 different species inhabit the gut of each person. These estimations are based on a variety of methods that take cultural differences into account (Lloyd-Price et al. 2016). These species are mostly found in the phyla Bacteroidetes, Actinobacteria, Firmicutes, and Proteobacteria. What is noteworthy is that during the course of an individual's lifetime, the relative proportions of different phyla can change dramatically amongst people and even within a similar individual (Odamaki et al. 2016; Flores et al. 2014). This underscores the uniqueness of each person's microbiome and its dynamic nature. Despite this individual variation, when we zoom out and examine microbiomes across populations worldwide, certain trends and patterns start to emerge.

Traditionally, it was widely believed that the healthy urinary bladder was devoid of any microorganisms. However, this long-held belief has been challenged by several research teams using metagenomic (DNA-based) and metaculturomic (culture-based) approaches. These studies have successfully refuted this conventional notion, revealing the presence of resident microbial communities within the urinary bladder, a phenomenon now referred to as the "urobiome". Interestingly, the composition of the urobiome varies between adult females and males, likely influenced by anatomical and hormonal distinctions between the sexes. More importantly, researchers have established a link between the urobiome and diseases of the lower urinary tract, such as recuperating urinary tract infections and overactive bladder. This discovery has opened up the exciting possibility that in the future, clinicians may be able to alleviate symptoms of these disorders by modifying the urobiome instead of solely targeting and eradicating suspected uropathogens (Brubaker et al. 2021).

The field of genomics, which originally focused on deciphering DNA sequences (the specific order of nucleotides in a segment of DNA), has rapidly evolved to encompass a more functional perspective. This broader approach involves the investigation of gene and protein expression patterns and their respective functions (Del Giacco and Cattaneo 2012). A significant portion of genomics research centres around the concept of reference genomes, which serve as comprehensive guides or blueprints for what is considered a "typical" individual of a given species. The development of these reference genomes has been a major focus of the genomics community, with substantial resources invested. For instance, creating the human reference genome alone required 13 years and a staggering $2.7 billion in funding. However, as the field of genomics has expanded, scientists have come to recognize the limitations posed by relying solely on a single reference genome. With the proliferation of sequencing experiments, it has become evident that such an approach cannot fully capture the genetic variation present within a species. To address this limitation and better encompass the diversity that can be missed by using a single reference, researchers have turned to the concept of a "pan-genome". This pan-genome offers a more thorough understanding of genetic diversity as it is a comprehensive compilation of all DNA sequences discovered within a species (Sherman and Salzberg 2020).

The application of metagenomics has played an important part in uncovering the incredible diversity of the gut microbiota. However, it has also brought to light the fact that a large percentage of the bacteria residing in the gut remains uncultured and thus difficult to study using traditional culture-based techniques. In response to this challenge, a methodology known as "culturomics" was developed. Culturomics represents a resurgence of culture-based techniques in microbiology aimed at cultivating and identifying previously unknown bacteria that are found in the human intestines. This approach involves the use of multiple culture conditions in combination with rapid and efficient methods for identifying bacteria. Culturomics has proven to be highly successful, facilitating the growth of thousands of new human-associated microbes. This breakthrough has provided researchers with fresh and interesting insights on the intricate relationships between hosts and the diverse array of bacteria that populate their bodies (Lagier et al. 2018).

Metagenomic sequencing has ushered in a transformative era in microbiology, revolutionizing the way we detect and characterize microbial species. Now that there are several software methods available for taxonomic categorization of this data, we can quickly identify the microbial species present in intricate biological and environmental specimens (Ye et al. 2019). Finding the microbial taxon or taxa in a sample has always been a fundamental difficulty in the field of microbiology. Finding the origin of an infection in an individual's blood sample or analysing the bacteria in soil from a sample taken from the environment (Jones et al. 2017; Pedersen et al. 2016), this task has traditionally relied on labour-intensive sequential testing of potential candidates (Pavia 2011; Venkatesan et al. 2013). The advent of genomic sequencing technologies has changed this landscape dramatically. Metagenomic sequencing allows us to directly probe how microbial colonies are made up in an unbiased manner. This breakthrough technology facilitates faster species identification, finding previously unknown species, and reduces our dependence on culturally specific methods, which were time-consuming and limited in scope (Knights et al. 2011; Loman et al. 2013; Mani 2020a, b). Furthermore, the potential applications of metagenomic sequencing extend far beyond basic research. It holds promise for enhancing diagnostics and has significant implications in public health settings (Chiu and Miller 2019; Miller et al. 2013). The potential to rapidly and accurately identify pathogens has been widely acknowledged. Ongoing efforts are dedicated to addressing the difficulties in using these techniques in clinical settings, ensuring that metagenomic sequencing can fulfil its potential in improving healthcare and microbiological research (Blauwkamp et al. 2019; Miller et al. 2019).

In contrast to metagenomics studies, which focus on the broader genetic makeup of a complicated microbial community, metatranscriptome studies delve into the complete set of genes actively expressed within such a community (Bashiardes et al. 2016). Undertaking transcriptome studies is a computationally demanding task that typically necessitates the use of several bioinformatics tools and techniques (Bharti and Grimm 2021; Ghurye et al. 2016). Among the crucial processes in a typical metatranscriptome analysis are the reconstruction of every one of the transcripts, annotation of these transcripts and the genes they are connected with, and quantification of their expression levels. These procedures also involve removing contaminated sequencing reads and editing low-quality data (Bharti and Grimm 2021; Wajid and Serpedin 2016; Koch et al. 2018). Within the realm of functional genomics, both metaproteomics and metatranscriptomics offer valuable insights into the extent and diversity of expression of gene within microbial communities. Metatranscriptomics, specifically the sequencing of total mRNA (referred to as RNAseq) from a microbial community (Ottesen et al. 2014), has gained significant popularity due to the ability to employ the use of the same sequencing techniques for RNA and DNA analysis. On the other hand, metaproteomics is designed to characterize the whole range of proteins produced by gene expression, offering an additional layer of information pertaining to post-translational modifications and the localization of proteins, which goes beyond what transcriptome measurements can provide. In general, a variety of techniques, such as label-free procedures, isobaric chemical labelling and metabolic labelling techniques, can be used to detect the relative numbers of proteins inside a sample (Li et al. 2012). These two complementary techniques, metatranscriptomics and metaproteomics, enable researchers to comprehensively explore the functional aspects of microbial communities' genetic and protein expression profiles.

Among the existing methodologies, network-based analysis stands out as the most biologically interpretable. Another promising approach is the use of a Bayesian network model, which can effectively model the probabilistic relationships among various multi-omics data and unveil the most likely network structure from this graphical model (Wang et al. 2013; McGeachie et al. 2016). Furthermore, constraint-based approaches like flux balance analysis are able to mimic metabolite flow and forecast microbial development; however their current usefulness is mostly restricted to tiny microbial communities (Chong and Xia 2017). In summary, while transcriptomics, proteomics, and metabolomics all capture aspects of gene expression, it is important to note that they may not exhibit precise quantitative alignment. Measuring transcript levels provides insight into protein production rates but does not accurately predict protein concentration or stability. Proteomics data is often a more reliable indicator of biological phenotypic characteristics than mRNA data, and protein abundance is frequently weak or absent in many biological systems (Maier et al. 2009; Vogel and Marcotte 2012).

1.2 Metagenomics in Microbial Characterization

Metagenomics is the comprehensive examination of both the genetic makeup and functional capabilities of entire nucleotide sequences found in a collective sample of organisms, typically microbes. This method entails analysing the genomes of microorganisms in an environmental sample (Handelsman et al. 1998). Traditional methods for obtaining pure cultures of bacteria are often limited, capturing only a tiny portion of the bacterial community prevalent in various environmental samples (Amann et al. 1995). Three longitudinal investigations using genomic, transcriptomic, metabolomic, and epigenomic data from microorganisms and humans were carried out by the Integrative Human Microbiome Project (iHMP). However, merely collecting such extensive datasets is insufficient. It is essential to embrace systems biology, personalized medicine, and cutting-edge computational analytic techniques for integrating multi-omics data between the microbiome and host in order to fully realize the promise of this data (Wang et al. 2019). While recent endeavours have made progress in cultivating previously unculturable bacteria (Zengler et al. 2002), it is still crucial to use a variety of strategies that are culture-independent to support these initiatives, particularly when dealing with the enormous number of undiscovered species in our natural world. Access to genomic information regarding uncultured bacteria holds the potential to expedite the development of culture media suitable for their growth (Beja et al. 2000). This advancement is a valuable addition to the ongoing efforts that have yielded significant progress in culturing technology (Rappe et al. 2002). Of particular interest is the Acidobacterium phylum, which has proven challenging to culture despite its widespread presence on Earth and abundance (Quaiser et al. 2003). There are eight separate groupings within this phylum (Barns et al. 1999; Hugenholtz et al. 1998). Although significant strides have been made in culturing members of this phylum (Sait et al. 2002), it is noteworthy that all cultured representatives belong to only three of these eight groups. Next-generation sequencing (NGS) methods have improved since the Human Microbiome Project (HMP) began in 2008, enabling thorough microbial investigation without the requirement for culturing (Turnbaugh et al. 2007; Mani 2020a, b). Shotgun metagenomics sequencing (MGS) and 16S ribosomal RNA (rRNA) gene sequencing (16-Sseq) are two extensively used techniques. The marker gene 16S rRNA gene, which has both conservative and hyper-variable regions, is used in 16-Sseq. This allows for the differentiation of various species and estimation of microbial taxonomic composition (Barko et al. 2018). Contrarily, MGS is more expensive due to the untargeted sequencing of every microorganism's genome that is present in the sample, although it is more accurate. When the 16S ribosomal RNA gene is sequenced, operational taxonomic units (OTUs) are created from the sequences by using similarity analysis. Based on a sample sequence inside each OTU, these OTUs are then annotated at multiple taxonomic levels, including phylum, class, order, family, genus, and species (Nguyen et al. 2016). However, when using reference databases for MGS, such as BLAST and SEED, the sequences are automatically categorized into taxonomic groups (Breitwieser et al. 2017). The fraction of sequencing reads attributed to a particular taxon compared to all reads in the sample may then be used to calculate the relative distribution of that taxon (Hua et al. 2015).

Metagenomics has offered valuable insights into the previously uncharacterized *Acidobacterium* subgroups, going beyond the information available from 16S rRNA gene sequences. This progress has been achieved through the sequencing of six fosmid clones, each approximately 35 kb in size (Quaiser et al. 2003), as well as 12 BAC clones, each 25 kb in size (Liles et al. 2003). These clones were derived from individuals of the uncultured soil-dwelling *Acidobacterium* genus. More specifically, two, five, and eleven of these clones come from members of *Acidobacterium* Group VI, Group V, and Group III. The examination of these particular clones offers a preliminary insight into the wide range of genetic variation found in the *Acidobacterium* phylum. The feasibility of sequencing the whole metagenome of a sample of the environment has increased as new sequencing technologies have emerged. However, in most cases, the complexity of microbial communities within these environments often surpasses the feasibility of obtaining a truly comprehensive metagenome sequence. Even in relatively simple communities, there exists microheterogeneity, which introduces a level of intricacy that necessitates simplification when reconstructing genomes.

The acid mine drainage biofilm community proved to be an ideal candidate for complete metagenome sequencing due to its relatively lower complexity. Upon analysing the 16S rRNA gene, it was determined that, biofilm was suitable since it included three bacterial species and two archaeal species. Marine ecosystems, in comparison, have substantially higher species diversity, with an average of between 100 and 200 species per millilitre of water (Curtis et al. 2002). This higher diversity significantly complicates the sequencing and assembly process. Moving up the complexity scale, soil ecosystems stand out with an estimated 4000 species/g of soil in terms of species richness (Curtis et al. 2002; Torsvik et al. 1990). The development of quicker and more affordable sequencing methods than what is now available is necessary to sequence the soil metagenome. Nevertheless, soil metagenomics continues to make strides by concentrating on specific biological inquiries, thereby allowing for the exploration of particular aspects of the community's functionality through genomic analysis.

1.2.1 Phage Metagenomics: Unravelling Viral Diversity in Microbial Communities

Microbial communities are primarily characterized by the prevalence of bacteria, and within these bacterial populations, bacteriophages exert significant influence. Bacteriophages are crucial in influencing the population dynamics and diversity of communities of microbes (Wommack and Colwell 2000). Much like their bacterial counterparts, the vast majority of bacteriophages remain largely unexplored in laboratory settings due to their staggering diversity (Rohwer 2003), and many of the bacteria they infect have not been successfully cultured. Furthermore, since there is no direct counterpart to the 16S rRNA gene, which is frequently employed in bacterial phylogeny research, the study of viral phylogeny requires full viral genome sequences. Recently, two notable research used metagenomic analysis to explore phage diversity in the marine environment and the human gut. These investigations have provided valuable insights into the intricate world of phages in these ecosystems.

Metagenomic analysis of the viral population discovered in human faeces was done in research (Breitbart et al. 2003). The method included assembling a library of random viral DNA fragments from virus particles recovered from a 500g sample of human faeces. The results of this investigation revealed that the viral community within this sample comprised approximately 1200 different genotypes. Surprisingly, data shows that the number of different viral genotypes likely outnumbers the bacterial species present in the human gut. This discovery underlines the noteworthy diversity and complexity of the viral populations residing in the ecology of the human gut.

1.2.2 Metagenomic Analysis of Microbial Communities in Acid Mine Drainage

Acid mine drainage is caused by iron oxidation by microorganisms, which causes acidity when pyrite dissolves in abandoned mining operations (Edwards et al. 2000). While one archaeon found in the mine called *Ferroplasma acidarmanus* fer1 has been successfully cultivated in isolation and had its genome sequenced, efforts to cultivate additional members of the community have been unsuccessful. DNA was taken straight from the biofilm in an effort to sequence the genetic material of the uncultured archaea and bacteria found there (Tyson et al. 2004). With 103,462 reads and a mean insert size of 3.2 kb, a small-insert repository was created, producing 76.2 Mbp of sequencing data. With this metagenomic sequence data in hand, a mission was embarked to elucidate the ecological function of the five prokaryotic organisms inhabiting the acid mine drainage environment (Tyson et al. 2004). Each genome sequence contained genes that most of these creatures would need in order for carbon fixation through the reductive acetyl coenzyme A pathway.

Furthermore, scientists deduced that these *Ferroplasma* species prefer a heterotrophic lifestyle by observing the existence of several genes in the genomes of *Ferroplasma* types I and II that showed notable similarities to amino acids and sugar transporters. Many of the hypotheses generated in the study of acid mine drainage will be subject to further evaluation through the development of microarrays and the innovation of novel techniques for isolating these uncultured prokaryotes (Tyson et al. 2004).

1.2.3 Metagenomic Exploration of Microbial Communities in the Sargasso Sea

It is a comprehensive sequencing effort that centred on the microbiota found in the Sargasso Sea, which has limited resources, a marine environment that has been extensively studied (Venter et al. 2004). From a pool of two million random sequences, they generated an astounding 1.6 billion base pairs of sequence data, which included roughly 1.0 billion base pairs of unique, nonredundant sequence information. The Sargasso Sea study lays the foundation for functional investigations aimed at unravelling how the genes responsible for phosphorus acquisition are strategically employed to enhance survival within a community facing a scarcity of phosphorus. To make sense of this voluminous dataset, the researchers employed a multi-step approach. They meticulously assembled overlapping sequences into larger scaffolds and then categorized them into provisional "organism bins". Three main factors—oligonucleotide frequencies, resemblances, and read depth to previously sequenced genomes—were used to classify the samples. Despite the enormous diversity of microorganisms that make up the Sargasso Sea, the assembling method only produced a few numbers of complete microbial genomes. The voluminous data collected from the Sargasso Sea project, however, provide a substantial contribution to the disciplines of metagenomics and the ecology of microbes and make up the largest genome dataset accessible for any population on Earth (Riesenfeld et al. 2004).

1.2.4 Challenges in Metagenomic Analysis: Navigating the Complexities of Microbial Genomic Data

The ideal phylogenetic reference point should be uniformly distributed among all species. Due to variances in the presence of ribosomal RNA (rrn) operons in microbial genomes, and ranging from 1 to 15, the 16S rRNA genes do not, however, comply to this requirement (Klappenbach et al. 2000). The amount of rrn operons has been hypothesized to be positively associated with the pace of microorganism development (Klappenbach et al. 2000). Consequently, slow-growing bacteria that are challenging to culture might be underrepresented in libraries of 16S rRNA generated via PCR. The genes of 16S rRNA are therefore less likely to appear frequently in metagenomic clones compared to those of their faster-growing counterparts. Though conceptually simple, building libraries of metagenomic data from environmental samples involves difficult technological issues. There are various challenges involved in forming and extracting valuable insights from metagenomic libraries: (a) The need to isolate and clone a substantial amount of DNA from a limited sample size, (b) the necessity to process a multitude of clones and sequences to derive meaningful and informative data, (c) the presence of population distributions of the lognormal type, which makes it challenging to accurately depict the minor species in a sample. Addressing these difficulties is an ongoing endeavour, with various studies published (Gillespie et al. 2004; Handelsman et al. 2003) that assess and refine methods for library construction.

Small-insert libraries offer a notable benefit compared to large-insert libraries due to their ability to use harsh methods for microbial lysis, which would otherwise overly fragment DNA when it comes to large inserts. Small-insert libraries are a useful tool for identifying novel metabolic activities encoded by individual genes, even if they might not be appropriate for capturing complicated processes involving many genes. Small-insert libraries are also important for reconstructing metagenomes (Tyson et al. 2004; Venter et al. 2004). Achieving the necessary library size to attain adequate metagenome coverage, even in the case of the simplest communities of microbes, poses a substantial screening challenge. The information obtained via sequencing is restricted by the gene annotations included in current databases and the capabilities of the available sequencing technology, despite the fact that brute-force sequencing approaches have produced useful insights into these libraries and their source communities (Tyson et al. 2004; Venter et al. 2004).

1.3 Metatranscriptomics in Microbial Characterization

Metatranscriptomics has recently gained prominence as a potent tool for learning more about the functional aspects of intricate microbiological groups, commonly referred to as microbiomes. By employing unbiased RNA-sequencing techniques, metatranscriptomics enables us to uncover valuable information about these communities. This includes not only identifying the various taxonomic components present but also shedding light on the active biochemical functions they perform within the community (Jiang et al. 2016).

1.3.1 Metatranscriptomic Analysis Uncovers Fundamental Metabolic Pathways and Unique Functions within Diverse Microbial Communities

The Sequence Read Archive (SRA) of the National Center for Biotechnology Information (NCBI) was used to get publicly accessible metatranscriptomic datasets (Leinonen et al. 2011). These datasets underwent a systematic data processing pipeline, encompassing five distinct metatranscriptomic studies: (a) 12 mouse large intestine samples were used to create the dataset, which had 30 million paired-end reads with 76 base pairs average length (Xiong et al. 2012); (b) a dataset containing 35 million single-end reads, each 101 base pairs in length, obtained from a kimchi sample on the 29th day of fermentation (Jung et al. 2013); (c) a dataset made up of 14 million 100 base pair long paired-end reads sourced from a bovine rumen sample (Poulsen et al. 2013); (d) a dataset consisting of 103 million paired-end reads, each with a length of 100 base pairs, acquired from a sample of deep sea (Baker et al. 2013); (e) a dataset encompassing 131 million paired-end reads, each with a length of 150 base pairs, derived from a permafrost-sourced sample. These datasets were all produced using Illumina sequencing technologies.

For each of the datasets, the Trinity RNA-Sequencing assembly algorithm was applied (Grabherr et al. 2011) to assemble putative mRNA reads. It had previously been determined that using this particular assembler would improve annotation and reduce the likelihood of misassemblies in metatranscriptomic data (Celaj et al. 2014). Notably, with percentages of 62% and 72%, respectively, the kimchi and deep-sea datasets showed the largest proportion of reads that were successfully assembled into contigs. The contigs in the kimchi dataset had an N50 length of 368 base pairs, which is probably a sign of the low diversity of the kimchi sample's microbiome.

1.3.2 Exploring Metatranscriptomic Insights into Bacterial Communities on Laundered Textiles

Modern washing machines are home to a plethora of bacteria. The most prevalent contamination pathways are worn clothing, water from the tap, and air (Novak Babič et al. 2020). Due to the humid, nutrient-rich, and warm environment within the machine, microorganisms like bacteria and fungi can congregate and grow there (Munk et al. 2001; Nix et al. 2015; Egert 2017). The negative consequences of such contamination include unsightly stains, malodour, and biofilm growth (Munk et al. 2001; Egert 2017). Resistant biofilm growth, in particular, may offer a concern to vulnerable individuals, since biofilms may serve as a reservoir for (possibly) harmful microbes that re-contaminate clothing after washing (Bloomfield et al. 2015; Gibson et al. 1999). The relative most common sequence types in home washing machines were found to be closely associated with potentially harmful microorganisms, such as *Brevundimonas vesicularis* or *Pseudomonas aeruginosa* inside of the detergent dispenser and *Moraxella osloensis* or *Acinetobacter parvus* inside the door seal, using the molecular technique of 16S rRNA gene pyrosequencing (Jacksch et al. 2020).

Studies on the laundry microbiota's metabolic activities are typically limited to certain capabilities, such as the generation and prevention of odours (Munk et al. 2001; Kubota et al. 2012; Stapleton et al. 2013), whereas the community of microbes composition of washing machines and laundry products has been investigated in this and other structural analyses (Jacksch et al. 2020; Schmithausen et al. 2019). Malodour can shorten the life cycle of a textile and is frequently linked to poor hygiene (Van Herreweghen et al. 2020). The metabolic potential at the moment of sampling may be calculated using next-generation RNA-sequencing (RNASeq), which quantifies almost all of the transcripts from the current cells. This approach can also aid in characterizing the functioning of a microbial community by providing a thorough understanding of the expression patterns of a complete microbiota in a single experiment (Kukurba and Montgomery 2015; Cardoso-Silva et al. 2014). Schematic representation of the use of multi-omics approach in tracing the information flow from DNA, RNA, protein (enzyme) to pathways is shown in Fig. 1.1 (Ebrahim et al. 2016).

Several bioinformatics tools are often needed for transcriptome research since they are computationally complex (Bharti and Grimm 2021; Ghurye et al. 2016). Common processes in a metatranscriptome investigation include the rebuilding of each of the transcripts, annotation of these transcripts and genes, and evaluation of their expression. Other crucial actions include quality cutting and the elimination of

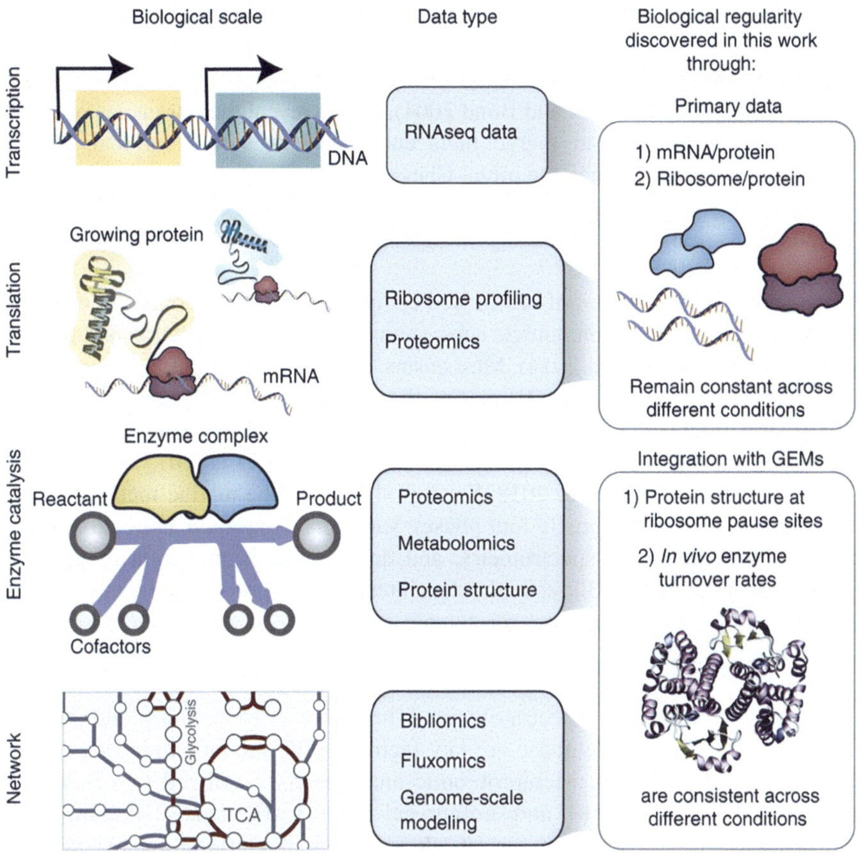

Fig. 1.1 A multi-scale multi-omics framework detects significant biological regularities in *E. coli*. Tracing the central dogma of biology (left column), we can link specific data types (middle column) to explain each of these biological processes. In this work, novel biological regularities that relate these processes are discovered through: (i) primary omics data (top box, right column) and (ii) integration with GEMs (bottom box, right column). (Adapted from Ebrahim et al. 2016)

tainted sequencing reads (Bharti and Grimm 2021; Wajid and Serpedin 2016; Koch et al. 2018). De novo assemblers, like Spades, built on de Bruijn graphs (rna mode) (Bushmanova et al. 2019), Trinity (Grabherr et al. 2011), Velvet/Oases (Schulz et al. 2012), or SOAPdenovo-trans (Xie et al. 2014), concentrate on the connection between substrings of a given length k (k-mers) in order to rebuild the transcriptome. If there is not a reference genome available, they can be used (Liu et al. 2019; Chen et al. 2017). Although each of these assemblers may create effective assemblies, there is a noticeable amount of heterogeneity when comparing several assembler programmes (Earl et al. 2011; Bradnam et al. 2013).

1.4 Metaproteomics in Microbial Characterization

The goal of metaproteomics is to characterize overall proteins extracted from populations of microbes (Wilmes and Bond 2004), and to reveal the intricacies of a particular ecosystem, we can merge metagenomics with the operational aspects (Franzosa et al. 2015). Since the initial Chesapeake Bay surroundings environmental metaproteomic investigation (Kan et al. 2015), additional studies have been conducted in various habitats using different methodologies (Matallana-Surget et al. 2018).

Researchers have commonly employed comparative metaproteomics to examine the fluctuations that arise in aquatic ecosystems based on both location and time of year, in situ (Georges et al. 2014), Mesocosms (Bryson et al. 2016), or Microcosms (Russo et al. 2016) approaches. Regarding the marine ecosystem analysis, metaproteomics is a quickly growing area that demands a high level of expertise in navigating difficult steps and making crucial decisions (Wilmes et al. 2015; Heyer et al. 2017; Matallana-Surget et al. 2018; Saito et al. 2019). The marine metaproteomic methodology comprises mostly four phases: sampling and protein extraction, separation of proteins, mass spectrometry, and detection and annotation of proteins (Wöhlbrand et al. 2013). Biases in the data interpretation process and methodological inconsistencies between metaproteomic studies persist due to the absence of standardized experimental protocols (Leary et al. 2013; Timmins-Schiffman et al. 2017).

The precision of mass spectra obtained through experimentation and the size of the protein research database are key factors in the identification of proteins (Wöhlbrand et al. 2013). Metaproteomic analyses have utilized two methods: gel-based (Wang et al. 2014) and shotgun gel-free (Morris et al. 2010; Saito et al. 2015) approaches, which demonstrate their complementary nature (Matallana-Surget et al. 2018). Taxonomic and functional annotation are also a part of metaproteomic data analysis. The two primary sources of information used to create protein search databases are metagenomic data and public protein repositories (Heyer et al. 2017). It can be difficult to utilize public protein repositories to find proteins like UniProtKB/SwissProt, UniRef, UniProtKB/TrEMBL, NCBI, or Ensembl because of the vastness of these databases, which reduces the total number of proteins found by expanding search space and overestimating false discovery rate (FDR) (Nesvizhskii 2010; Tanca et al. 2013; Timmins-Schiffman et al. 2017). A large increase in the overall number of detected proteins is made possible by the two-round searching of databases technique, which involves scanning experimental mass spectrometry (MS) against a tuned database made up of the protein sequences found following an initial error-tolerant search. Recent metaproteomic studies have heavily relied on this strategy (Serrano-Villar et al. 2016). In order to create the protein search database for metagenomic data, both assembled (Teeling et al. 2012) and non-assembled (Tanca et al. 2016) sequencing reads were used.

1.4.1 Metaproteome of the Microbiota of Human Gut

The microbiome of the human gut, an intricate ecosystem vital for digestion, significantly impacts our health through its diverse bacteria, most of which are owned by four main phyla: Bacteroides, Firmicutes, Proteobacteria, and Actinobacteria (Frank et al. 2007). According to metagenomic data, it is home to over 1000 different bacterial species and has over nine million genes (Li et al. 2014; Qin et al. 2010). Although shared genes exist, merely a few hundred bacteria species make up each person's gut microflora, leading to variations in composition and abundance (Qin et al. 2010; Zhang et al. 2018). Changes in the microbiome of gut, reflected in the metaproteome, are associated with various diseases (Gilbert et al. 2016).

1.4.2 Analysing the Microbiota Using a Metaproteomic Method

Metaproteomic analysis of the gut microbiome involves studying bacteria from different sources. A typical workflow, as depicted in Fig. 1.2, includes protein extraction, database search, preparation, LC-MS/MS analysis, and taxonomic/functional analysis. Identifying various peptides and proteins within microbial communities requires extensive databases encompassing proteins expressed by all gut microorganisms (Pan and Chen 2020). Building an ideal human gut microbiome database is crucial but challenging due to the limited annotation of 40–80% of the >9 million genes found in this microbiome (Heintz-Buschart and Wilmes 2018; Heyer et al. 2017).

Furthermore, the exploration of post-translational modifications (PTMs) is gaining traction in metaproteomics. According to recent research, several PTMs, including methylation, phosphorylation, deamination, acetylation, hydroxylation, and nitrosylation, have different distributions in different microbial taxonomic and functional groups. These PTMs play a role in bacterial adaptation to diverse environmental conditions (Zhang et al. 2016).

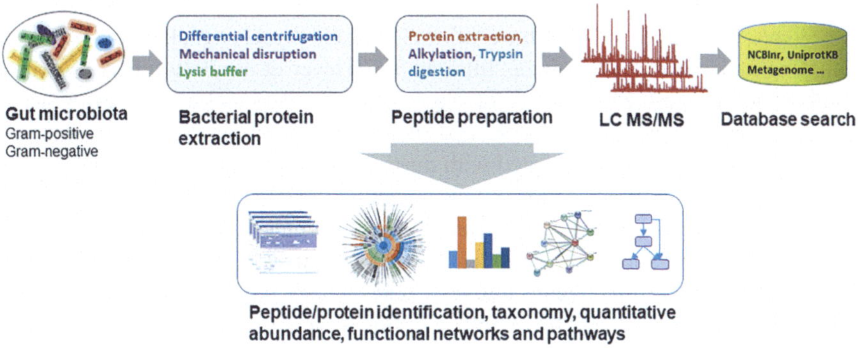

Fig. 1.2 A concise description of the experimental workflow for metaproteomic analysis of the human gut microbiome. (Adapted with permission from Pan and Chen 2020)

1.4.3 Current Challenges in Metaproteomics

Several new developments in mass spectrometry are currently reshaping the landscape of single-species proteomics. These innovations hold great promise for potential future applications in metaproteomics. Firstly, the incorporation of ion mobility as a new technique for ion filtration and analysis offers the potential to further fractionate peptide mixtures. This can effectively reduce peptide co-elution, enhancing the specificity of analyses. It is imperative to integrate functional information pertaining to the samples' unique circumstances and the biological environment. However, making these improvements may necessitate a reconsideration of our current representation of the microbiome, our knowledge of the numerous metabolic routes used by the most common taxonomic components, as well as its overall metabolism.

In current events, proteomics has undergone significant advancements that now permit the exploration of cellular diversity within populations at the individual cell level, introducing a novel parameter of understanding (Ctortecka et al. 2022). The utilization of non-isobaric mass tags has made it feasible to perform advantageous multiplexing of single-cell analyses using a data-independent acquisition mode (Derks et al. 2022). Currently, application of single-cell metaproteomics faces certain challenges due to the tandem mass spectrometers' sensitivity restrictions, compounded by the fact that metaproteomics aims to investigate microorganisms that are approximately 1000 times smaller in size compared to mammalian cells.

1.4.4 Future Trends and Applications

In the upcoming years, there will probably be more ambitious metaproteomics projects, handling, analysing, and comparing a larger number of samples. Metaproteomics should work to strengthen the analytical pipeline's resilience and evaluate their techniques across several centres. Rapid technology is high-resolution tandem mass spectrometry (HRMS). For each sample, it typically takes 1–2 h of mass spectrometry to find substantial alterations in microbial community structure or function. Nowadays, compared to the 24-h workflow suggested 3 years ago, the time between sampling and outcomes can be as little as 5 h (Heyer et al. 2019). This provides ample room to optimize the different stages of the process and achieve even faster results through enhanced protocols and increased computing power. In order to reduce the time and cost associated with mass spectrometry, it may be beneficial to label samples for multiplex analysis. One potent method that yields a detailed inventory of all the microorganisms in a sample is metaproteomics. It also offers functional information, such as the generation of toxins, resistance to antibiotics, and additional microbial pathogenic markers, and enables the measurement of these bacteria at various levels of taxonomic classification. It can detect hosts' markers indicating inflammation or defense efficiency and determine their abundance, useful for diagnosis and personalized medicine. Various studies have already demonstrated the informative nature of faecal, oral, and pulmonary metaproteomics

(Hardouin et al. 2021; Henry et al. 2022; Young et al. 2015). The analysis of quality of water, diversity of soil, biodiversity, and ecosystem changes can be simplified and expedited through the use of mass spectrometry, making it an ideal tool for monitoring animal health and environmental conditions. Recent studies have shown that new opportunities are emerging for the systematic application of metaproteomics in different sectors, including biobanking of microflora (Zhang et al. 2022) and sweeteners' effects on microbiota of human gut models *in vivo* (Sun et al. 2022).

1.5 Metabolomics in Microbial Characterization

In 1995, *Haemophilus influenzae*, a free-living bacterium, had its first whole-genome sequencing published (Fleischmann et al. 1995). Completing the sequences of genomes from different creatures is now feasible due to developments in sequencing technologies. Currently, NCBI-GenBank has over 1600 prokaryotes with completed genome sequences. These prokaryotes are classified into several evolutionary groupings within the Bacteria and Archaea domains. Furthermore, the Joint Genome Institute has made available more than 500 finished microbial genomes as of July 2011. After the introduction of genomics, transcriptomics and proteomics emerged as powerful methods to evaluate cellular activities at the translation and transcription levels. Several new "omics" approaches were developed, resulting in the creation of high-throughput analytical techniques. These techniques have played a key role in the creation of metabolomics, a crucial instrument for evaluating the function of genes (Villas-Boas et al. 2005; Hollywood et al. 2006). Examining the metabolome—the whole collection of metabolites generated by an organism—reflects the genome, containing enzymatic pathways and networks. These pathways, along with the metabolites present in the body, are affected by a number of variables, including developmental processes and changes in the environment. By using metabolomics to monitor these factors and their impact on the body, it is possible to get a more precise image of the physiological status of the organism (Bundy et al. 2005; Garcia et al. 2008; Nicholson and Lindon 2008).

Ancient cultures have long engaged in the monitoring of a system's constituent parts, a process known as metabolite profiling. For instance, clinicians used to diagnose illnesses by monitoring alterations in a patient's bodily fluids, including their urine and saliva (Nicholson and Lindon 2008). With the advent of modern analytical technologies, many biomarkers for diseases have been identified through the analysis of metabolites. These include conditions such as coronary heart disease, cancer, mental disorders, and diabetes (Ohdoi et al. 2003; Li et al. 2008; Brindle et al. 2002). Metabolomics has also proven beneficial to other fields of medicine, such as toxicity, nutrition (nutrigenomics), *in vitro* fertilization, and oncology (Nicholson and Lindon 2008; Seli et al. 2008; Spratlin et al. 2009). The metabolome analysis's high sensitivity for detection has allowed the plant sciences field to pursue a wide range of metabolomics research topics. These include recognizing particular compounds in medicinal plants, tracking compound production in transgenic crops, and differentiating wild types from their transgenic counterparts in fruits and vegetables

(Ketchum et al. 2003; Fiehn 2003; Schaneberg et al. 2003; Burns et al. 2003; Stobiecki et al. 2003). Prokaryotes are the most fully sequenced creatures, and microbial genomics has been at the forefront of sequencing technology (Kyrpides 2009). However, metabolomics in microorganisms is relatively new in comparison. Nonetheless, microbial metabolomics has made significant progress and contributions.

1.5.1 Development in the Field of Metabolomics

Metabolomics has leveraged knowledge from other "omics" to make studying and analysing the metabolome feasible.

1.5.1.1 Enhancing Analytical Tools

The improvement of analytical tools is among the most significant results of the post-genomic age, which has contributed to the advancement of metabolomics research. This need arose from the requirement to examine numerous intracellular metabolites in functional genomics. Two frequently utilized technologies in metabolome investigations are MS and nuclear magnetic resonance (NMR) spectroscopy (Nicholson et al. 2005). Both approaches can handle several metabolites in an individual experiment without needing to pre-select specific analytes. These advancements also enable the measurement of molecular concentrations and the identification of the structures of metabolites (Nicholson and Lindon 2008). Both of these approaches are therefore accepted by many scientific domains as the industry standard analytical methods for research in metabolomics (Spratlin et al. 2009; Villas-Boas et al. 2005; Breitling et al. 2008).

1.5.1.2 Access to Genomic Sequences

After sequencing the genomes of individual microbes, sequencing technologies with high-throughput capabilities enable community member gene assessments on a broad scale to advance systems biology. However, the majority of microorganisms observed in nature live in intricate, interconnected communities and are difficult to isolate or cultivate in a lab. Environmental genomics, or genomic analysis of entire populations, reveals previously unrecognized microbial diversity. Billions of base pairs have been contributed to the sequencing database as a result of biodiverse metagenomic initiatives and consortia, which are essential for understanding community diversity and the complexities of microbial members living in various ecological niches. It has also been done using metagenomic sequencing to investigate microbial communities connected to people. For instance, the whole genetic diversity of gut microbes was defined and their gene functions predicted using a whole-genome shotgun (WGS) technique. This makes it possible for environmental biology to reconstruct system-wide networks in communities of microbes using information from metagenomics and other meta-"omics" (Raes and Bork 2008; Walker 2010; Vogel et al. 2009).

1.5.1.3 Standardization, Models, and Data Integration

The hierarchical structure of the various "omics" provides an integrative strategy that serves as the cornerstone of systems biology (Vemuri and Aristidou 2005). There are numerous advantages, but the most important is that data integration and correlation generate insights not available through other methodologies (Hollywood et al. 2006). Numerous databases and network analysis software, as well as annotation and statistical tools, have been developed (Breitling et al. 2008; Smith et al. 2005; Rodionov et al. 2010). Among these are those relevant to metabolomics (for example, Entrez Gene, Transport DB, UniProtKB, the Kyoto Encyclopedia of Genes and Genomics (KEGG), and Clusters of Orthologous Groups (COG)) (Gill et al. 2006; Feist et al. 2009; Raes and Bork 2008). Significant progress has been made to develop standard interchangeable norms in proteomics and genomics to facilitate the effective integration of data sets (Kyrpides 2009; Sintchenko et al. 2007). Additionally, new models are being developed to handle particular difficulties brought on by metabolomics (Hollywood et al. 2006). These informatics methods can be useful for metabolic engineering in different microorganisms and for reconstructing biochemical networks (Caspi et al. 2010).

1.5.2 Applications

Metabolic profiling of microorganisms is a pivotal tool within the realm of systems microbiology, enabling the amalgamation of biological data to enhance our comprehension of microbial interplay and the inner workings of cells.

1.5.2.1 Human Ecosystem

The inception of the HMP in 2007 marked a significant shift in the perception of microbes within the realm of human biology. Prior to this, bacteria were predominantly seen as unwelcome guests that disrupted the systems they inhabited. While various metagenomics projects had already hinted at the vital function's microbes performed, this acknowledgment remained largely confined to the field of environmental microbiology. As an expansion of the Human Genome Project (HGP), the HMP presents a significant chance for the integration of intricate systems encompassing the structure of microbial communities (microbiota), genetic content (metagenomics), gene expression (metatranscriptomics and metaproteomics), and metabolic processes (meta-metabolomics) (Turnbaugh et al. 2007). This initiative comprehensively inventoried five distinct sites within the human body: oral cavity, GIT, vaginal region, skin, and nasal cavity (Turnbaugh et al. 2007; Peterson et al. 2009; Grice and Segre 2011). Simultaneously, endeavours to sequence new reference genomes have played a vital role in enhancing our comprehension of these microbial communities. By the conclusion of 2010, researchers had successfully sequenced 500 of the 3000 pertinent microbial genomes (Pennisi 2010).

In the realm of gastrointestinal ecology, the microorganisms, encompassing bacteria, archaea, and fungi, that inhabit the intestinal tract have been recognized as pivotal constituents (Li et al. 2008; Ventura et al. 2009; Walker 2010). The human gut, home to an estimated 10^{14} microorganisms, encompasses over a thousand distinct species, a significant portion of which remains uncultivable (Gill et al. 2006; Atarashi et al. 2011; Kassinen et al. 2007). These advantages encompass protection against harmful pathogens, the bestowal of immunity, the replenishment of the epithelial cells present in the gut, the extraction of otherwise inaccessible nutrients, and the production of energy through anaerobic metabolism of substances (Nicholson et al. 2005; Ventura et al. 2009; Turnbaugh et al. 2007). A systems biology approach can help us gain a comprehensive knowledge of the interaction between the microbiome of the gut and the host.

The interaction between an individual's gut microbiome and drug response holds significant implications for the field of personalized medicine and the development of innovative drug discovery strategies. In the years to come, the evaluation of an individual's gut microbiome could potentially become a complementary approach alongside the assessment of single nucleotide polymorphisms (SNPs) to anticipate and mitigate adverse effects resulting from drug metabolism, such as harmful drug interactions (Nicholson et al. 2005). It is noteworthy that the Consortium for Metabonomic Toxicology (COMET) has previously acknowledged the significance of this issue and made it one of its main goals (Nicholson and Lindon 2008; Lindon et al. 2005).

1.5.2.2 Environmental Ecosystem

In the realm of ecological systems, microbial communities play a pivotal role. Traditionally, our comprehension of these communities was constrained to microorganisms that could be cultivated (Tringe et al. 2005). However new discoveries in DNA sequencing technology and creative approaches have revolutionized the research and identification of environmental bacteria, allowing for the examination of entire populations, including those microbes that have escaped domestication (Torsvik and Ovreas 2002; Rusch et al. 2007; Raes et al. 2007; Wilmes et al. 2010). The term "microbial ecosystem" describes the community of microbes and their cohabitants, which might include other living things like plants and animals that are present in a particular area. This system is characterized by its dynamic nature, which is impacted by a variety of factors, including environmental conditions such as temperature and moisture, chemical composition such as acidity and concentration of nutrients, and physical structure of the environment, such as whether it is solid or liquid. Consequently, the elements and relationships within these communities might vary from basic to intricate (in controlled conditions with just one or two species) to complexity (as seen in environments like soil, oceans, rhizospheres, and wastewater) (Tringe et al. 2005; Jiao et al. 2010). The interactions between community members are generally categorized as symbiotic, mutualistic, or competitive.

Within microbial ecosystems, certain dynamics lead to specific outcomes, and one of these influential dynamics is the metabolic activity of individual members. This is notably evident when we examine the microbiota in the human gut, where the size of the community and the interactions among its members have significant impacts not only on the host but also on one another, as evidenced in studies (Turnbaugh et al. 2007; Ley et al. 2006; Manichanh et al. 2006). Consequently, metabolomics emerges as a pivotal tool for comprehending the composition of these populations and their intricate interactions within microbial ecosystems. It not only capitalizes on findings and advancements from other "omics" disciplines but also provides valuable insights into the overall functions of these communities, as exemplified by selected illustrations in the subsequent sections.

The connection between microorganisms and persistent dissolved organic matter remains enigmatic. Given the intricate nature of the biogeochemical interplay at hand, it is probable that numerous facets are at play in this interaction. To shed light on this complex phenomenon, one can establish a conceptual framework and formulate hypotheses by drawing upon insights and data derived from the field of systems biology. Within this context, microbial metabolomics emerges as a pivotal bridge, facilitating a deeper comprehension of these processes, as the cycling of recalcitrant carbon intricately intertwines with microbial activities (Jiao et al. 2010).

1.5.3 Microbial Metabolomics: Navigating Challenges and Forging Collaborative Paths in Systems Microbiology

Being a relatively recent field, microbial metabolomics has faced its share of challenges and setbacks. The hope is that the insights gained from other "omics" disciplines will help overcome these obstacles. Indeed, issues have been identified with standardization, range variability, metabolite flux measurement, metabolic annotation, coverage depth, and managing sizable databases and informatics. Discussions of potential remedies for these issues have taken place at conferences, working groups, and academic publications (Hollywood et al. 2006; Kyrpides 2009; Sintchenko et al. 2007; Nicholson et al. 2005).

Effective interdisciplinary collaboration is crucial for advancing systems biology. Innovative methods to handle the complexity of studying pathogen and host interactions are provided by the HMP's successful model and the recent launch of the National Institute of Allergy and Infectious Disease (NIAID)-sponsored Systems Biology Programme for Infectious Disease Research (Aderem et al. 2011). The existence of metabolomics societies with members from many fields of study is evidence of the benefits of cooperation and information exchange. It will take time to transition from technology- and computation-driven research to a thorough understanding of the data and their implications because systems biology symbolizes a significant shift in approach and uses an integrated methodology that is very different from traditional research methods. More than 500 articles on metabolomics are published yearly, demonstrating the growing body of research in this field in academia, business, and government. Hence, microbial metabolomics is poised to gain recognition and make a considerable impact in the realm of systems microbiology.

1.6 Conclusion and Future Perspectives

Our understanding of microbial communities has been completely transformed by metagenomics, analysing genetic material from various environments, revealing the vast microbial diversity and functional capabilities. It is especially crucial in studying unculturable microorganisms like the Acidobacterium phylum, shedding light on their genetic diversity. Metagenomics has also exposed the diversity and impact of bacteriophages in ecosystems, such as the human gut and oceans. Despite its progress, metagenomics faces challenges including the need for better reference points for phylogenetic analysis, the complexities of constructing metagenomic libraries, and the difficulties in achieving adequate coverage of complex microbial communities. Metatranscriptomics reveals diverse microbial functions within microbiomes, emphasizing functional differences despite similar taxonomy. In laundered textiles, RNA-sequencing uncovered post-washing harmful bacteria, raising hygiene concerns. The analysis of metatranscriptomic data has shown that microbial communities are not solely defined by their taxonomic composition; their functional capabilities can vary widely even when taxonomically similar. This highlights the complex interplay between taxa and their contributions to ecosystem functions. Metagenomics and metatranscriptomics have transformed our understanding of microbial communities, from uncovering hidden diversity to unravelling the functional intricacies of these complex ecosystems. These approaches continue to advance our knowledge of the microbial world and its relevance to various fields, including environmental science, health, and biotechnology.

Metaproteomics has evolved significantly, focusing on the analysis of microbial proteins within ecosystems. Researchers have applied comparative metaproteomics to investigate aquatic environments, though standardization remains an issue. Advances in mass spectrometry have enhanced specificity and data integration. However, single-cell metaproteomics, while promising, encounters sensitivity challenges. It reflects genetic function and helps identify biomarkers for various diseases. It has significantly contributed to understanding microbial metabolism. Metaproteomics projects are expected to handle more samples and become more robust. High-resolution tandem mass spectrometry accelerates the process, enabling faster results. Multiplexing and the application of metaproteomics in monitoring animal health and environmental conditions are promising future trends.

Microbial metabolomics, a crucial component of systems microbiology, has evolved alongside genomics, transcriptomics, and proteomics. It offers insights into metabolic pathways encoded within genomes and has applications in various fields. Metabolomics leverages analytical tools like NMR spectroscopy and mass spectrometry. Access to genomic sequences aids in understanding microbial communities. The topic of microbial metabolomics is promising and is quickly developing for advancing systems microbiology. Collaborative efforts and interdisciplinary approaches are essential to overcome challenges and harness the full potential of metabolomics in understanding microbial systems. Challenges remain, including orphan reactions and non-essential processes, which metabolomics can address. With substantial genomic, transcriptomic, epigenomic, and metabolomic data from hosts and microorganisms, the iHMP carried out three longitudinal experiments. Yet, collecting data alone is not enough. To unlock its potential, we must employ systems biology, personalized medicine, and advanced computational analysis to combine data from multi-omics between the host and microbiome.

Acknowledgements The authors are thankful to the principal of Gargi College for providing the infrastructural support.

Conflict of Interest None.

References

Aderem A, Adkins JN, Ansong C, Galagan J, Kaiser S, Korth MJ, Law GL, McDermott JG, Proll SC, Rosenberger C, Schoolnik G, Katze MG (2011) A systems biology approach to infectious disease research: innovating the pathogen-host research paradigm. MBio 2(1):e00325

Amann RI, Ludwig W, Schleifer KH (1995) Phylogenetic identification and in situ detection of individual microbial cells without cultivation. Microbiol Rev 59:143–169

Atarashi K, Tanoue T, Shima T, Imaoka A, Kuwahara T, Momose Y, Cheng G, Yamasaki S, Saito T, Ohba Y, Taniguchi T, Takeda K, Hori S, Ivanov II, Umesaki Y, Itoh K, Honda K (2011) Induction of colonic regulatory T cells by indigenous Clostridium species. Science 331(6015):337–341

Baker BJ, Sheik CS, Taylor CA, Jain S, Bhasi A, Cavalcoli JD et al (2013) Community transcriptomic assembly reveals microbes that contribute to deep-sea carbon and nitrogen cycling. ISME J 7:1962–1973

Barko PC, McMichael MA, Swanson KS, Williams DA (2018) The gastrointestinal microbiome: a review. J Vet Intern Med 32:9–25

Barns SM, Takala SL, Kuske CR (1999) Wide distribution and diversity of members of the bacterial kingdom Acidobacterium in the environment. Appl Environ Microbiol 65:1731–1737

Bashiardes S, Zilberman-Schapira G, Elinav E (2016) Use of metatranscriptomics in microbiome research. Bioinform Biol Insights 10:19–25. https://doi.org/10.4137/BBI.S34610

Beja O, Aravind L, Koonin EV, Suzuki MT, Hadd A et al (2000) Bacterial rhodopsin: evidence for a new type of phototrophy in the sea. Science 289:1902–1906

Bharti R, Grimm DG (2021) Current challenges and best-practice protocols for microbiome analysis. Brief Bioinform 22:178–193. https://doi.org/10.1093/bib/bbz155

Blauwkamp TA, Thair S, Rosen MJ, Blair L, Lindner MS, Vilfan ID, Kawli T, Christians FC, Venkatasubrahmanyam S, Wall GD et al (2019) Analytical and clinical validation of a microbial cell-free DNA sequencing test for infectious disease. Nat Microbiol 4:663–674

Bloomfield SF, Exner M, Goroncy-Bermes P, Hartemann P, Heeg P, Ilschner C, Krämer I, Merkens W, Oltmanns P, Rotter M et al (2015) Lesser-known or hidden reservoirs of infection and implications for adequate prevention strategies: where to look and what to look for. GMS Hyg Infect Control 10:Doc04. https://doi.org/10.3205/dgkh000247

Bradnam KR, Fass JN, Alexandrov A, Baranay P, Bechner M, Birol I, Boisvert S, Chapman JA, Chapuis G, Chikhi R et al (2013) Assemblathon 2: evaluating de novo methods of genome assembly in three vertebrate species. Gigascience 2:10. https://doi.org/10.1186/2047-217X-2-10

Breitbart M, Hewson I, Felts B, Mahaffy JM, Nulton J et al (2003) Metagenomic analyses of an uncultured viral community from human feces. J Bacteriol 185:6220–6223

Breitling R, Vitkup D, Barrett MP (2008) New surveyor tools for charting microbial metabolic maps. Nat Rev Microbiol 6(2):156–161

Breitwieser FP, Lu J, Salzberg SL (2017) A review of methods and databases for metagenomic classification and assembly. Brief Bioinform 20:1125. https://doi.org/10.1093/bib/bbx120

Brindle JT, Antti H, Holmes E, Tranter G, Nicholson JK, Bethell HW, Clarke S, Schofield PM, McKilligin E, Mosedale DE, Grainger DJ (2002) Rapid and noninvasive diagnosis of the presence and severity of coronary heart disease using 1H-NMR-based metabonomics. Nat Med 8(12):1439–1444

Brubaker L, Putonti C, Dong Q, Wolfe AJ (2021) The human urobiome. Mamm Genome 32(4):232–238. https://doi.org/10.1007/s00335-021-09862-8. Epub 2021 Mar 2

Bryson S, Li Z, Pett-Ridge J, Hettich RL, Mayali X, Pan C et al (2016) Proteomic stable isotope probing reveals taxonomically distinct patterns in amino acid assimilation by coastal marine bacterioplankton. mSystems 1:e27–e15

Bundy JG, Willey TL, Castell RS, Ellar DJ, Brindle KM (2005) Discrimination of pathogenic clinical isolates and laboratory strains of Bacillus cereus by NMR-based metabolomic profiling. FEMS Microbiol Lett 242(1):127–136

Burns J, Fraser PD, Bramley PM (2003) Identification and quantification of carotenoids, tocopherols and chlorophylls in commonly consumed fruits and vegetables. Phytochemistry 62(6):939–947

Bushmanova E, Antipov D, Lapidus A, Prjibelski AD (2019) rnaSPAdes: a de novo transcriptome assembler and its application to RNA-Seq data. Gigascience 8. https://doi.org/10.1093/gigascience/giz100

Cardoso-Silva CB, Costa EA, Mancini MC, Balsalobre TW, Canesin LE, Pinto LR, Carneiro MS, Garcia AA, de Souza AP, Vicentini R (2014) De novo assembly and transcriptome analysis of contrasting sugarcane varieties. PLoS One 9:e88462. https://doi.org/10.1371/journal.pone.0088462

Caspi R, Altman T, Dale JM, Dreher K, Fulcher CA, Gilham F, Kaipa P, Karthikeyan AS, Kothari A, Krummenacker M, Latendresse M, Mueller LA, Paley S, Popescu L, Pujar A, Shearer AG, Zhang P, Karp PD (2010) The MetaCyc database of metabolic pathways and enzymes and the BioCyc collection of pathway/genome databases. Nucleic Acids Res 38(Database issue):D473–D479

Celaj A, Markle J, Danska J, Parkinson J (2014) Comparison of assembly algorithms for improving rate of metatranscriptomic functional annotation. Microbiome 2:39

Chen Q, Lan C, Zhao L, Wang J, Chen B, Chen Y-PP (2017) Recent advances in sequence assembly: principles and applications. Brief Funct Genomics 16:361–378. https://doi.org/10.1093/bfgp/elx006

Chiu CY, Miller SA (2019) Clinical metagenomics. Nat Rev Genet 20:341–355

Cho I, Blaser MJ (2012) The human microbiome: at the interface of health and disease. Nat Rev Genet 13:260–270

Chong J, Xia J (2017) Computational approaches for integrative analysis of the metabolome and microbiome. Metabolites 7(4):62

Ctortecka C, Stejskal K, Krssakova G, Mendjan S, Mechtler K (2022) Quantitative accuracy and precision in multiplexed single-cell proteomics. Anal Chem 94:2434–2443

Curtis TP, Sloan WT, Scannell JW (2002) Estimating prokaryotic diversity and its limits. Proc Natl Acad Sci U S A 99:10494–10499

Del Giacco L, Cattaneo C (2012) Introduction to genomics. Methods Mol Biol 823:79–88. https://doi.org/10.1007/978-1-60327-216-2_6

Derks J, Leduc A, Wallmann G, Huffman RG, Willetts M, Khan S et al (2022) Increasing the throughput of sensitive proteomics by plexDIA. Nat Biotechnol 41:50–59

Earl D, Bradnam K, St John J, Darling A, Lin D, Fass J, Yu HOK, Buffalo V, Zerbino DR, Diekhans M et al (2011) Assemblathon 1: a competitive assessment of de novo short read assembly methods. Genome Res 21:2224–2241. https://doi.org/10.1101/gr.126599.111

Ebrahim A, Brunk E, Tan J, O'Brien EJ, Kim D, Szubin R, Lerman JA, Lechner A, Sastry A, Bordbar A, Feist AM, Palsson BO (2016) Multi-omic data integration enables discovery of hidden biological regularities. Nat Commun 7:13091. https://doi.org/10.1038/ncomms13091

Edwards KJ, Bond PL, Druschel GK, McGuire MM, Hamers RJ et al (2000) Geochemical and biological aspects of sulfide mineral dissolution: lessons from Iron Mountain, California. Chem Geol 169:383–397

Egert M (2017) The BE-microbiome-communities with relevance for laundry and home care. SOFW J 143:44–48

Feist AM, Herrgard MJ, Thiele I, Reed JL, Palsson BO (2009) Reconstruction of biochemical networks in microorganisms. Nat Rev Microbiol 7(2):129–143

Fiehn O (2003) Metabolic networks of Cucurbita maxima phloem. Phytochemistry 62(6):875–886

Fleischmann RD, Adams MD, White O, Clayton RA, Kirkness EF, Kerlavage AR, Bult CJ, Tomb JF, Dougherty BA, Merrick JM et al (1995) Whole-genome random sequencing and assembly of Haemophilus influenzae Rd. Science 269(5223):496–512

Flores GE, Caporaso JG, Henley JB, Rideout JR, Domogala D, Chase J et al (2014) Temporal variability is a personalized feature of the human microbiome. Genome Biol 15:531

Frank DN, St Amand AL, Feldman RA, Boedeker EC, Harpaz N, Pace NR (2007) Molecular-phylogenetic characterization of microbial community imbalances in human inflammatory bowel diseases. Proc Natl Acad Sci U S A 104:13780–13785

Franzosa EA, Hsu T, Sirota-Madi A, Shafquat A, Abu-Ali G, Morgan XC et al (2015) Sequencing and beyond: integrating molecular 'omics' for microbial community profiling. Nat Rev Microbiol 13:360–372. https://doi.org/10.1038/nrmicro3451

Garcia DE, Baidoo EE, Benke PI, Pingitore F, Tang YJ, Villa S, Keasling JD (2008) Separation and mass spectrometry in microbial metabolomics. Curr Opin Microbiol 11(3):233–239

Georges AA, El-Swais H, Craig SE, Li WK, Walsh DA (2014) Metaproteomic analysis of a winter to spring succession in coastal northwest Atlantic Ocean microbial plankton. ISME J 8:1301–1313. https://doi.org/10.1038/ismej.2013.234

Ghurye JS, Cepeda-Espinoza V, Pop M (2016) Metagenomic assembly: overview, challenges and applications. Yale J Biol Med 89:353–362

Gibson LL, Rose JB, Haas CN (1999) Use of quantitative microbial risk assessment for evaluation of the benefits of laundry sanitation. Am J Infect Control 27:S34–S39. https://doi.org/10.1016/S0196-6553(99)70040-4

Gilbert JA, Quinn RA, Debelius J, Xu ZZ, Morton J, Garg N, Jansson JK, Dorrestein PC, Knight R (2016) Microbiome-wide association studies link dynamic microbial consortia to disease. Nature 535:94–103

Gilbert JA, Blaser MJ, Caporaso JG et al (2018) Current understanding of the human microbiome. Nat Med 24:392–400

Gill SR, Pop M, Deboy RT, Eckburg PB, Turnbaugh PJ, Samuel BS, Gordon JI, Relman DA, Fraser-Liggett CM, Nelson KE (2006) Metagenomic analysis of the human distal gut microbiome. Science 312(5778):1355–1359

Gillespie DE, Rondon MR, Handelsman J (2004) Metagenomic libraries from uncultured microorganisms. In: Molecular microbial ecology. Bios Sci. Publ. In press

Grabherr MG, Haas BJ, Yassour M, Levin JZ, Thompson DA, Amit I, Adiconis X, Fan L, Raychowdhury R, Zeng Q et al (2011) Full-length transcriptome assembly from RNA-Seq data without a reference genome. Nat Biotechnol 29:644–652. https://doi.org/10.1038/nbt.1883

Grice EA, Segre JA (2011) The skin microbiome. Nat Rev Microbiol 9(4):244–253

Gupta A, Singh V, Mani I (2022) Dysbiosis of human microbiome and infectious diseases. Prog Mol Biol Transl Sci 192(1):33–51. https://doi.org/10.1016/bs.pmbts.2022.06.016

Handelsman J, Rondon MR, Brady SF, Clardy J, Goodman RM (1998) Molecular biological access to the chemistry of unknown soil microbes: a new frontier for natural products. Chem Biol 5:R245–R249

Handelsman J, Liles MR, Mann DA, Riesenfeld CS, Goodman RM (2003) Cloning the metagenome: culture-independent access to the diversity and functions of the uncultivated microbial world. In: Wren B, Dorrell N (eds) Functional microbial genomics. Academic, New York, pp 241–255

Hardouin P, Chiron R, Marchandin H, Armengaud J, Grenga L (2021) Metaproteomics to decipher CF host–microbiota interactions: overview, challenges and future perspectives. Genes (Basel) 12:892

Heintz-Buschart A, Wilmes P (2018) Human gut microbiome: function matters. Trends Microbiol 26:563–574

Henry C, Bassignani A, Berland M, Langella O, Sokol H, Juste C (2022) Modern metaproteomics: a unique tool to characterize the active microbiome in health and diseases, and pave the road towards new biomarkers-example of Crohn's disease and ulcerative colitis flare-ups. Cells 11:1340

Heyer R, Schallert K, Zoun R, Becher B, Saake G, Benndorf D (2017) Challenges and perspectives of metaproteomic data analysis. J Biotechnol 261:24–36. https://doi.org/10.1016/j.jbiotec.2017.06.1201

Heyer R, Schallert K, Budel A, Zoun R, Dorl S, Behne A et al (2019) A robust and universal metaproteomics workflow for research studies and routine diagnostics within 24 h using phenol extraction, FASP digest, and the MetaProteomeAnalyzer. Front Microbiol 10:1883

Hollywood K, Brison DR, Goodacre R (2006) Metabolomics: current technologies and future trends. Proteomics 6(17):4716–4723

Hua X, Song L, Yu G et al (2015) MicrobiomeGWAS: a tool for identifying host genetic variants associated with microbiome composition. https://doi.org/10.1101/031187

Hugenholtz P, Goebel BM, Pace NR (1998) Impact of culture-independent studies on the emerging phylogenetic view of bacterial diversity. J Bacteriol 180:4765–4774

Jacksch S, Kaiser D, Weis S, Weide M, Ratering S, Schnell S, Egert M (2020) Influence of sampling site and other environmental factors on the bacterial community composition of domestic washing machines. Microorganisms 8:30. https://doi.org/10.3390/microorganisms8010030

Jiang Y, Xiong X, Danska J et al (2016) Metatranscriptomic analysis of diverse microbial communities reveals core metabolic pathways and microbiome-specific functionality. Microbiome 4:2. https://doi.org/10.1186/s40168-015-0146-x

Jiao N, Herndl GJ, Hansell DA, Benner R, Kattner G, Wilhelm SW, Kirchman DL, Weinbauer MG, Luo T, Chen F, Azam F (2010) Microbial production of recalcitrant dissolved organic matter: long-term carbon storage in the global ocean. Nat Rev Microbiol 8(8):593–599

Jones S, Baizan-Edge A, MacFarlane S, Torrance L (2017) Viral diagnostics in plants using next generation sequencing: computational analysis in practice. Front Plant Sci 8:1770

Jung JY, Lee SH, Jin HM, Hahn Y, Madsen EL, Jeon CO (2013) Metatranscriptomic analysis of lactic acid bacterial gene expression during kimchi fermentation. Int J Food Microbiol 163:171–179

Kan J, Hanson TE, Ginter JM, Wang K, Chen F (2015) Metaproteomic analysis of Chesapeake Bay microbial communities. Saline Syst. 1:7. https://doi.org/10.1186/1746-1448-1-7.

Kassinen A, Krogius-Kurikka L, Makivuokko H, Rinttila T, Paulin L, Corander J, Malinen E, Apajalahti J, Palva A (2007) The fecal microbiota of irritable bowel syndrome patients differs significantly from that of healthy subjects. Gastroenterology 133(1):24–33

Ketchum RE, Rithner CD, Qiu D, Kim YS, Williams RM, Croteau RB (2003) Taxus metabolomics: methyl jasmonate preferentially induces production of taxoids oxygenated at C-13 in Taxus x media cell cultures. Phytochemistry 62(6):901–909

Klappenbach JA, Dunbar JM, Schmidt TM (2000) rRNA operon copy number reflects ecological strategies of bacteria. Appl Environ Microbiol 66:1328–1333

Knights D, Kuczynski J, Charlson ES, Zaneveld J, Mozer MC, Collman RG, Bushman FD, Knight R, Kelley ST (2011) Bayesian community-wide culture-independent microbial source tracking. Nat Methods 8:761–763

Koch CM, Chiu SF, Akbarpour M, Bharat A, Ridge KM, Bartom ET, Winter DR (2018) A beginner's guide to analysis of RNA sequencing data. Am J Respir Cell Mol Biol 59:145–157. https://doi.org/10.1165/rcmb.2017-0430TR

Kubota H, Mitani A, Niwano Y, Takeuchi K, Tanaka A, Yamaguchi N, Kawamura Y, Hitomi J (2012) Moraxella species are primarily responsible for generating malodor in laundry. Appl Environ Microbiol 78:3317–3324. https://doi.org/10.1128/AEM.07816-11

Kukurba KR, Montgomery SB (2015) RNA sequencing and analysis. Cold Spring Harb Protoc 2015:951–969. https://doi.org/10.1101/pdb.top084970

Kyrpides NC (2009) Fifteen years of microbial genomics: meeting the challenges and fulfilling the dream. Nat Biotechnol 27(7):627–632

Lagier JC, Dubourg G, Million M, Cadoret F, Bilen M, Fenollar F, Levasseur A, Rolain JM, Fournier PE, Raoult D (2018) Culturing the human microbiota and culturomics. Nat Rev Microbiol 16:540–550. https://doi.org/10.1038/s41579-018-0041-0

Leary DH, Hervey IVWJ, Deschamps JR, Kusterbeck AW, Vora GJ (2013) Which metaproteome? The impact of protein extraction bias on metaproteomic analyses. Mol Cell Probes 27:193–199. https://doi.org/10.1016/j.mcp.2013.06.003

Leinonen R, Sugawara H, Shumway M (2011) The sequence read archive. Nucleic Acids Res 39:D19–D21

Ley RE, Turnbaugh PJ, Klein S, Gordon JI (2006) Microbial ecology: human gut microbes associated with obesity. Nature 444(7122):1022–1023

Li M, Wang B, Zhang M, Rantalainen M, Wang S, Zhou H, Zhang Y, Shen J, Pang X, Wei H, Chen Y, Lu H, Zuo J, Su M, Qiu Y, Jia W, Xiao C, Smith LM, Yang S, Holmes E, Tang H, Zhao G, Nicholson JK, Li L, Zhao L (2008) Symbiotic gut microbes modulate human metabolic phenotypes. Proc Natl Acad Sci U S A 105(6):2117–2122

Li Z, Adams RM, Chourey K, Hurst GB et al (2012) Systematic comparison of label-free, metabolic labeling, and isobaric chemical labeling for quantitative proteomics on LTQ Orbitrap Velos. J Proteome Res 11:1582–1590

Li J, Jia H, Cai X, Zhong H, Feng Q et al (2014) MetaHIT Consortium. An integrated catalog of reference genes in the human gut microbiome. Nat Biotechnol 32(8):834–841. https://doi.org/10.1038/nbt.2942

Liles MR, Manske BF, Bintrim SB, Handelsman J, Goodman RM (2003) A census of rRNA genes and linked genomic sequences within a soil metagenomic library. Appl Environ Microbiol 69:2684–2691

Lindon JC, Keun HC, Ebbels TM, Pearce JM, Holmes E, Nicholson JK (2005) The Consortium for Metabonomic Toxicology (COMET): aims, activities and achievements. Pharmacogenomics 6(7):691–669

Liu J, Yu T, Mu Z, Li G (2019) TransLiG: a de novo transcriptome assembler that uses line graph iteration. Genome Biol 20:81. https://doi.org/10.1186/s13059-019-1690-7

Lloyd-Price J, Abu-Ali G, Huttenhower C (2016) The healthy human microbiome. Genome Med 8:51

Loman NJ, Constantinidou C, Christner M, Rohde H, Chan JZ-M, Quick J, Weir JC, Quince C, Smith GP, Betley JR et al (2013) A culture-independent sequence-based metagenomics approach to the investigation of an outbreak of Shiga-toxigenic *Escherichia coli* O104:H4. JAMA 309:1502–1510

Maier T, Guell M, Serrano L (2009) Correlation of mRNA and protein in complex biological samples. FEBS Lett 583:3966–3973

Mani I (2020a) Metagenomics approach for bioremediation: challenges and perspectives. In: Pandey VC, Singh V (eds) Bioremediation of pollutants. Elsevier, pp 275–285., ISBN: 9780128190258. https://doi.org/10.1016/B978-0-12-819025-8.00012-0

Mani I (2020b) Current status and challenges of DNA sequencing. In: Singh V (ed) Advances in synthetic biology. Springer, Singapore. https://doi.org/10.1007/978-981-15-0081-7_5

Manichanh C, Rigottier-Gois L, Bonnaud E, Gloux K, Pelletier E, Frangeul L, Nalin R, Jarrin C, Chardon P, Marteau P, Roca J, Dore J (2006) Reduced diversity of faecal microbiota in Crohn's disease revealed by a metagenomic approach. Gut 55(2):205–211

Marchesi JR, Ravel J (2015) The vocabulary of microbiome research: a proposal. Microbiome 3:31

Matallana-Surget S, Jagtap PD, Griffin TJ, Beraud M, Wattiez R (2018) Comparative metaproteomics to study environmental changes. In: Nagarajan M (ed) Metagenomics: perspectives, methods, and applications. Elsevier Inc., Amsterdam, pp 327–363. https://doi.org/10.1016/b978-0-08-102268-9.00017-3

McGeachie MJ, Sordillo JE, Gibson T et al (2016) Longitudinal prediction of the infant gut microbiome with dynamic Bayesian networks. Sci Rep 6:20359

Miller RR, Montoya V, Gardy JL, Patrick DM, Tang P (2013) Metagenomics for pathogen detection in public health. Genome Med 5:81

Miller S, Naccache SN, Samayoa E, Messacar K, Arevalo S, Federman S, Stryke D, Pham E, Fung B, Bolosky WJ et al (2019) Laboratory validation of a clinical metagenomic sequencing assay for pathogen detection in cerebrospinal fluid. Genome Res 29:831–842
Morris RM, Nunn BL, Frazar C, Goodlett DR, Ting YS, Rocap G (2010) Comparative metaproteomics reveals ocean-scale shifts in microbial nutrient utilization and energy transduction. ISME J 4:673–685. https://doi.org/10.1038/ismej.2010.4
Munk S, Johansen C, Stahnke LH, Adler-Nissen J (2001) Microbial survival and odor in laundry. J Surfact Deterg 4:385–394. https://doi.org/10.1007/s11743-001-0192-2
Nesvizhskii AI (2010) A survey of computational methods and error rate estimation procedures for peptide and protein identification in shotgun proteomics. J Proteome 73:2092–2123. https://doi.org/10.1016/j.jprot.2010.08.009
Nguyen N-P, Warnow T, Pop M, White B (2016) A perspective on 16S rRNA operational taxonomic unit clustering using sequence similarity. NPJ Biofilms Microbiomes 2:16004
Nicholson JK, Lindon JC (2008) Metabonomics. Nature 455(23):1054–1056
Nicholson JK, Holmes E, Wilson ID (2005) Gut microorganisms, mammalian metabolism and personalized health care. Nat Rev Microbiol 3(5):431–438
Nix ID, Frontzek A, Bockmühl DP (2015) Characterization of microbial communities in household washing machines. Tenside Surfactants Deterg 52:432–440. https://doi.org/10.3139/113.110394
Novak Babič M, Gostinčar C, Gunde-Cimerman N (2020) Microorganisms populating the water-related indoor biome. Appl Microbiol Biotechnol 104:6443–6462. https://doi.org/10.1007/s00253-020-10719-4
Odamaki T, Kato K, Sugahara H, Hashikura N, Takahashi S, Xiao J-Z et al (2016) Age-related changes in gut microbiota composition from newborn to centenarian: a cross-sectional study. BMC Microbiol 16:90
Ohdoi C, Nyhan WL, Kuhara T (2003) Chemical diagnosis of Lesch-Nyhan syndrome using gas chromatography-mass spectrometry detection. J Chromatogr B Anal Technol Biomed Life Sci 792(1):123–130
Ottesen EA, Young CR, Gifford SM, Eppley JM et al (2014) Multispecies diel transcriptional oscillations in open ocean heterotrophic bacterial assemblages. Science 345:207–212
Pan S, Chen R (2020) Metaproteomic analysis of human gut microbiome in digestive and metabolic diseases. Adv Clin Chem 97:1–12
Pavia AT (2011) Viral infections of the lower respiratory tract: old viruses, new viruses, and the role of diagnosis. Clin Infect Dis 52(Suppl 4):S284–S289
Pedersen HK, Gudmundsdottir V, Nielsen HB, Hyotylainen T, Nielsen T, Jensen BAH, Forslund K, Hildebrand F, Prifti E, Falony G et al (2016) Human gut microbes impact host serum metabolome and insulin sensitivity. Nature 535:376–381
Pennisi E (2010) Body's hardworking microbes get some overdue respect. Science 330(6011):1619
Peterson J, Garges S, Giovanni M, McInnes P, Wang L et al (2009) The NIH human microbiome project. Genome Res 19(12):2317–2323
Poulsen M, Schwab C, Jensen BB, Engberg RM, Spang A, Canibe N et al (2013) Methylotrophic methanogenic Thermoplasmata implicated in reduced methane emissions from bovine rumen. Nat Commun 4:1428
Qin J, Li R, Raes J, Arumugam M, Burgdorf KS et al (2010) A human gut microbial gene catalogue established by metagenomic sequencing. Nature 464(7285):59–65
Quaiser A, Ochsenreiter T, Lanz C, Schuster SC, Treusch AH et al (2003) Acidobacteria form a coherent but highly diverse group within the bacterial domain: evidence from environmental genomics. Mol Microbiol 50:563–575
Raes J, Bork P (2008) Molecular eco-systems biology: towards an understanding of community function. Nat Rev Microbiol 6(9):693–699
Raes J, Foerstner KU, Bork P (2007) Get the most out of your metagenome: computational analysis of environmental sequence data. Curr Opin Microbiol 10(5):490–498
Rappe MS, Connon SA, Vergin KL, Giovannoni SJ (2002) Cultivation of the ubiquitous SAR11 marine bacterioplankton clade. Nature 418:630–633

Riesenfeld CS, Schloss PD, Handelsman J (2004) Metagenomics: genomic analysis of microbial communities. Annu Rev Genet 38:525–552. https://doi.org/10.1146/annurev.genet.38.072902.091216

Rodionov DA, Yang C, Li X, Rodionova IA, Wang Y, Obraztsova AY, Zagnitko OP, Overbeek R, Romine MF, Reed S, Fredrickson JK, Nealson KH, Osterman AL (2010) Genomic encyclopedia of sugar utilization pathways in the Shewanella genus. BMC Genomics 11:494

Rohwer F (2003) Global phage diversity. Cell 113:141

Rusch DB, Halpern AL, Sutton G, Heidelberg KB, Williamson S et al (2007) The Sorcerer II Global Ocean Sampling expedition: northwest Atlantic through eastern tropical Pacific. PLoS Biol 5(3):e77

Russo A, Couto N, Beckerman A, Pandhal J (2016) A metaproteomic analysis of the response of a freshwater microbial community under nutrient enrichment. Front Microbiol 7:1172. https://doi.org/10.3389/fmicb.2016.01172

Sait M, Hugenholtz P, Janssen PH (2002) Cultivation of globally distributed soil bacteria from phylogenetic lineages previously only detected in cultivation-independent surveys. Environ Microbiol 4:654–666

Saito MA, Alexander D, Anton FP, Matthew RM, Michael SR, Giacomo RD et al (2015) Needles in the blue sea: sub-species specificity in targeted protein biomarker analyses within the vast oceanic microbial metaproteome. Proteomics 15:3521–3531. https://doi.org/10.1002/pmic.201400630

Saito MA, Bertrand EM, Duffy ME, Gaylord DA, Held NA, Hervey IVWJ (2019) Progress and challenges in ocean metaproteomics and proposed best practices for data sharing. J Proteome Re 31:1461–1476. https://doi.org/10.1021/acs.jproteome.8b00761

Schaneberg BT, Crockett S, Bedir E, Khan IA (2003) The role of chemical fingerprinting: application to Ephedra. Phytochemistry 62(6):911–918

Schmithausen RM, Sib E, Exner M, Hack S, Rösing C, Ciorba P, Bierbaum G, Savin M, Bloomfield SF, Kaase M et al (2019) The washing machine as a reservoir for transmission of extended-spectrum-beta-lactamase (CTX-M-15)-producing Klebsiella oxytoca ST201 to newborns. Appl Environ 85:e01435-19. https://doi.org/10.1128/AEM.01435-19

Schulz MH, Zerbino DR, Vingron M, Birney E (2012) Oases: robust de novo RNA-seq assembly across the dynamic range of expression levels. Bioinformatics 28:1086–1092. https://doi.org/10.1093/bioinformatics/bts094

Seli E, Botros L, Sakkas D, Burns DH (2008) Noninvasive metabolomic profiling of embryo culture media using proton nuclear magnetic resonance correlates with reproductive potential of embryos in women undergoing in vitro fertilization. Fertil Steril 90(6):2183–2189

Serrano-Villar S, Rojo D, Martínez-Martínez M, Deusch S, Vázquez-Castellanos JF, Sainz T, Vera M, Moreno S, Estrada V, Gosalbes MJ, Latorre A, Margolles A, Seifert J, Barbas C, Moya A, Ferrer M (2016) HIV infection results in metabolic alterations in the gut microbiota different from those induced by other diseases. Sci Rep. 6:26192. https://doi.org/10.1038/srep26192.

Sherman RM, Salzberg SL (2020) Pan-genomics in the human genome era. Nat Rev Genet 21(4):243–254. https://doi.org/10.1038/s41576-020-0210-7. Epub 2020 Feb 7

Shreiner AB, Kao JY, Young VB (2015) The gut microbiome in health and in disease. Curr Opin Gastroenterol 31:69–75

Sintchenko V, Iredell JR, Gilbert GL (2007) Pathogen profiling for disease management and surveillance. Nat Rev Microbiol 5(6):464–470

Smith CA, O'Maille G, Want EJ, Qin C, Trauger SA, Brandon TR, Custodio DE, Abagyan R, Siuzdak G (2005) METLIN: a metabolite mass spectral database. Ther Drug Monit 27(6):747–751

Spratlin JL, Serkova NJ, Eckhardt SG (2009) Clinical applications of metabolomics in oncology: a review. Clin Cancer Res 15(2):431–440

Stapleton K, Hill K, Day K, Perry JD, Dean JR (2013) The potential impact of washing machines on laundry malodour generation. Lett Appl Microbiol 56:299–306. https://doi.org/10.1111/lam.12050

Stobiecki M, Matysiak-Kata I, Franski R, Skala J, Szopa J (2003) Monitoring changes in anthocyanin and steroid alkaloid glycoside content in lines of transgenic potato plants using liquid chromatography/mass spectrometry. Phytochemistry 62(6):959–969

Sun Z, Wang W, Li L, Zhang X, Ning Z, Mayne J et al (2022) Comprehensive assessment of functional effects of commonly used sugar substitute sweeteners on ex vivo human gut microbiome. Microbiol Spectr 10:e0041222

Tanca A, Palomba A, Deligios M, Cubeddu T, Fraumene C, Biosa G et al (2013) Evaluating the impact of different sequence databases on metaproteome analysis: insights from a lab-assembled microbial mixture. PLoS One 8:e82981. https://doi.org/10.1371/journal.pone.0082981

Tanca A, Palomba A, Fraumene C, Pagnozzi D, Manghina V, Deligios M et al (2016) The impact of sequence database choice on metaproteomic results in gut microbiota studies. Microbiome 4:51

Teeling H, Fuchs BM, Becher D, Klockow C, Gardebrecht A, Bennke CM et al (2012) Substrate-controlled succession of marine bacterioplankton populations induced by a phytoplankton bloom. Science 336:608–611. https://doi.org/10.1126/science.1218344

Timmins-Schiffman E, May DH, Mikan M, Riffle M, Frazar C, Harvey HR et al (2017) Critical decisions in metaproteomics: achieving high confidence protein annotations in a sea of unknowns. ISME J 11:309–314. https://doi.org/10.1038/ismej.2016.132

Torsvik V, Ovreas L (2002) Microbial diversity and function in soil: from genes to ecosystems. Curr Opin Microbiol 5(3):240–245

Torsvik V, Goksoyr J, Daae FL (1990) High diversity in DNA of soil bacteria. Appl Environ Microbiol 56:782–787

Tringe SG, von Mering C, Kobayashi A, Salamov AA, Chen K, Chang HW, Podar M, Short JM, Mathur EJ, Detter JC, Bork P, Hugenholtz P, Rubin EM (2005) Comparative metagenomics of microbial communities. Science 308(5721):554–557

Turnbaugh PJ, Ley RE, Hamady M, Fraser-Liggett CM, Knight R, Gordon JI (2007) The human microbiome project. Nature 449(7164):804–810

Tyson GW, Chapman J, Hugenholtz P, Allen EE, Ram RJ et al (2004) Community structure and metabolism through reconstruction of microbial genomes from the environment. Nature 428:37–43

Van Herreweghen F, Amberg C, Marques R, Callewaert C (2020) Biological and chemical processes that lead to textile malodour development. Microorganisms 8:1709. https://doi.org/10.3390/microorganisms8111709

Vemuri GN, Aristidou AA (2005) Metabolic engineering in the -omics era: elucidating and modulating regulatory networks. Microbiol Mol Biol Rev 69(2):197–216

Venkatesan A, Tunkel AR, Bloch KC, Lauring AS, Sejvar J, Bitnun A, Stahl J-P, Mailles A, Drebot M, Rupprecht CE et al (2013) Case definitions, diagnostic algorithms, and priorities in encephalitis: consensus statement of the international encephalitis consortium. Clin Infect Dis 57:1114–1128

Venter JC, Remington K, Heidelberg JF, Halpern AL, Rusch D et al (2004) Environmental genome shotgun sequencing of the Sargasso Sea. Science 304:66–74

Ventura M, O'Flaherty S, Claesson MJ, Turroni F, Klaenhammer TR, van Sinderen D, O'Toole PW (2009) Genome-scale analyses of health-promoting bacteria: probiogenomics. Nat Rev Microbiol 7(1):61–71

Villas-Boas SG, Mas S, Akesson M, Smedsgaard J, Nielsen J (2005) Mass spectrometry in metabolome analysis. Mass Spectrom Rev 24(5):613–646

Vogel C, Marcotte EM (2012) Insights into the regulation of protein abundance from proteomic and transcriptomic analyses. Nat Rev Genet 13:227–232

Vogel TM, Simonet P, Jansson JK, Hirsch PR, Tiedje JM, van Elsas JK, Bailey MJ, Nalin R (2009) TerraGenome: a consortium for the sequencing of a soil metagenome. Nat Rev Microbiol 7:652

Wajid B, Serpedin E (2016) Do it yourself guide to genome assembly. Brief Funct Genomics 15:1–9. https://doi.org/10.1093/bfgp/elu042

Walker A (2010) A glut from the gut: metagenomics takes a giant step forward. Nat Rev Microbiol 8(5):315

Walter J, Ley R (2011) The human gut microbiome: ecology and recent evolutionary changes. Ann Rev Microbiol 65:411–429
Wang W, Baladandayuthapani V, Morris JS et al (2013) iBAG: integrative Bayesian analysis of high-dimensional multiplatform genomics data. Bioinformatics 29:149–159
Wang DZ, Xie ZX, Zhang SF (2014) Marine metaproteomics: current status and future directions. J Proteome 97:27–35. https://doi.org/10.1016/j.jprot.2013.08.024
Wang Q, Wang K, Wu W, Giannoulatou E, Ho JWK, Li L (2019) Host and microbiome multi-omics integration: applications and methodologies. Biophys Rev 11(1):55–65. https://doi.org/10.1007/s12551-018-0491-7. Epub 2019 Jan 9
Wilmes P, Bond PL (2004) The application of two-dimensional polyacrylamide gel electrophoresis and downstream analyses to a mixed community of prokaryotic microorganisms. Environ Microbiol 6:911–920. https://doi.org/10.1111/j.1462-2920.2004.00687.x
Wilmes P, Bowen BP, Thomas BC, Mueller RS, Denef VJ, Verberkmoes NC, Hettich RL, Northen TR, Banfield JF (2010) Metabolome-proteome differentiation coupled to microbial divergence. MBio 1(5)
Wilmes P, Heintz-Buschart A, Bonf PL (2015) A decade of metaproteomics: where we stand and what the future holds. Proteomics 15:3409–3417. https://doi.org/10.1002/pmic.201500183
Wöhlbrand L, Trautwein K, Rabus R (2013) Proteomic tools for environmental microbiology—a roadmap from sample preparation to protein identification and quantification. Proteomics 13:2700–2730. https://doi.org/10.1002/pmic.201300175
Wommack KE, Colwell RR (2000) Virioplankton: viruses in aquatic ecosystems. Microbiol Mol Biol Rev 64:69–114
Xie Y, Wu G, Tang J, Luo R, Patterson J, Liu S, Huang W, He G, Gu S, Li S et al (2014) SOAPdenovo-Trans: de novo transcriptome assembly with short RNA-Seq reads. Bioinformatics 30:1660–1666. https://doi.org/10.1093/bioinformatics/btu077
Xiong X, Frank DN, Robertson CE, Hung SS, Markle J, Canty AJ et al (2012) Generation and analysis of a mouse intestinal metatranscriptome through Illumina based RNA-sequencing. PLoS One 7:e36009
Ye SH, Siddle KJ, Park DJ, Sabeti PC (2019) Benchmarking metagenomics tools for taxonomic classification. Cell 178(4):779–794. https://doi.org/10.1016/j.cell.2019.07.010
Young JC, Pan C, Adams RM, Brooks B, Banfield JF, Morowitz MJ et al (2015) Metaproteomics reveals functional shifts in microbial and human proteins during a preterm infant gut colonization case. Proteomics 15:3463–3473
Zengler K, Toledo G, Rappe M, Elkins J, Mathur EJ et al (2002) Cultivating the uncultured. Proc Natl Acad Sci U S A 99:15681–15686
Zhang W, Sun J, Cao H, Tian R, Cai L, Ding W, Qian PY (2016) Post-translational modifications are enriched within protein functional groups important to bacterial adaptation within a deep-sea hydrothermal vent environment. Microbiome 4:49
Zhang X, Deeke SA, Ning Z, Starr AE, Butcher J, Li J, Mayne J, Cheng K, Liao B, Li L, Singleton R, Mack D, Stintzi A, Figeys D (2018) Metaproteomics reveals associations between microbiome and intestinal extracellular vesicle proteins in pediatric inflammatory bowel disease. Nat Commun 9:2873
Zhang X, Walker K, Mayne J, Li L, Ning Z, Stintzi A et al (2022) Evaluating live microbiota biobanking using an ex vivo microbiome assay and metaproteomics. Gut Microbes 14:2035658

Technology for Studying Multi-omics 2

Truong Pham Thanh, Yen-Vy Nguyen Thi,
and Dinh-Toi Chu

2.1 Introduction

Multi-omics, a cross-disciplinary domain, concentrates on the architecture, role, development, charting, and modification of genomes which integrate data from multiple "omics" fields comprising genomics, transcriptomics, proteomics, metabolomics, and others in order to provide a holistic overview of the molecular mechanisms behind the biological processes and diseases. This approach can reveal how the changes at the genetic level can affect gene expression and metabolic processes, leading to changes in phenotype or disease state (Papatheodorou et al. 2017). To understand thoroughly how each omics layer contributes to the bigger picture, each of the omics layers has been explained in the table below (Table 2.1).

Instead of a singular biomolecule type being the sole object of concentration, the ethos of multi-omics epitomizes the dynamic and interconnected essence of biological systems. This is accomplished by meticulously analyzing the genomic, transcriptomic, proteomic, and metabolomic networks by using numerous methods, for instance, the Bayesian inference (Vaske et al. 2010; Mo et al. 2018; Argelaguet et al. 2018; Bhattacharyya et al. 2022; Kirk et al. 2023; Maity et al. 2020), meta-analysis (Kale et al. 2023) as well as machine learning and deep learning (Waqas et al. 2023; Benkirane 2022; Tabakhi et al. 2023) to investigate how these networks engage in mutual interactions and exert reciprocal influence on each other (Hasin et al. 2017).

T. P. Thanh · Y.-V. N. Thi · D.-T. Chu (✉)
Center for Biomedicine and Community Health, International School, Vietnam National University, Hanoi, Vietnam

Faculty of Applied Sciences, International School, Vietnam National University, Hanoi, Vietnam
e-mail: toicd@vnu.edu.vn

I. Mani, V. Singh (eds.), *Multi-Omics Analysis of the Human Microbiome*,
https://doi.org/10.1007/978-981-97-1844-3_2

Table 2.1 An apprehensive overview of Omics layers

Omics layer	Description	References
Genomics	The comprehensive study of the complete set of genes or to be known as the genome of an organism involves the procedure of DNA sequencing and analysis	ENCODE Project Consortium (2007, 2012), Thurman et al. (2012), Zhao et al. (2023), Fresca and Manzoni (2021)
Transcriptomics	The study of RNA transcripts produced by the genome under experimental circumstances or in a known or experimental cell	Dong and Chen (2013), Downing and Angelopoulos (2023)
Proteomics	The study of the structure and functions of proteins—The vital parts of living organisms, as the main components of the physiological metabolic pathways of cells	Deshmukh et al. (2015), Kwok et al. (2008)
Metabolomics	The study of chemical processes of metabolites, or to be known as the study of the unique chemical fingerprints that specific cell processes leave behind	Li and Luo (2022), Pretorius et al. (2021)

This comprehensive approach is critical for understanding the emergent properties of life, which arise from interactions between fields that have traditionally been studied in isolation and then combine the results (Huang et al. 2017; Moco 2022; Panguluri et al. 2010; Song et al. 2011). By integrating biological data layers ranging from genes to proteins to metabolites, multi-omics paints a more comprehensive tableau of the system's dynamics and the connections that underpin health and disease (Hasin et al. 2017; Zhou et al. 2019; Graw et al. 2021; Zhou et al. 2021). This integrated perspective has illuminated the origins of complex diseases, identified novel diagnostic markers, and proposed new therapeutic targets (Subramanian et al. 2020; Ahmed 2020; Lorenzon et al. 2018). There is no doubt that multi-omics has played an instrumental role in deciphering the complexities of how molecules, pathways, cells, organs, and organisms function collectively (Huang et al. 2017; Kloehn et al. 2020; Budamagunta et al. 2021). In order to provide a stronger grasp into key aspects of the field, Table 2.2 provides real-life examples of multi-omics studies and their applications.

Evidently, the merits of pursuing research in the multi-omics approach are manifold as research in the field has unveiled a cornucopia of applications with substantial real-life implications (Subramanian et al. 2020). And to conduct as well as achieve these findings, technology integration is essential; the integration of diverse technologies and methodologies offers vast potential for discoveries and a deeper understanding of biological systems.

The integration of large-scale database analysis tools has opened new frontiers in multi-omics research, promising transformative discoveries and novel insights (Gonzales et al. 2023). These technologies have revolutionized our ability to

Table 2.2 Some real-life applications of multi-omics studies

Application	Description	References
Precision medicine	By integrating the multi-omics method, researchers and professionals can now provide personalized treatment for patients since the multi-omics method aids in tailoring therapies to an individual's unique molecular profile, enhancing treatment efficacy and minimizing adverse effects	Hasin et al. (2017), Everad John (2022), Bonnechère (2022), Raufaste-Cazavieille et al. (2022), Hu and Jia (2021), Ota and Fujio (2021)
Disease biomarker discovery	By comprehensively analyzing multiple biomolecular classes, multi-omics studies contribute to the identification of novel disease biomarkers. These molecular signatures serve as indicators for early disease detection, prognosis, and response to treatment	Camacho et al. (2018), Park et al. (2022), Kerdidani et al. (2022), Chikwambi et al. (2023)
Drug development	Multi-omics research facilitates the elucidation of complex drug responses and mechanisms of action. By integrating genomics, proteomics, and metabolomics data, this approach enables a comprehensive understanding of drug-target interactions, drug metabolism, and toxicity profiles	Hasin et al. (2017), Liu and Mei (2023), Guo et al. (2023)
Microbiome studies	Multi-omics techniques have revolutionized the exploration of the human microbiome. By integrating metagenomics, metatranscriptomics, metaproteomics, and metabolomics, researchers gain insights into microbial communities' structure, function, and impact on human health and disease	Bolyen et al. (2019), Franzosa et al. (2014), Couvillion et al. (2023)
Systems biology	Multi-omics research forms the cornerstone of systems biology, enabling a holistic understanding of biological systems at various levels of complexity. It aids in deciphering complex biological networks, elucidating emergent properties, and advancing our knowledge of fundamental biological processes	Pinu et al. (2019), Baharin et al. (2023), Liang et al. (2022)

analyze complex datasets, uncover hidden relationships, and predict biological outcomes (Gonzales et al. 2023). Additionally, the technology also provides researchers with a powerful platform for visualizing and interpreting multi-omics data, capturing the intricate interactions within biological systems. Some of the popular technologies have been reviewed and compiled in the following (Table 2.3).

Based on the observable high-level technologies and the rapid growth of them, researchers might well foresee further advancements in the multi-omics research field as these technology solutions progress, generating forth fresh discoveries and advances.

Table 2.3 Applied technology in omics layers

Technology	Omics layer	Description	References
Next-generation sequencing (NGS)	Genomics	This is a technology that allows quick and constant sequencing in DNA or RNA samples supporting various applications, for instance, gene expression profiling, as well as the identification of epigenetic markers	Reiter et al. (2010), GTEx Consortium (2017)
RNA-Seq	Transcriptomics	Deep-sequencing technologies are exploited in this advancement for transcriptome profiling, providing a highly accurate measurement of transcription levels and their isoforms compared to other methods	Lowe et al. (2017), Wang et al. (2009), Rao et al. (2021), Andersson et al. (2020)
Mass spectrometry (MS)	Proteomics	A powerful solution for the identification and quantification of proteins; in recent years, the tool is widely used in proteomics for protein identification in sophisticated samples	Doncheva et al. (2019), Aslam et al. (2017), Doncheva et al. (2018), Schoof et al. (2021)
Nuclear magnetic resonance spectroscopy (NMR)	Metabolomics	One of the state-of-the-art techniques for metabolomics research, NMR, or to be known as a physical phenomenon based on the magnetic properties of the atomic nuclei	Xia et al. (2015), Johnson et al. (2016), Würtz et al. (2017), Tzoulaki et al. (2014)

2.2 Genomics

Genomics, an advanced methodology within the field of biomedicine, focuses on the study of entire genomes, including all genes and their interactions, in order to gain a deeper understanding of the genetic information encoded within an organism. Genomics research takes a holistic approach by examining the complete set of genes, along with noncoding regions, regulatory elements, and other components of the genome. This broader perspective allows researchers to explore the intricate interplay between genes, their regulation, and the complex networks that influence gene expression and function (Dorman 2013). To further elucidate this point, a table (Table 2.4) was included, which illustrates and adds depth to it through various aspects.

Along with the competency and advancement in methodology, genomic technology has also played a vital role in this revolution. The growth and refinements witnessed in NGS started with the development of pyrosequencing and was first officially available in 2005 as the 454/Roche platform (Akintunde et al. 2023). NGS utilizes a massive parallel sequencing approach, compared to the rather low throughput and costly method of High-Throughput Sequencing (HTS), and second

Table 2.4 Implementation of genomic research on varied species

Species	Application	References
Humans	The investigation of the spatial arrangement of the genomic sequence is applied in comprehending the functionality of biological processes such as gene transcription, replication, repair, and regulation medicine for human healthcare, agriculture, forestry, food safety, environmental studies	Gong et al. (2021), Couvillion et al. (2023), Hien et al. (2021), Bae et al. (2022)
Non-human model organisms	In non-human model organisms, genomics has been implemented in creating computational interactome networks for protein interaction studies, optimizing reduced representation sequencing for population genomics, as well as determining sex in the case when absence of field records for conservation studies	Dong et al. (2021), Davison and Neiman (1825), Christiansen et al. (2021), Onley et al. (2020)
Plants	The enhancement of crop improvement is facilitated through the application of innovative technologies such as next-generation sequencing (NGS) and omics techniques such as genomics, transcriptomics, and proteomics	Yang et al. (2021), Schwartz et al. (2020), Sahu et al. (2020)

generation sequencing. It can be observed that Next-Generation Sequencing had a major impact on how people conduct, analyze, and research as it elevated the speed, expenditure, and accuracy remarkably. Millions or even billions of reads can be produced in a single run within a few hours or days (Akintunde et al. 2023; Mardis 2008; Bean et al. 2020). This amplification in technology has not merely enhanced genomics research but also established a firm foundation for multi-omics studies since to conduct this integrative work, comprehensive genomics data is essential. Together with NGS, the growth in the multi-modal learning method reinforced the significance of genomics in multi-omics studies as the method utilizes genomics data in parallel with other layers in multi-omics, which is the key to have a more complete view of biological systems and to uncover the characteristics that cannot be observed by investigating single omics (Gonzales et al. 2023).

Furthermore, the notable advancement in single-cell multi-omics in recent years also unlocked new paths to discover in the tissues as well as organisms as it provides the researcher a powerful way to profile the multi-omics of a single cell, unveiling exceptional amount of information on the cell functions as well as its diversity; this type of data can shed light on the root of complex pathologies and so on (Gonzales et al. 2023; Cano-Gamez and Trynka 2020). Apart from that, computational methods for interconnecting the multi-omics data are critical for extracting meaningful information from these complex datasets. Machine learning and deep learning techniques have shown great promise in this regard (Lin et al. 2022; Bréhélin 2023; Adornetto and Greco 2023; Huang et al. 2021). With training algorithms fine-tuned on genomic data, researchers can forecast gene regulation, identify genomic variations, and even anticipate the consequences of specific genetic changes. The exponential progression and upheaval of innovations and undertakings transpiring globally over the preceding 10 years have engendered the burgeoning utilization of genomics to scrutinize pan-omics transversely over manifold taxa (Hasin et al.

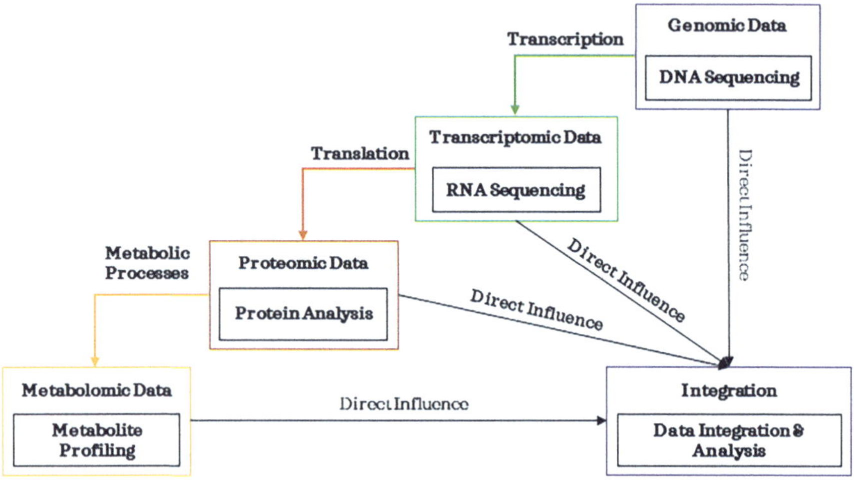

Fig. 2.1 The hierarchical flow of biological information in multi-omics research

2017). The amalgamation of genomics with supplementary omics spheres constitutes the cornerstone for progressions in customized therapeutics, morbid exploration, genetic governance, and agrarian bioresearch (Roy et al. 2018; Rosario et al. 2020). Figure 2.1 proffers perceptivity into the genomics dramas in multi-omics probing, illuminating the applications inside each class.

2.3 Transcriptomics

Transcriptomics is a scientific discipline dedicated to the examination and interpretation of the entirety of RNA transcripts within a given biological specimen at a specific time and environment providing the researcher profound understanding of the state of the cell, which enables them to understand not solely the factors of the genome or the components of cells or tissues but also a crucial understanding of the patterns of gene expression, the processes involved in RNA manipulation, and the regulatory systems at play within a cellular or organismal context (Lowe et al. 2017). Of the omics techniques, transcriptomics is uniquely valuable for multi-omics investigations for its depth in RNA analysis; the characteristic of this research method provides an additional layer to the multi-omics research as supporting detail for other omics data granted that genomics provides visions upon the unchanging traits of the cell. Transcriptomics, in contrast, unveils the active alternations in gene expression under given circumstances, which can also be used as validation and interpretation data for other omics analyses enriching the extensiveness of multi-omics studies (Xu et al. 2019; Williams et al. 2022; Zhang et al. 2022).

The RNA-seq (high-throughput RNA sequencing) technology revolutionized the field with the power to bring unprecedented accuracy and resolution in the data of gene expression quantification (Valenti et al. 2021; Chen et al. 2022; Maheendran et al. 2022); these upticks have amplified the range of transcriptome analysis, enabling its utilization in miscellaneous areas of probing, embracing disease check-ups, life-form categorizations, and grasping organic computations (Lowe et al. 2017). The subsequent list (Table 2.5) stipulates a summing up of some of the ongoing upticks in the research and utilization of transcriptome analysis. Transcriptomics technology serves as a potent instrument for investigating species classification through the lens of gene expression, offering groundbreaking perspectives on the evolutionary ties among distinct species and illuminating their genetic congruities and disparities. In the realm of botany and agriculture, transcriptomics plays a pivotal role in investigating plant growth and development, the impact of environmental variables on plants, and their responses to pathogenic entities (Cha et al. 2022), thereby informing the development of strategies aimed at enhancing agricultural productivity and sustainability (Saeed et al. 2003; Joshi et al. 2022). Transcriptomics also finds its application in the examination of microorganisms, specifically concerning environmental surveillance. Through the analysis of the transcriptome of microbial populations, investigators can track alterations in the environment and their subsequent effects on these communities (Caspi et al. 2016).

Within the sphere of human and animal healthcare, transcriptomics assumes a pivotal function in the identification of diseases (Jiang et al. 2023). Not only so, but transcriptomics methodology integrated with artificial intelligence has also been implemented in the field of dentistry and oral healthcare as the saliva biomarkers of patients have been exploited to harness personalized and precision dentistry (Paqué et al. 2022). In addition, transcriptomics technology has also been used in stem cell research as the foundation for tissue engineering and regenerative medicine (Jain

Table 2.5 Significant advancements in transcriptomic technology

Advancement	Description	References
Single-cell RNA sequencing (scRNA-seq)	The technology was developed to enable the users to analyze the gene expressions at a single-cell level, which provides a better data resolution as well as a clearer insight into the specific cell functions	Vieth et al. (2019), Olsen and Baryawno (2018), Haque et al. (2017)
Long-read sequencing technologies	A technology platform that allows high and comprehensive transcriptome profiling as it provides the capability of long-read sequencing	Marx (2023), Amarasinghe et al. (2020), Mantere et al. (2019)
Spatial transcriptomics	In this technology, histological imaging and RNA are first combined to map with the transcriptome in the specific tissue, proving a highly holistic viewpoint of the biology within the tissue as well as the pathology of diseases	Williams et al. (2022), Marx (2021), Yue et al. (2023)
RNA-seq in combination with CRISPR	This technology helps researchers to have a better understanding of the role of interested genes in numerous pathology states as well as biological processes by using the gene modifying technology to functionally validate the gene expression changes	Burgess (2017), Selvakumar et al. (2022), Alqahtani et al. (2021)

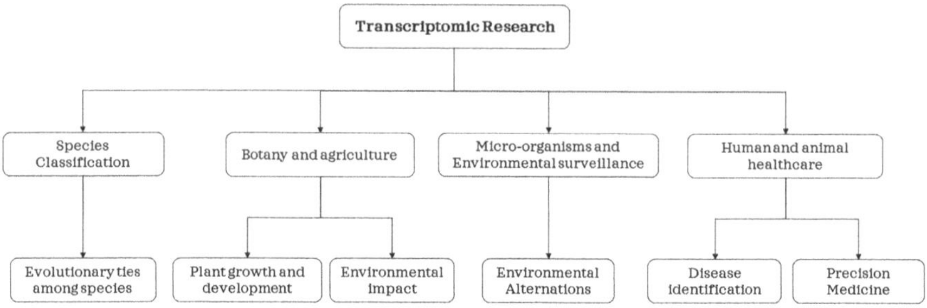

Fig. 2.2 The multidisciplinary impact of transcriptomic research

et al. 2022), opening the way to treat some incurable diseases (Jasim et al. 2022). Depicted in Fig. 2.2 is a schematic representation of the multifaceted applications of transcriptomic research across a range of scientific disciplines. The central node signifies the domain of transcriptomic research, which bifurcates into four key applications: taxonomic classification of species, botanical and agricultural studies, environmental surveillance through microorganisms, and applications in human and animal health. Each of these primary nodes further diverges into specific areas of impact, demonstrating the extensive scope and profound influence of transcriptomic research in enhancing our comprehension of intricate biological systems.

2.4 Proteomics

Proteomics focuses on uncovering vibrant insights into the realm of proteins, empowering the ability to investigate which molecules "carry out" instructions encoded in the DNA and regulated at the transcription phase. Considered the "hard worker" of the cell responsible for the conduct of most of the biological functions, having a holistic observation of the proteome provides a critical understanding of the functionality of the desired biological systems, which supports the layer of proteomics and transcriptomics by shedding light on protein levels, alternations, relationships, and spatialization as this information, solely by DNA or RNA analysis, cannot be achieved (Zhang et al. 2010; Beck et al. 2006; Devireddy et al. 2021).

The advent of high-throughput methodologies, exemplified by mass spectrometry, has marked a significant turning point in the field of proteomics, enabling a comprehensive examination of protein components within a wide range of biological systems (Beck et al. 2006; Devireddy et al. 2021; Smolka et al. 2002). No longer confined to studying proteins in isolation, proteomics offers a system-wide perspective into the dynamic interplay between proteins, their diverse modifications and interactions, and how these complex relationships ultimately govern biological structure and function (Zhang et al. 2010; Beck et al. 2006; Devireddy et al. 2021). Not solely the advancement of MS (Mass Spectrometry) contributed to the growth of the field but notable progress was also seen in the integrative artificial intelligent technologies; for instance, Deep Visual Proteomics was a great breakthrough that enabled a significantly more comprehensive and accurate analysis as the tool granted researchers insights upon the interconnection between protein abundance and the complex or subcellular phenotypes (Mund et al. 2022; Rosenberger et al. 2022). Numerous approaches exist for the characterization of the proteome, which includes both gel-reliant and gel-independent methods (Beck et al. 2006; Devireddy et al. 2021; Smolka et al. 2002). Progress in protein/peptide marking methods, sample preparation procedures, and bioinformatics tools has expedited the extensive analysis of post-translational modifications, protein interactions, and protein positioning within cells (Beck et al. 2006; Devireddy et al. 2021; Smolka et al. 2002).

Proteomics allows researchers to navigate the intricate protein networks underlying health and disease (Zhang et al. 2010; Devireddy et al. 2021). By pinpointing differences in protein expression and interaction patterns between normal and pathological states, proteomics holds promise for uncovering novel diagnostic markers and therapeutic targets (Zhang et al. 2010; Devireddy et al. 2021). The proteome is highly dynamic and continuously remodeled in response to endogenous and exogenous stimuli (Zhang et al. 2010; Devireddy et al. 2021). Hence, proteomics serves as a lens through which we can observe how cells modify their protein ensemble in response to environmental shifts and disturbances in homeostasis (Zhang et al. 2010; Devireddy et al. 2021). In Ghosh and Xu's research conducted in 2014, proteomics has been used to find out stress in plants when given different environmental conditions; the analysis revealed changes in the abundance of proteins involved in various biological processes (Ghosh and Xu 2014; Singh et al. 2022). By leveraging powerful technologies to probe the dynamic protein landscape of the cell, proteomics promises to yield fundamental insights into cellular organization and regulation in both health and disease (Zhang et al. 2010; Beck et al. 2006). The ensuing chart presents an assortment of seminal research endeavors utilizing proteomics within multi-omics explorations across diverse biological taxa, concentrating specifically on *Homo sapiens*' microbiota (Zhang et al. 2010; Beck et al. 2006).

2.5 Metabolomics

Metabolomics represents a subdiscipline of molecular biology, concentrating on the comprehensive examination of metabolites—the diminutive biochemical entities generated during metabolic processes within cells and organisms (Patti et al. 2012; Chong et al. 2018). This field is an integral component of systems biology, an expansive discipline that investigates the interplay among biological systems (Patti et al. 2012; Fang et al. 2019). The metabolomics approach can be either targeted, quantifying a select few specific metabolites, or untargeted, striving to quantify an extensive array of metabolites from a given sample (Patti et al. 2012; Chong et al. 2018; Fang et al. 2019). The utility of metabolomics spans a multitude of sectors, encompassing medical and pharmaceutical research, food studies, and environmental investigations (Patti et al. 2012; Chong et al. 2018; Fang et al. 2019). To provide a better overview of the application of the field, a table has been made available to demonstrate some of the most recent applications of the field in numerous species (Table 2.6).

Table 2.6 The application of metabolomics in various species

Species	Application	References
Plants	Metabolomics in plant research has been used to develop and investigate pesticide impacts on crops as well as to understand plant responses to environmental-related stress or characterize plant extracts, and aid in ethnobotany-driven drug discoveries, highlighting its capability in enhancing the knowledge of plant biology and the utilization in farming and medicine	Shahid et al. (2023), Mashabela et al. (2023), Difonzo et al. (2022), Makunga et al. (2022)
Bacteria and viruses	The application of metabolomics was witnessed in investigating how the infection alters the metabolite profile of mice that have been colonized with a specific bacterium. Furthermore, another application of it was shown in the study of the effects of toxic plants on the microbiota and metabolism of grass-eating animals. Also, a multiplex approach has been developed for the simultaneous identification of multiple respiratory tract viruses and bacteria, including SARS-CoV-2	Spring et al. (2023), Xu et al. (2022), Brinkmann et al. (2022), Mote et al. (2022)
Non-human model organisms	Metabolomics has been exploited to conduct research on the effect of an olive tree extract on the metabolism of a fungal pathogen; its role is further affirmed as it was considered a new era in the diagnosis or prognosis of B-cell non-Hodgkin's lymphoma, highlighting its potential in veterinary medicine. It is also notable that an *N*-acyl glycine modification methodology has been developed for targeted metabolomics analysis and has been applied to study the progress of diabetes in mice, demonstrating its utility in animal and human disease research	Papadopoulou et al. (2023), Xiang et al. (2023), Alfaifi et al. (2023)

In recent years, remarkable transformation with significant strides witnessed and made in both methodological research and practical applications of this field (Wishart et al. 2018; Noecker et al. 2016; Gillet et al. 2012). The deciphering of data generated from metabolomics studies has been refined, a feat accomplished through the advent of bespoke software tools designed specifically for this purpose (Wishart et al. 2018; Noecker et al. 2016; Gillet et al. 2012). However, the pursuit of a holistic software solution capable of performing all requisite steps for the analysis of combined Nuclear Magnetic Resonance (NMR) and Mass Spectrometry (MS) data sets remains a challenge (Wishart et al. 2018; Gillet et al. 2012). This is primarily attributable to the distinct pre-processing tools and algorithms necessitated by each technique (Wishart et al. 2018; Gillet et al. 2012).

In the endeavor to amalgamate NMR and MS data sets, two primary strategies have been conceived (Noecker et al. 2016; Gillet et al. 2012). The first approach entails an autonomous analysis of samples employing each methodology, followed by a comparative evaluation of the resultant data sets to identify recurrent metabolic modifications as detected by each procedure (Noecker et al. 2016; Gillet et al. 2012). The second strategy, in contrast, involves the fusion of each data set into a singular statistical model employing a multiblock analysis, an approach that has demonstrated superior predictive prowess and resolving power compared to either method in isolation (Noecker et al. 2016; Gillet et al. 2012).

Among the recent advancements in metabolomics, the development of dynamic nuclear polarization (DNP) for NMR metabolomics investigations is noteworthy (Jannin et al. 2019; Dey et al. 2020). This innovation holds the potential to significantly augment sensitivity and facilitate the detection of metabolites present in low abundance (Jannin et al. 2019; Dey et al. 2020). Additionally, the field of metabolomics has been effectively utilized to characterize individual reactions to pharmaceutical therapies or other medical procedures, achieved by observing changes in the concentrations of metabolites within biological fluids, including blood and urine (Holmes et al. 2008; Beger et al. 2016).

2.6 Conclusion, Limitations, and Future Directions

In summary, the exploration of multi-omics, which includes genomics, transcriptomics, and proteomics, has granted us an unparalleled comprehension of the intricate and dynamic molecular processes that drive biological events. The amalgamation of these varied yet synergistic disciplines through a cross-disciplinary approach has enabled a comprehensive assessment of biological systems, resulting in pioneering discoveries in the realm of biomedicine.

Nonetheless, this review is not devoid of its constraints. The intricacy and enormity of the data produced by multi-omics investigations present considerable obstacles concerning data governance, scrutiny, and elucidation. Moreover, the substantial expense of multi-omics methodologies and the requirement for specialized proficiency to manage and decipher the data may restrict their broad utilization.

Despite these hurdles, the prospect of multi-omics research is incredibly promising. The ongoing progress in technologies such as deep learning/machine learning and R technology for data visualizations herald transformative findings and fresh perspectives. Moreover, the employment of multi-omics in precision medicine, the identification of disease biomarkers, drug creation, microbiome research, and systems biology can radically alter our comprehension of health and disease, leading to enhanced diagnostic and treatment approaches.

As we progress, it is vital to tackle the current constraints and persist in refining and broadening our multi-omics techniques. This will permit us to probe further into the complexities of biological systems, resulting in transformative innovations that will augment our comprehension of life in health and disease.

Acknowledgments Figures in this chapter were created by PowerPoint—Microsoft 365.

References

Adornetto C, Greco G (2023) A new deep learning and xai-based algorithm for features selection in genomics. arxiv:2303.16914[q-bio.GN,cs.AI,cs.LG]

Ahmed Z (2020) Practicing precision medicine with intelligently integrative clinical and multi-omics data analysis. Hum Genomics 14(1):35
Akintunde O, Tucker T, Carabetta V (2023) The evolution of next-generation sequencing technologies. arXiv pre-print server
Alfaifi A et al (2023) Metabolomics: a new era in the diagnosis or prognosis of B-cell non-Hodgkin's lymphoma. Diagnostics 13(5):861
Alqahtani FM et al (2021) Combining genome-wide gene expression analysis (RNA-seq) and a gene editing platform (CRISPR-Cas9) to uncover the selectively pro-oxidant activity of aurone compounds against *Candida albicans*. Front Microbiol 12:708267
Amarasinghe SL et al (2020) Opportunities and challenges in long-read sequencing data analysis. Genome Biol 21(1):30
Andersson A et al (2020) Single-cell and spatial transcriptomics enables probabilistic inference of cell type topography. Commun Biol 3(1):565
Argelaguet R et al (2018) Multi-omics factor analysis—a framework for unsupervised integration of multi-omics data sets. Mol Syst Biol 14(6):e8124
Aslam B et al (2017) Proteomics: technologies and their applications. J Chromatogr Sci 55(2):182–196
Bae EB et al (2022) Network analysis for estimating standardization trends in genomics using MEDLINE. BMC Med Res Methodol 22(1):263
Baharin A, Ting T-Y, Goh H-H (2023) Omics approaches in uncovering molecular evolution and physiology of botanical carnivory. Plan Theory 12(2):408
Bean L et al (2020) Diagnostic gene sequencing panels: from design to report—a technical standard of the American College of Medical Genetics and Genomics (ACMG). Genet Med 22(3):453–461
Beck HC et al (2006) Quantitative proteomic analysis of post-translational modifications of human histones. Mol Cell Proteomics 5(7):1314–1325
Beger RD et al (2016) Metabolomics enables precision medicine: "A White Paper, Community Perspective". Metabolomics 12(9):149
Benkirane H et al (2022) CustOmics: a versatile deep-learning based strategy for multi-omics integration. arXiv pre-print server
Bhattacharyya R, Henderson N, Baladandayuthapani V (2022) Functional integrative bayesian analysis of high-dimensional multiplatform genomic data. arXiv pre-print server
Bolyen E et al (2019) Reproducible, interactive, scalable and extensible microbiome data science using QIIME 2. Nat Biotechnol 37(8):852–857
Bonnechère B (2022) Integrating rehabilomics into the multi-omics approach in the management of multiple sclerosis: the way for precision medicine? Genes 14(1):63
Bréhélin L (2023) Advancing regulatory genomics with machine learning. arXiv pre-print server
Brinkmann A et al (2022) RespiCoV: Simultaneous identification of Severe Acute Respiratory Syndrome Coronavirus 2 (SARS-CoV-2) and 46 respiratory tract viruses and bacteria by amplicon-based Oxford-Nanopore MinION sequencing. PLoS One 17(3):e0264855
Budamagunta V, Foster TC, Zhou D (2021) Cellular senescence in lymphoid organs and immunosenescence. Aging 13(15):19920–19941
Burgess DJ (2017) Combining CRISPR perturbations and RNA-seq. Nat Rev Genet 18(2):67–67
Camacho DM et al (2018) Next-generation machine learning for biological networks. Cell 173(7):1581–1592
Cano-Gamez E, Trynka G (2020) From GWAS to function: using functional genomics to identify the mechanisms underlying complex diseases. Front Genet 11:424
Caspi R et al (2016) The MetaCyc database of metabolic pathways and enzymes and the BioCyc collection of pathway/genome databases. Nucleic Acids Res 44(D1):D471–D480
Cha O-K, Yang S, Lee H (2022) Transcriptomics using the enriched Arabidopsis shoot apex reveals developmental priming genes involved in plastic plant growth under salt stress conditions. Plan Theory 11(19):2546
Chen A et al (2022) Global spatial transcriptome of macaque brain at single-cell resolution. Cold Spring Harbor Laboratory, Cold Spring Harbor, NY

Chikwambi Z et al (2023) Multi-omics data integration approach identifies potential biomarkers for prostate cancer. Cold Spring Harbor Laboratory, Cold Spring Harbor, NY

Chong J et al (2018) MetaboAnalyst 4.0: towards more transparent and integrative metabolomics analysis. Nucleic Acids Res 46(W1):W486–W494

Christiansen H et al (2021) Facilitating population genomics of non-model organisms through optimized experimental design for reduced representation sequencing. BMC Genomics 22(1):625

Couvillion SP et al (2023) Interrogating the role of the milk microbiome in mastitis in the multi -omics era. Front Microbiol 14:1105675

Davison A, Neiman M (1825) Mobilizing molluscan models and genomes in biology. Philos Trans R Soc B Biol Sci 2021(376):20200163

Deshmukh AS et al (2015) Deep proteomics of mouse skeletal muscle enables quantitation of protein isoforms, metabolic pathways, and transcription factors*. Mol Cell Proteomics 14(4):841–853

Devireddy AR et al (2021) Integration of reactive oxygen species and hormone signaling during abiotic stress. Plant J 105(2):459–476

Dey A et al (2020) Hyperpolarized NMR metabolomics at natural ^{13}C abundance. Anal Chem 92(22):14867–14871

Difonzo G et al (2022) Metabolomics approach to characterize green olive leaf extracts classified based on variety and season. Plan Theory 11(23):3321

Doncheva NT et al (2018) Cytoscape stringApp: network analysis and visualization of proteomics data. Cold Spring Harbor Laboratory, Cold Spring Harbor, NY

Doncheva NT et al (2019) Cytoscape stringApp: network analysis and visualization of proteomics data. J Proteome Res 18(2):623–632

Dong Z, Chen Y (2013) Transcriptomics: advances and approaches. Sci China Life Sci 56(10):960–967

Dong C et al (2021) INFIMA leverages multi-omics model organism data to identify effector genes of human GWAS variants. Genome Biol 22(1):241

Dorman CJ (2013) Genome architecture and global gene regulation in bacteria: making progress towards a unified model? Nat Rev Microbiol 11(5):349–355

Downing T, Angelopoulos N (2023) A primer on correlation-based dimension reduction methods for multi-omics analysis. arXiv pre-print server

ENCODE Project Consortium (2012) An integrated encyclopedia of DNA elements in the human genome. Nature 489(7414):57–74

ENCODE Project Consortium et al (2007) Identification and analysis of functional elements in 1% of the human genome by the ENCODE pilot project. Nature 447(7146):799–816

Everad John A (2022) Editorial: multi-omics studies and applications in precision medicine. Front Genet 13:1034283

Fang C, Fernie AR, Luo J (2019) Exploring the diversity of plant metabolism. Trends Plant Sci 24(1):83–98

Franzosa EA et al (2014) Relating the metatranscriptome and metagenome of the human gut. Proc Natl Acad Sci 111(22):E2329–E2338

Fresca S, Manzoni A (2021) Real-time simulation of parameter-dependent fluid flows through deep learning-based reduced order models. Fluids 6(7):259

Ghosh D, Xu J (2014) Abiotic stress responses in plant roots: a proteomics perspective. Front Plant Sci 5:6

Gillet LC et al (2012) Targeted data extraction of the MS/MS spectra generated by data-independent acquisition: a new concept for consistent and accurate proteome analysis. Mol Cell Proteomics 11(6):O111.016717

Gong H et al (2021) Application of Hi-C and other omics data analysis in human cancer and cell differentiation research. Comput Struct Biotechnol J 19:2070–2083

Gonzales S et al (2023) SOX7: novel autistic gene identified by analysis of multi-omics data. Cold Spring Harbor Laboratory, Cold Spring Harbor, NY

Graw S et al (2021) Multi-omics data integration considerations and study design for biological systems and disease. Mol Omics 17(2):170–185

GTEx Consortium et al (2017) Genetic effects on gene expression across human tissues. Nature 550(7675):204–213
Guo S et al (2023) Editorial: Computational and systematic analysis of multi-omics data f or drug discovery and development. Front Med 10:1146896
Haque A et al (2017) A practical guide to single-cell RNA-sequencing for biomedical research and clinical applications. Genome Med 9(1):75
Hasin Y, Seldin M, Lusis A (2017) Multi-omics approaches to disease. Genome Biol 18:1
Hien LTT et al (2021) Genomics and big data: research, development and applications. Vietnam J Biotechnol 19(3):393–410
Holmes E, Wilson ID, Nicholson JK (2008) Metabolic phenotyping in health and disease. Cell 134(5):714–717
Hu C, Jia W (2021) Multi-omics profiling: the way towards precision medicine in metabolic diseases. J Mol Cell Biol 13(8):mjab051
Huang S, Chaudhary K, Garmire LX (2017) More is better: recent progress in multi-omics data integration methods. Front Genet 8:84
Huang K et al (2021) Machine learning applications for therapeutic tasks with genomics data. arxiv:2105.01171[cs.LG,q-bio.GN,q-bio.QM]
Jain S et al (2022) Stem cells research prospects towards precision medicine. Bratisl Med J 123(11):795–805
Jannin S et al (2019) Application and methodology of dissolution dynamic nuclear polarization in physical, chemical and biological contexts. J Magn Reson 305:41–50
Jasim SA et al (2022) (2022) Shining the light on clinical application of mesenchymal stem cell therapy in autoimmune diseases. Stem Cell Res Ther 13:101. https://doi.org/10.1186/s13287-022-02782-7
Jiang L et al (2023) Applications of genomics and transcriptomics in precision medicine for myopia control or prevention. Biomol Ther 13(3):494
Johnson CH, Ivanisevic J, Siuzdak G (2016) Metabolomics: beyond biomarkers and towards mechanisms. Nat Rev Mol Cell Biol 17(7):451–459
Joshi S, Chinnusamy V, Joshi R (2022) Root system architecture and omics approaches for belowground abiotic stress tolerance in plants. Agriculture 12(10):1677
Kale A et al (2023) MetaExplorer: facilitating reasoning with epistemic uncertainty in meta-analysis. ACM
Kerdidani D et al (2022) Rebooting regulatory T cell and dendritic cell function in immune-mediated inflammatory diseases: biomarker and therapy discovery under a multi-omics lens. Biomedicines 10(9):2140
Kirk PDW, Pagani F, Richardson S (2023) Bayesian outcome-guided multi-view mixture models with applications in molecular precision medicine. arXiv pre-print server
Kloehn J et al (2020) Multi-omics analysis delineates the distinct functions of sub-cellular acetyl-CoA pools in Toxoplasma gondii. BMC Biol 18(1):67
Kwok MCM et al (2008) Proteomics of photoreceptor outer segments identifies a subset of SNARE and Rab proteins implicated in membrane vesicle trafficking and fusion. Mol Cell Proteomics 7(6):1053–1066
Li Y, Luo Y (2022) Metabolomics of aging and Alzheimer's disease: from single-omics to multi-omics. arXiv pre-print server
Liang X et al (2022) Gut microbiome, cognitive function and brain structure: a multi-omics integration analysis. Transl Neurodegen 11(1):49
Lin X et al (2022) Clustering of single-cell multi-omics data with a multimodal deep learning method. Nat Commun 13(1)
Liu X-Y, Mei X-Y (2023) Prediction of drug sensitivity based on multi-omics data using deep learning and similarity network fusion approaches. Front Bioeng Biotechnol 11:1156372
Lorenzon R et al (2018) Clinical and multi-omics cross-phenotyping of patients with autoimmune and autoinflammatory diseases: the observational TRANSIMMUNOM protocol. BMJ Open 8(8):e021037
Lowe R et al (2017) Transcriptomics technologies. PLOS Comput Biol 13(5):e1005457

Maheendran R et al (2022) Transcriptomics and its role in crop improvement. p. 179–194
Maity AK et al (2020) Bayesian structural equation modeling in multiple omics data integration with application to circadian genes. arxiv:2112.03330[stat.ME,stat.AP]
Makunga N, Kai G, Sieniawska E (2022) Editorial: metabolomics as a tool in ethnobotany-driven drug discoveries. Front Pharmacol 13:1066875
Mantere T, Kersten S, Hoischen A (2019) Long-read sequencing emerging in medical genetics. Front Genet 10:426
Mardis ER (2008) Next-generation DNA sequencing methods. Annu Rev Genomics Hum Genet 9(1):387–402
Marx V (2021) Method of the year: spatially resolved transcriptomics. Nat Methods 18(1):9–14
Marx V (2023) Method of the year: long-read sequencing. Nat Methods 20(1):6–11
Mashabela MD, Masamba P, Kappo AP (2023) Applications of metabolomics for the elucidation of abiotic stress tolerance in plants: a special focus on osmotic stress and heavy metal toxicity. Plan Theory 12(2):269
Mo Q et al (2018) A fully Bayesian latent variable model for integrative clustering analysis of multi-type omics data. Biostatistics 19(1):71–86
Moco S (2022) Studying metabolism by NMR-based metabolomics. Front Mol Biosci 9:882487
Mote RS et al (2022) Integrative interactomics applied to bovine fescue toxicosis. Sci Rep 12(1):4899
Mund A et al (2022) Deep visual proteomics defines single-cell identity and heterogeneity. Nat Biotechnol 40(8):1231–1240
Noecker C et al (2016) Metabolic model-based integration of microbiome taxonomic and metabolomic profiles elucidates mechanistic links between ecological and metabolic variation. mSystems 1(1)
Olsen TK, Baryawno N (2018) Introduction to single-cell RNA sequencing. Curr Protoc Mol Biol 122(1):e57
Onley I, Austin J, Mitchell K (2020) Sex determination of non-model organisms in the absence of field records using Diversity Arrays Technology (DArT) data. Authorea, Inc.
Ota M, Fujio K (2021) Multi-omics approach to precision medicine for immune-mediated diseases. Inflamm Regen 41(1):23
Panguluri SK et al (2010) Genomic profiling of messenger RNAs and microRNAs reveals potential mechanisms of TWEAK-induced skeletal muscle wasting in mice. PLoS One 5(1):e8760
Papadopoulou E-A et al (2023) GC/EI/MS and ^{1}H NMR metabolomics reveal the effect of an olive tree endophytic *Bacillus* sp. lipopeptide extract on the metabolism of *Colletotrichum acutatum*. Metabolites 13(4):462
Papatheodorou I et al (2017) Expression atlas: gene and protein expression across multiple studies and organisms. Nucleic Acids Res 46(D1):D246–D251
Paqué PN et al (2022) Salivary biomarkers as key to monitor personalized oral healthcare and precision dentistry: a scoping review. Front Oral Health 3:1003679
Park M-K et al (2022) Deep-learning algorithm and concomitant biomarker identification for NSCLC prediction using multi-omics data integration. Biomol Ther 12(12):1839
Patti GJ, Yanes O, Siuzdak G (2012) Metabolomics: the apogee of the omics trilogy. Nat Rev Mol Cell Biol 13(4):263–269
Pinu FR et al (2019) Systems biology and multi-omics integration: viewpoints from the metabolomics research community. Meta 9(4):76
Pretorius CJ et al (2021) Metabolomics for biomarker discovery: key signatory metabolic profiles for the identification and discrimination of oat cultivars. Meta 11(3):165
Rao A et al (2021) Exploring tissue architecture using spatial transcriptomics. Nature 596(7871):211–220
Raufaste-Cazavieille V, Santiago R, Droit A (2022) Multi-omics analysis: paving the path toward achieving precision medicine in cancer treatment and immuno-oncology. Front Mol Biosci 9:962743
Reiter NJ et al (2010) Structure of a bacterial ribonuclease P holoenzyme in complex with tRNA. Nature 468(7325):784–789

Rosario D et al (2020) Systems biology approaches to understand the host-microbiome interactions in neurodegenerative diseases. Front Neurosci 14:716

Rosenberger FA et al (2022) Spatial single-cell mass spectrometry defines zonation of the hepatocyte proteome. Cold Spring Harbor Laboratory, Cold Spring Harbor, NY

Roy KR et al (2018) Multiplexed precision genome editing with trackable genomic barcodes in yeast. Nat Biotechnol 36(6):512–520

Saeed AI et al (2003) TM4: a free, open-source system for microarray data management and analysis. BioTechniques 34(2):374–378

Sahu PK et al (2020) Next generation sequencing based forward genetic approaches for identification and mapping of causal mutations in crop plants: a comprehensive review. Plan Theory 9(10):1355

Schoof EM et al (2021) Quantitative single-cell proteomics as a tool to characterize cellular hierarchies. Nat Commun 12(1):3341

Schwartz SH et al (2020) Carbon dots for efficient small interfering RNA delivery and gene silencing in plants. Plant Physiol 184(2):647–657

Selvakumar SC et al (2022) CRISPR/Cas9 and next generation sequencing in the personalized treatment of cancer. Mol Cancer 21(1)

Shahid M, Singh UB, Khan MS (2023) Metabolomics-based mechanistic insights into revealing the adverse effects of pesticides on plants: an interactive review. Metabolites 13(2):246

Singh PK et al (2022) Genomic and proteomic responses to drought stress and biotechnological interventions for enhanced drought tolerance in plants. Curr Plant Biol 29:100239

Smolka M, Zhou H, Aebersold R (2002) Quantitative protein profiling using two-dimensional gel electrophoresis, isotope-coded affinity tag labeling, and mass spectrometry. Mol Cell Proteomics 1(1):19–29

Song L et al (2011) Open chromatin defined by DNaseI and FAIRE identifies regulatory elements that shape cell-type identity. Genome Res 21(10):1757–1767

Spring J et al (2023) Retroviral infection and commensal bacteria dependently alter the metabolomic profile in a sterile organ. Viruses 15(2):386

Subramanian I et al (2020) Multi-omics data integration, interpretation, and its application. Bioinform Biol Insights 14:117793221989905

Tabakhi S et al (2023) Multimodal learning for multi-omics: a survey. World Sci Annu Rev Artif Intell. https://doi.org/10.1142/S2811032322500047

Thurman RE et al (2012) The accessible chromatin landscape of the human genome. Nature 489(7414):75–82

Tzoulaki I et al (2014) Design and analysis of metabolomics studies in epidemiologic research: a primer on -omic technologies. Am J Epidemiol 180(2):129–139

Valenti F et al (2021) Precision medicine and melanoma: multi-omics approaches to monitoring the immunotherapy response. Int J Mol Sci 22(8):3837

Vaske CJ et al (2010) Inference of patient-specific pathway activities from multi-dimensional cancer genomics data using PARADIGM. Bioinformatics 26(12):i237–i245

Vieth B et al (2019) A systematic evaluation of single cell RNA-seq analysis pipelines. Nat Commun 10(1):4667

Wang Z, Gerstein M, Snyder M (2009) RNA-Seq: a revolutionary tool for transcriptomics. Nat Rev Genet 10(1):57–63

Waqas A et al (2023) Multimodal data integration for oncology in the era of deep neural networks: a review. arXiv pre-print server

Williams CG et al (2022) An introduction to spatial transcriptomics for biomedical research. Genome Med 14(1):68

Wishart DS et al (2018) HMDB 4.0: the human metabolome database for 2018. Nucleic Acids Res 46(D1):D608–D617

Würtz P et al (2017) Quantitative serum nuclear magnetic resonance metabolomics in large-scale epidemiology: a primer on -omic technologies. Am J Epidemiol 186(9):1084–1096

Xia J et al (2015) MetaboAnalyst 3.0—making metabolomics more meaningful. Nucleic Acids Res 43(W1):W251–W257

Xiang L et al (2023) Derivatization of N-acyl glycines by 3-nitrophenylhydrazine for targeted metabolomics analysis and their application to the study of diabetes progression in mice. Anal Chem 95(4):2183–2191

Xu C et al (2019) Probabilistic harmonization and annotation of single-cell transcriptomics data with deep generative models. Cold Spring Harbor Laboratory, Cold Spring Harbor, NY

Xu B et al (2022) A holistic genome dataset of bacteria, archaea and viruses of the Pearl River estuary. Sci Data 9(1):49

Yang Y et al (2021) Applications of multi-omics technologies for crop improvement. Front Plant Sci 12:563953

Yue L et al (2023) A guidebook of spatial transcriptomic technologies, data resources and analysis approaches. Comput Struct Biotechnol J 21:940–955

Zhang W, Li F, Nie L (2010) Integrating multiple 'omics' analysis for microbial biology: application and methodologies. Microbiology 156(2):287–301

Zhang L et al (2022) Clinical and translational values of spatial transcriptomics. Signal Transduct Target Ther 7(1)

Zhao Z et al (2023) Tutorial on survival modelling with omics data. arXiv pre-print server

Zhou W et al (2019) Longitudinal multi-omics of host–microbe dynamics in prediabetes. Nature 569(7758):663–671

Zhou Y et al (2021) AlzGPS: a genome-wide positioning systems platform to catalyze multi-omics for Alzheimer's drug discovery. Alzheimers Res Ther 13(1)

Next Generation Sequencing in Metagenomics and Metatranscriptomics

3

Ajay Kumar and Anuj Yadav

3.1 Introduction

Next-generation sequencing (NGS) represents a revolutionary technology used to decode the DNA or RNA sequences (refer to Table 3.1), serving as a crucial instrument in exploring genetic variations associated with diseases and various biological processes.

Its debut in the commercial arena dates back to 2005, initially dubbed "massively parallel sequencing" due to its capacity to concurrently sequence numerous DNA strands (Ansorge 2009), in stark contrast to the traditional Sanger sequencing, which sequences one strand at a time via capillary electrophoresis (CE). Both of these methodologies possess their unique merits in the contemporary genetic analysis landscape. Sanger sequencing excels in examining a limited number of gene targets and samples, achieving results within a day. Furthermore, it stands as the pinnacle in sequencing technology, often utilized to validate NGS findings. In contrast, NGS empowers the simultaneous exploration of hundreds to thousands of genes across multiple samples. It enables the in-depth analysis of diverse genomic traits in a single sequencing operation, spanning single nucleotide variants (SNVs), copy number and structural alterations, and even RNA fusions. NGS offers exceptional throughput per run, facilitating swift and cost-effective research endeavors. Additionally, it presents advantages such as reduced sample input requirements, heightened precision, and the ability to detect variants at lower allele frequencies when compared to Sanger sequencing. The rapidity, capacity, and accuracy of NGS have triggered a transformative revolution in genetic analysis, paving the way for novel discoveries in genomic and clinical research, reproductive health, and applications in environmental, agricultural, and forensic sciences (Auger et al. 2014).

A. Kumar (✉) · A. Yadav
Department of Biotechnology, Faculty of Engineering and Technology, Rama University, Kanpur, Uttar Pradesh, India

I. Mani, V. Singh (eds.), *Multi-Omics Analysis of the Human Microbiome*,
https://doi.org/10.1007/978-981-97-1844-3_3

Table 3.1 Comparison between DNA and RNA sequencing

S. No	DNA sequencing	RNA sequencing
1	This includes high-throughput DNA sequencing	This is accomplished by RNA high-throughput sequencing
2	Deep sequencing of DNA	RNA sequencing in depth
3	The simultaneous sequencing of multiple DNA strands	The gene expression and identify various RNA strands, simultaneously
4	This encompasses whole exome sequencing, targeted sequencing, and whole genome sequencing methods	Whole transcriptome sequencing and mRNA sequencing examine all messenger RNA in a sample, revealing gene expression patterns and RNA processes
5	Massive parallel sequencing examines both DNA strands, while exome sequencing specifically targets the mRNA coding region, representing only 1–3% of the entire DNA	The collected RNA is initially reverse transcribed into cDNA and subsequently amplified before analysis
6	Targeted sequencing focuses on specific subsets of genes to analyze particular regions of interest within the DNA	cDNA is produced when reverse transcriptase acts on mRNA
7	When the required DNA quantity is limited, PCR products are utilized to quantify the DNA accurately	This method involves expressing the desired protein by utilizing its corresponding cDNA

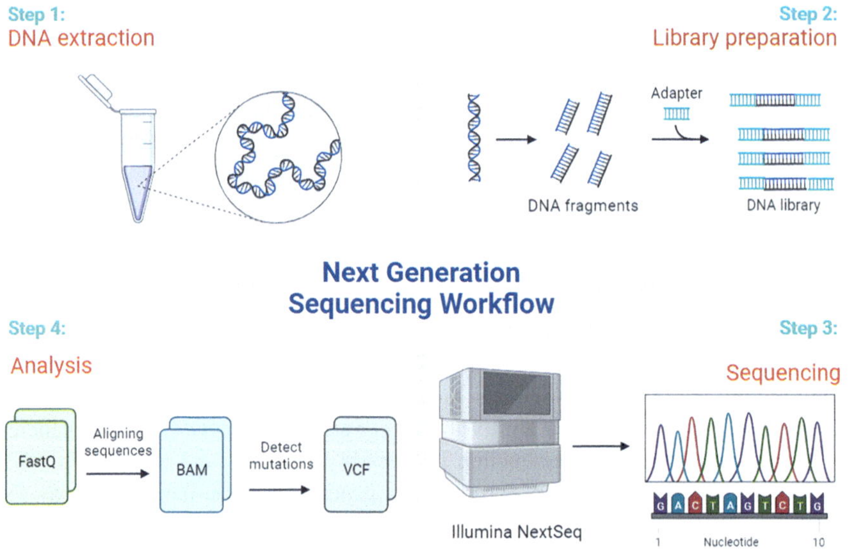

Fig. 3.1 NGS workflow steps

3.2 NGS Technologies

Regardless of the instrument technology employed, a standard NGS experiment follows consistent steps, as depicted in Fig. 3.1.

3.3 Construct Library

Initiating the sequencing process begins with the creation of what's termed a sequencing "library" derived from the sample under investigation. In the initial phase, the DNA (or cDNA) sample undergoes a transformation into concise double-stranded fragments, typically ranging from 100 to 800 base pairs in length. The approach to DNA fragmentation varies based on specific research goals, employing methods like physical shearing, enzymatic digestion, or PCR-based amplification techniques targeting specific genetic regions. Following this fragmentation process, the resulting DNA fragments are paired with technology-specific adaptor sequences, leading to the creation of a unique fragment library. It is noteworthy that some of these adaptors might also incorporate unique molecular "barcodes" designed to individually tag each sample (Zhao et al. 2021). This barcode system facilitates the concurrent sequencing of multiple samples, enabling researchers to amalgamate them and perform analysis within a single sequencing run. To illustrate, barcodes numbered 1 through 20 could be employed to uniquely label 20 distinct samples, which can then be simultaneously analyzed in a consolidated sequencing process. This methodology, commonly referred to as "pooling" or "multiplexing," presents several advantages, including efficiency gains in terms of time and resources during sequencing experiments, as well as the effective standardization of workflow procedures, given that grouped samples are handled collectively.

Apart from the previously mentioned fragment libraries, it is crucial to highlight two other specialized library preparation techniques: paired-end libraries and mate-pair libraries. Paired-end libraries add a novel aspect to sequencing by enabling the sequencing of DNA fragments from both ends, diverging from the conventional single-direction sequencing approach. The creation of paired-end libraries closely mirrors that of standard fragment libraries, with a key difference lying in the presence of adaptor tags at both ends of the DNA insert. These tags enable sequencing from both the forward and reverse directions. Such an approach not only streamlines the alignment of sequence reads but also enhances the ability to detect genomic rearrangements, repetitive sequence elements, and various structural features, including RNA gene fusions and splice variants. Nevertheless, it is crucial to acknowledge that advancements in contemporary library preparation methods and analytical tools have considerably broadened the scope of single-direction sequencing, making it capable of discerning these genomic attributes as well.

Mate-pair libraries require a more complex preparation process compared to fragment or paired-end libraries. These libraries entail the incorporation of significantly larger-sized DNA inserts, frequently surpassing 2 kilobases (kb) and potentially extending up to 30 kb in size. Sequencing mate-pair libraries produces a pair of reads that are distantly located from each other and display opposite orientations. This distinctive characteristic offers invaluable insights into tasks such as de novo genome assembly, the identification of extensive structural variations, and the detection of complex genomic rearrangements. Mate-pair sequencing leverages the spatial information between these two sequencing reads, making it an indispensable tool in genomic research, particularly in scenarios where comprehensive structural insights are essential (Sandhu et al. 2019).

3.4 Clonal Amplification

To prepare for sequencing, it is vital to affix the DNA library to a stable substrate and undergo clonal amplification. This step amplifies each target, boosting the detectable signal during the sequencing process (Fujii et al. 2020). During this amplification phase, every unique DNA molecule in the library becomes attached to the surface of a bead or a flow cell. Subsequently, it undergoes replication through PCR, creating a group of identical clones. When utilizing Ion Torrent technology, a specific procedure known as "templating" comes into play to introduce library molecules onto the beads, facilitating the subsequent sequencing steps.

3.5 Sequence Library

The entirety of the DNA within the library undergoes sequencing simultaneously by means of a specialized sequencing instrument. While each Next-Generation Sequencing (NGS) technology possesses its distinct attributes, they all share a common methodology known as "sequencing by synthesis." In this approach, individual bases are read as they elongate along a polymerized DNA strand. This procedure unfolds in a cyclical fashion, marked by systematic steps: the creation of DNA bases on a single-stranded DNA template, the identification of the integrated base, and the subsequent removal of reaction components to commence the cycle again. The majority of sequencing instruments utilize optical detection methods to confirm the incorporation of nucleotides during DNA synthesis (Oniciuc et al. 2018). However, Ion Torrent instruments deviate from this convention by employing electrical detection. These instruments detect the release of hydrogen ions, a natural phenomenon that happens when nucleotides are incorporated into the growing DNA strand during synthesis. This distinctive electrical detection mechanism sets Ion Torrent technology apart from other sequencing approaches and contributes to its efficiency in deciphering DNA sequences.

3.6 Analyze Data

Every NGS experiment produces substantial volumes of intricate data comprising short DNA reads. While different technology platforms have their unique algorithms and data analysis tools, they follow a similar analysis "pipeline" and employ common metrics to assess the quality of NGS datasets. The analysis process can be categorized into three stages: primary, secondary, and tertiary analysis (Aguiar-Pulido et al. 2016). Primary analysis encompasses the conversion of raw signals from instrument detectors into digitized data or base calls, collected throughout each sequencing cycle. The outcome of primary analysis includes files containing base calls organized into sequencing reads (FASTQ files) along with their corresponding quality scores (Phred quality score). Secondary analysis consists of read filtering and trimming based on quality, followed by the alignment of reads to a reference genome or the assembly of reads for novel genomes. This phase concludes with variant calling, producing a BAM file containing aligned reads. Tertiary analysis, the most intricate stage, involves interpreting results and extracting meaningful insights from the data. Given the intricacy of NGS data and associated algorithms, NGS analysis is typically conducted by specialized bioinformatics experts (Araujo et al. 2018). To assist individuals without specialized bioinformatics training, platforms such as Ion Torrent have developed user-friendly, intuitive software. This software simplifies the analysis process and does not necessitate programming skills to yield results.

3.7 Advantages of Next-Generation Sequencing

NGS has transformed the realms of metagenomics and metatranscriptomics by facilitating high-throughput sequencing of DNA and RNA extracted from intricate microbial communities. NGS technologies, such as Illumina sequencing, allow for the parallel sequencing of millions of DNA or RNA fragments in a single run (Nakazato et al. 2013). This method empowers researchers to acquire thorough profiles of microbial diversity and functional potential within a given sample.

3.8 Metagenomics

Metagenomics represents a branch of molecular biology dedicated to studying genetic material directly extracted from environmental samples like soil, water, air, and the human body, without the necessity of culturing or isolating individual organisms. This approach enables scientists to explore the combined genetic content of all microorganisms within a specific sample, offering valuable insights into the diversity, composition, and functional capabilities of microbial communities. The metagenomics process involves several steps, as illustrated in Fig. 3.2.

Samples from diverse environments, including soil, oceans, intestines, and extreme habitats like hot springs or deep-sea hydrothermal vents, are collected. Genetic material, predominantly DNA, is extracted from these samples, originating from a broad spectrum of microorganisms such as bacteria, archaea, viruses, and fungi. The extracted DNA is subsequently sequenced utilizing high-throughput sequencing technologies. This procedure yields extensive genetic data, representing a mixture of DNA from all the microorganisms found in the sample.

The sequenced data is analyzed using bioinformatics tools and techniques. This involves several steps, including sequence assembly (putting together short DNA fragments), annotation (identifying genes and their functions), and comparing sequences to databases to identify known and novel genetic elements (Ayling et al. 2020). Metagenomics allows researchers to infer the potential functions of the microbial community by identifying genes associated with specific biological processes or metabolic pathways. This information helps us understand the roles microorganisms play in their environments. Through the analysis of specific genetic

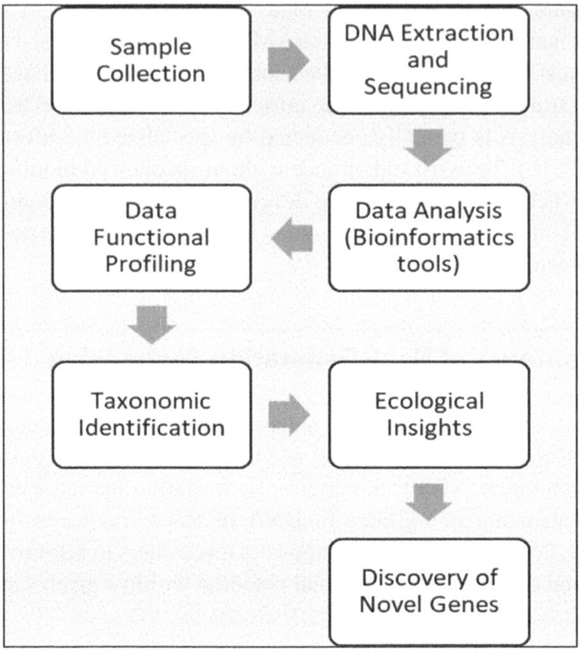

Fig. 3.2 Schematic workflows of metagenomics next generation sequencing

markers, such as the 16S rRNA gene for bacteria, researchers can discern the taxonomic composition of the microbial community. This approach reveals the types of microorganisms present in the sample.

Metagenomics provides insights into the interactions, relationships, and dynamics of microbial communities within their ecosystems. This can include understanding how microorganisms respond to changes in their environment or how they contribute to nutrient cycling and ecosystem stability. Metagenomics has the potential to uncover novel genes, enzymes, and metabolic pathways that have not been previously identified in cultured microorganisms. These discoveries can have applications in various fields, including biotechnology, medicine, and environmental management. Overall, metagenomics has revolutionized our understanding of microbial diversity and function, enabling us to explore the previously hidden world of microorganisms and their roles in shaping our planet's ecosystems and influencing human health (Bhargava et al. 2019b).

3.9 Sample Collection and DNA Extraction

Collecting samples and extracting DNA are pivotal stages in metagenomics, serving as the cornerstone for exploring the genetic diversity and functional capabilities of microbial communities across different environments (Bhargava et al. 2019a). Here's a breakdown of these steps:

Sample Collection: Choose the appropriate sampling site that represents the environment of interest. This could be a soil sample, water sample, gut microbiome sample, etc. Use aseptic techniques to minimize contamination from external sources. Sterilized equipment should be used to collect samples. Collect samples from different locations within the environment to capture the overall diversity of microorganisms present. Samples should be stored and transported properly to maintain the integrity of the genetic material. For instance, samples might be frozen or preserved in suitable solutions to prevent degradation.

DNA Extraction: Once the samples are collected, the next step is to extract the genetic material (DNA) from the collected microbial cells. The goal is to obtain a high-quality DNA sample that represents the genetic diversity of the microbial community. The selection of a DNA extraction method is contingent upon the sample type (such as soil, water, or feces) and the anticipated microbial content (including bacteria, archaea, or viruses). Common DNA extraction methods involve breaking open the microbial cells to release their genetic material and then purifying the DNA. These methods might involve the use of chemical and physical methods to lyse cells and separate DNA from other cellular components. Some DNA extraction methods might incorporate bead beating, enzymatic digestion, or mechanical disruption to break cell walls and membranes. It is important to note that different microorganisms might have varying cell wall structures or protective layers that require specific extraction protocols.

Quality Control: Throughout the DNA extraction process, quality control steps are essential to ensure the extracted DNA is free from contaminants and is of sufficient quality for downstream analysis. Measuring DNA concentration and purity using spectrophotometry or fluorometry helps assess the amount of DNA recovered and its quality. Checking the DNA fragment size can ensure that the DNA is intact and not significantly degraded.

Storage: Extracted DNA should be stored appropriately to prevent degradation. DNA is typically stored at minimal temperatures (e.g., −20 °C or −80 °C) to maintain its stability. Successful sample collection and DNA extraction are crucial for obtaining accurate and representative genetic information from microbial communities. Adopting appropriate techniques and meticulous handling during these stages is crucial to guarantee that subsequent steps, including DNA sequencing and bioinformatics analysis, yield meaningful insights into the microbial world within the collected samples (Chiu and Miller 2019).

3.10 Library Preparation for Metagenomic Sequencing

Library preparation is a critical step in metagenomic sequencing that involves converting the extracted DNA from a sample into a format suitable for high-throughput sequencing. The library preparation process includes several key steps:

DNA Fragmentation: The extracted DNA is often too long for sequencing. Therefore, it needs to be fragmented into smaller pieces. This can be achieved using physical methods (such as sonication) or enzymatic methods (such as restriction enzyme digestion).

End Repair and A-Tailing: The fragmented DNA ends are often uneven and may have overhangs or gaps. End repair enzymes are used to blunt-end the DNA fragments and remove damaged ends. Following this, an "A" base is affixed to the 3′ ends of the DNA fragments, a procedure referred to as A-tailing. This prepares the fragments for the next step of adapter ligation.

Adapter Ligation: Sequencing adapters are short DNA molecules with sequences that match the sequencing platforms. These adapters contain sequences necessary for attachment to the sequencing instrument and for priming the sequencing reactions. Adapters are ligated to the A-tailed DNA fragments. They provide the priming sites for sequencing and help identify the samples during data analysis.

Library Amplification (PCR): To create a sufficient amount of DNA for sequencing, the library fragments with attached adapters are amplified using PCR (polymerase chain reaction). During this step, primers specific to the adapter sequences are used to amplify the DNA fragments. It is crucial to optimize the number of PCR cycles to prevent over-amplification, which might lead to bias or errors in the results.

Library Size Selection: The amplified library contains fragments of varying lengths. To ensure optimal sequencing and avoid issues related to fragment size, a size selection step might be performed. This can involve gel electrophoresis, bead-based size selection, or other methods to isolate fragments of the desired size range.

Quality Control (QC): Throughout the library preparation process, quality control steps are important to ensure that the library is of high quality and ready for sequencing. QC can include assessing the fragment size distribution using gel electrophoresis or capillary electrophoresis and quantifying the library using methods like qPCR or fluorometry.

Pooling and Sequencing: Once the libraries are prepared and quality-checked, they can be pooled together with other libraries for sequencing. Pooling allows multiple samples to be sequenced simultaneously, which is efficient and cost-effective. The pooled libraries are then loaded onto a high-throughput sequencing platform (e.g., Illumina, PacBio, Oxford Nanopore) for sequencing.

Library preparation stands as a pivotal step in metagenomics, given that the quality of the library can profoundly influence the sequencing data and the subsequent bioinformatics analysis. Proper library preparation ensures that the genetic information from the original microbial community is faithfully represented in the sequenced data (Gaulke et al. 2021).

3.11 NGS Platforms for Metagenomics

Numerous Next-Generation Sequencing (NGS) platforms find common use in metagenomic sequencing owing to their high-throughput capabilities. These platforms enable the efficient and cost-effective sequencing of intricate microbial communities (Wang et al. 2019). Here are some of the prominent NGS platforms used in metagenomics; Illumina platforms, such as the MiSeq, HiSeq, and NovaSeq, are widely used for metagenomic studies. These platforms employ a sequencing-by-synthesis technique, where fluorescently labeled nucleotides are incorporated into growing DNA strands. The emitted fluorescence is then detected to deduce the sequence. Illumina platforms offer high accuracy and depth of sequencing, making them suitable for detecting rare microbial species and assessing microbial diversity. Short read lengths (typically up to 300 bases) are a limitation for assembling longer DNA fragments, but this can be overcome through bioinformatics techniques (Zerbino 2010).

PacBio's unique Single Molecule, Real-Time (SMRT) sequencing technology offers longer reads than Illumina, making it exceptionally valuable for deciphering complex metagenomes and enhancing our understanding of diverse microbial communities. Long reads are beneficial for resolving complex microbial populations, identifying novel species, and reconstructing complete or near-complete genomes from metagenomic data. However, PacBio sequencing can be more error-prone than Illumina sequencing. Oxford Nanopore Technologies' (ONT) MinION and PromethION sequencers use nanopore-based sequencing technology, where DNA strands pass through nanopores, and the resulting electrical signals are used to determine the sequence. Nanopore sequencing offers long reads, sometimes spanning tens of thousands of bases, which is advantageous for metagenomics to resolve complex genomes and identify structural variants. While ONT reads have higher error rates compared to Illumina, improvements in base-calling algorithms have been made over time.

Ion Torrent sequencing, developed by Thermo Fisher Scientific, detects changes in pH caused by the release of protons during DNA synthesis. It offers relatively short reads and is suitable for smaller-scale metagenomic studies. While Ion Torrent platforms provide rapid turnaround times and cost-effectiveness, they may have limitations in accurately resolving complex microbial communities. Each NGS platform has its own strengths and limitations. The selection of a sequencing platform relies on research goals, the intricacy of the microbial community, financial limitations, and the computational resources at hand. In many cases, researchers use a combination of platforms to leverage the advantages of each for a comprehensive analysis of metagenomic samples. Additionally, advances in technology continue to improve the accuracy, read lengths, and throughput of these platforms, further enhancing their utility in metagenomics.

3.12 Data Analysis in Metagenomics

Certainly, data analysis in metagenomics involves several key steps to make sense of the raw sequencing data and extract meaningful biological insights from complex microbial communities (Thomas et al. 2012). Here is an overview of the data analysis process, including quality control, read alignment, taxonomic classification, functional annotation, and pathway analysis; Quality control encompasses evaluating the quality of raw sequencing data, ensuring precision for subsequent analysis.

This includes checking for issues like base-calling errors, adapter contamination, and low-quality reads. Quality control tools (e.g., FastQC) generate quality reports and identify problematic regions in the data. Read trimming involves removing low-quality bases and adapter sequences to improve the accuracy of downstream analyses. Tools like Trimmomatic or Cutadapt are commonly used for read trimming. Read alignment involves mapping the trimmed reads to reference databases to identify the organisms present in the metagenomic sample. Common reference databases for taxonomic classification include the NCBI nt/nr databases or specialized databases like SILVA (for 16S rRNA sequences) or Green Genes. Alignment tools such as Bowtie2, BWA, or DIAMOND are used to align reads to the reference databases. Taxonomic classification algorithms assign reads to different taxonomic levels (kingdom, phylum, genus, etc.) based on sequence similarity (Sayers et al. 2021).

Functional annotation involves identifying the functions and biological processes associated with the genes present in the metagenomic sample. Tools like HMMER or InterProScan are used to search for protein domains and functional motifs within the predicted genes. Pathway analysis aims to identify metabolic pathways and biological functions that are enriched in the metagenomic sample. This can involve mapping the predicted genes to pathway databases like KEGG or MetaCyc and using statistical tests to identify significantly enriched pathways. Metagenomic data analysis is a multifaceted process, demanding the integration of bioinformatics tools, databases, and computational resources. It is essential to acknowledge the constraints and biases inherent in each step, employing suitable statistical methods to derive meaningful insights from the data. Additionally, the field of metagenomics is continuously evolving, and new tools and techniques are being developed to enhance the accuracy and depth of analysis (Huang et al. 2020).

3.13 Applications of Metagenomics in Environmental and Human Microbiome Studies

Metagenomics has found numerous applications in both environmental and human microbiome studies, providing insights into microbial diversity, function, interactions, and their impact on ecosystems and human health (see Fig. 3.3).

Here are some key applications in each field:

Environmental Microbiome Studies: Metagenomics allows researchers to comprehensively assess the microbial diversity within various ecosystems, such as soil, oceans, freshwater bodies, and extreme environments. This approach aids in comprehending the role of microorganisms in maintaining ecosystem stability, facilitating nutrient cycling, and contributing to overall ecosystem health (Oulas et al. 2015). Metagenomics can identify microorganisms and genes that have the potential to degrade pollutants and contaminants in the environment (Michealsamy et al. 2021). This information can be harnessed for bioremediation strategies to clean up polluted sites (Driscoll et al. 2017).

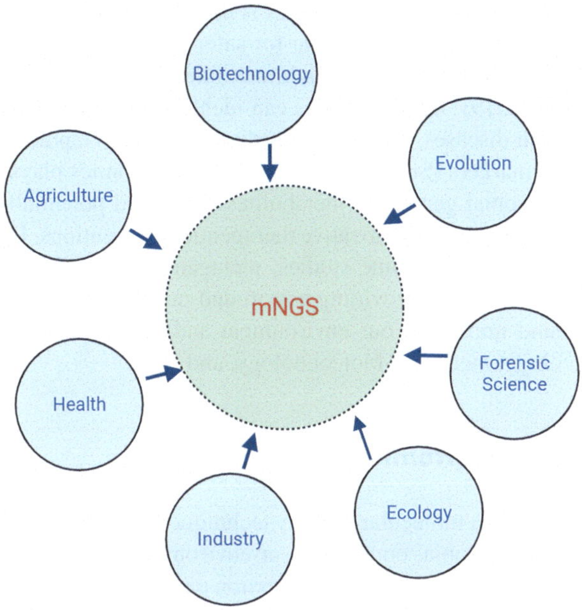

Fig. 3.3 Application of metagenomic next generation sequencing (mNGS) technology in different fields

Metagenomics, by delving into the functional genes within microbial communities, offers valuable insights into the biogeochemical cycling of elements such as carbon, nitrogen, and sulfur. This knowledge enhances our understanding of how microbes contribute to nutrient transformations and global elemental cycles. Metagenomics has the potential to discover previously unknown microbial species and genes with unique functions that could be of biotechnological interest. Metagenomic studies can reveal how microbial communities respond to environmental disturbances, such as climate change, pollution, and habitat alteration (Gökdemir et al. 2022). This information contributes to predicting and managing ecosystem responses.

Human Microbiome Studies: Metagenomics has significantly advanced our comprehension of the gut microbiome's impact on human health. It has revealed associations between microbiome composition and various health conditions, including obesity, inflammatory bowel disease, diabetes, and mental health disorders (Diao et al. 2022). Metagenomics can help identify how the gut microbiome interacts with drugs, affecting drug metabolism, efficacy, and side effects. This has implications for personalized medicine and drug development (Ramachandran and Wilson 2020). Analyzing oral and skin microbiomes through metagenomics plays a pivotal role in unraveling the significance of these microbial communities in oral health, various skin conditions, and diseases (Godlewska et al. 2020). Studying the human microbiome reveals how microbial communities interact with the host's immune system, affecting both immune responses and disease susceptibility. Metagenomics is used to characterize microbial communities in fecal transplants, allowing for safer and more targeted fecal microbiota transplantation for treating conditions like *Clostridium difficile* infection (Takeuchi et al. 2019). Metagenomics can identify microbial signatures associated with certain diseases, potentially leading to the development of diagnostic and prognostic markers (Suzuki et al. 2017). Metagenomics plays a vital role in pinpointing microbial genes and metabolites that could potentially become targets for the development of innovative therapeutic interventions. In both environmental and human microbiome studies, metagenomics has paved the way for fresh avenues of research, providing a profound insight into the microbial world and its profound impact on our environment and health. It continues to shape fields like ecology, medicine, biotechnology, and more.

3.14 Metatranscriptomics

Metatranscriptomics is a molecular biology technique focused on studying all RNA molecules within a microbial community or environment at a specific moment in time. It provides insights into the gene expression patterns of the microorganisms in the community, revealing which genes are actively transcribed and their corresponding functions. Metatranscriptomics extends our understanding beyond the potential functions of genes (provided by metagenomics) to the actual activities of the microbial community. The process of metatranscriptomics includes the following steps (Yang and Kim 2015) (see Fig. 3.4).

RNA Extraction: Similar to metagenomics, the first step in metatranscriptomics is to extract RNA from the microbial community of interest. RNA molecules encompass messenger RNA (mRNA), ribosomal RNA (rRNA), transfer RNA (tRNA), and various noncoding RNAs (Jang et al. 2020).

cDNA Synthesis: The isolated RNA is subsequently converted into complementary DNA (cDNA) through the process of reverse transcription, employing reverse transcriptase enzymes. This process converts RNA templates into DNA molecules, allowing for subsequent DNA-based analyses (Zhao et al. 2021).

Library Preparation: The cDNA is then prepared into a sequencing library, which involves adding adapters and performing PCR amplification. These adapters enable the cDNA fragments to be sequenced using NGS platforms. The cDNA library is sequenced using NGS platforms, generating a dataset of short sequence reads representing the RNA molecules present in the sample (Kopylova et al. 2012).

During the data analysis phase, raw sequencing reads undergo quality control and trimming processes to eliminate low-quality bases, adapters, and other artifacts. The trimmed reads are aligned to reference genomes or transcriptomes to determine which genes are being transcribed. This allows for quantification of gene expression levels. The identified transcripts are annotated by comparing them to existing databases of known genes and functional annotations.

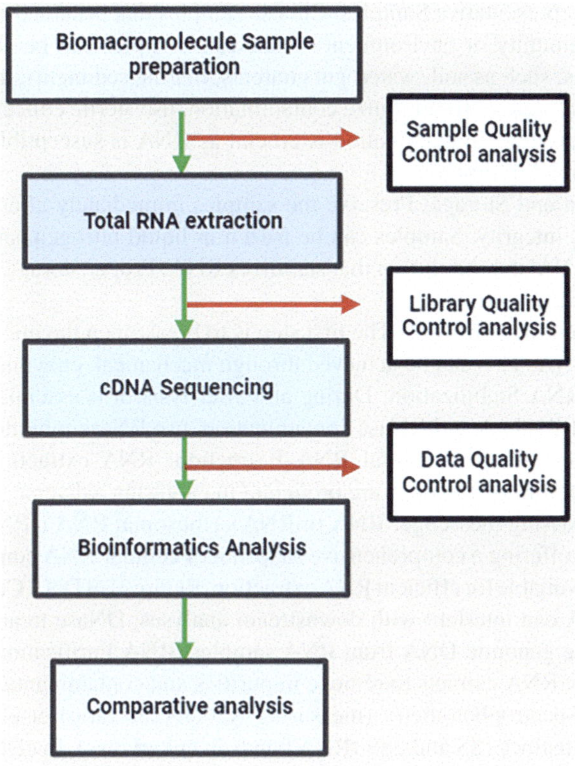

Fig. 3.4 Schematic workflow of metatranscriptomic next generation sequencing

Statistical methods are applied to compare gene expression levels between various conditions or samples, aiming to identify genes that exhibit differential expression. By analyzing gene expression patterns, researchers can gain insights into the activities and responses of microbial communities to various environmental conditions or perturbations. This information helps in understanding the functional roles of microorganisms within ecosystems. Metatranscriptomics offers a dynamic snapshot of gene expression within microbial communities, providing a comprehensive insight into how microorganisms adapt to evolving environments, engage in interactions, and contribute to diverse biological processes. It has applications in environmental studies, host-microbiome interactions, disease pathogenesis, biotechnological processes, and more.

3.15 Gathering Samples and Isolating RNA

Sample collection and RNA extraction are crucial steps in metatranscriptomics that ensure the preservation of RNA molecules and their representation in subsequent analyses (Mukherjee and Sudhakara Reddy 2020).

Selecting Representative Samples: Choose samples that accurately represent the microbial community or environment of interest. Samples can be collected from various sources, such as soil, water, gut contents, marine sediments, and more.

Aseptic Techniques: To minimize contamination, use sterile collection tools and containers. Preventing contamination is crucial as RNA is susceptible to degradation and microbial infiltration.

Preservation and Storage: Preserve the samples immediately after collection to maintain RNA integrity. Samples can be frozen in liquid nitrogen, snap-frozen, or preserved in RNAlater, a solution that stabilizes RNA. Proper storage ensures RNA remains intact until extraction.

RNA Extraction; Cell Lysis: The first step is to break open the microbial cells to release their RNA. This can be achieved through mechanical, enzymatic, or chemical methods. RNA Stabilization; During and after lysis: it is essential to maintain the integrity of RNA. Avoid RNase contamination, use RNase inhibitors, and work in RNA-safe environments. Total RNA Extraction: RNA extraction techniques designed for total RNA recovery are proficient in capturing a diverse range of RNA types, encompassing messenger RNA (mRNA), ribosomal RNA (rRNA), and non-coding RNAs, offering a comprehensive snapshot of cellular RNA content. Kits and protocols are available for efficient RNA extraction. Removal of DNA Contamination: Genomic DNA can interfere with downstream analyses. DNase treatment is often used to remove genomic DNA from RNA samples. RNA Purification and Quality Control: Purify RNA extracts to remove impurities and contaminants. Assess RNA quality using spectrophotometry (measuring A260/A280 ratio) or electrophoresis (checking for distinct 18S and 28S rRNA bands in eukaryotes). In cDNA synthesis, reverse transcription transforms RNA into complementary DNA (cDNA), providing a foundational material for subsequent library preparation and sequencing applications. The cDNA is prepared into a sequencing library, including adapter ligation and amplification steps. These steps vary depending on the sequencing platform used. Sample collection and RNA extraction need to be performed meticulously to ensure the accuracy and reliability of metatranscriptomic data. Contamination, degradation, or biases at these stages can lead to inaccurate gene expression profiling. Careful handling, adherence to protocols, and using appropriate quality control measures are crucial for successful metatranscriptomics studies (Jang et al. 2020).

3.16 Library Preparation for Metatranscriptomic Sequencing

Library preparation for metatranscriptomic sequencing involves converting the extracted RNA molecules into a format suitable for high-throughput sequencing. This procedure encompasses several pivotal steps aimed at guaranteeing the precise representation of the transcriptome and enabling the subsequent analysis of gene expression patterns (Gann et al. 2021). The first step is to fragment the RNA molecules into smaller pieces. Fragmentation is typically achieved by using enzymes that specifically cleave the RNA at random locations. The goal is to create fragments that are of suitable size for sequencing. The fragmented RNA acts as a template for reverse transcription, enabling the synthesis of complementary DNA (cDNA). This process utilizes reverse transcriptase enzymes and random primers to generate cDNA from the RNA fragments. After cDNA synthesis, a second strand of DNA is synthesized to create double-stranded cDNA. This step is performed using DNA polymerase and other enzymes. The double-stranded cDNA fragments undergo end repair to ensure blunt ends and are then adenylated (A-tailed) by adding an "A" base to the 3′ ends. These modifications ready the fragments for adapter ligation. Sequencing adapters, containing essential sequences for binding to the sequencing platform and indices for sample identification in multiplexed sequencing runs, are ligated to the A-tailed cDNA fragments. The ligated cDNA fragments are amplified using PCR to generate sufficient quantities of the library for sequencing. PCR primers with sequences matching the adapter regions are used for amplification. Due to the varied lengths of cDNA fragments, size selection is frequently carried out to eliminate fragments that are either excessively short or long, ensuring accurate sequencing results. This can be achieved using gel electrophoresis, bead-based methods, or automated platforms. Throughout the library preparation process, quality control steps are crucial to ensure that the library is of high quality and accurately represents the transcriptome. Quantify the library using qPCR or fluorometry, and assess the fragment size distribution using gel electrophoresis or capillary electrophoresis. Once the libraries have passed QC, they can be pooled together with other libraries for sequencing. Pooling allows multiple samples to be sequenced simultaneously, which is cost-effective. Subsequently, the combined libraries undergo sequencing using a high-throughput platform, such as Illumina, to generate comprehensive and detailed sequence data. Library preparation for metatranscriptomic sequencing is a crucial step that determines the accuracy and quality of downstream transcriptomic analysis. Proper techniques, careful handling, and quality control measures ensure that the library accurately represents the expression profile of the RNA molecules in the sample (Sanchez-Flores and Abreu-Goodger 2014).

3.17 NGS Platforms for Metatranscriptomics

Metatranscriptomic sequencing commonly employs various Next-Generation Sequencing (NGS) platforms, each presenting unique advantages and capabilities. These platforms serve as valuable tools for exploring gene expression patterns within microbial communities (Gorni 2013). Here are some of the prominent NGS platforms used in metatranscriptomics; Illumina platforms, such as the HiSeq and NovaSeq, are widely used for metatranscriptomic studies due to their high throughput and accuracy. These platforms produce short-read sequences with extensive coverage, facilitating the quantification of gene expression levels and the identification of differentially expressed genes. Illumina sequencing is particularly adept at studying the transcriptional dynamics of microbial communities. In contrast, PacBio's Single Molecule, Real-Time (SMRT) sequencing technology generates longer reads compared to Illumina, providing a valuable advantage in the context of metatranscriptomics. Longer reads help in resolving complex transcriptomes and identifying full-length transcripts. This is particularly useful for studying alternative splicing and operon structures in microbial communities. ONT's nanopore sequencing technology offers the advantage of producing long reads in real time. This allows for direct RNA sequencing without the need for amplification. Long reads provided by ONT sequencing are beneficial for capturing full-length transcripts, isoform diversity, and identifying rare transcripts. Ion Torrent sequencing offers rapid turnaround times and can be used for smaller-scale metatranscriptomic studies. Ion Torrent platforms, although generating shorter reads in comparison to Illumina, remain valuable for gaining insights into gene expression profiles within microbial communities. PacBio's Iso-Seq protocol is specifically designed for studying full-length transcripts. It generates long reads that span full transcript isoforms, enabling accurate identification of different isoforms and splice variants. ONT's innovative RNA sequencing technology, known for its direct RNA sequencing capability, eliminates the need for cDNA conversion. This groundbreaking approach allows for the direct sequencing of RNA molecules, setting a new standard in the field. This approach provides a direct view of RNA modifications and full-length transcripts, which can be valuable for studying transcript structure and post-transcriptional modifications.

3.18 Data Analysis

Data analysis in metatranscriptomics involves a series of computational steps to transform raw sequencing data into meaningful biological insights about gene expression patterns in microbial communities (Bashiardes et al. 2016). Below is an outline of the data analysis procedure, encompassing quality control, read alignment, quantification, gene expression analysis, and differential expression analysis. Quality control aims to ensure the reliability of the raw sequencing data. It involves checking for read quality and removing low-quality bases, adapter sequences, and other artifacts that can affect downstream analyses. Tools like FastQC are used to assess read quality, and programs like Trimmomatic or Cutadapt are employed for read trimming. Aligned reads indicate which parts of the genome/transcriptome the reads originate from. Alignment tools compare the sequenced reads to reference sequences (genomes or transcriptomes) to identify their origin. Alignment can be performed using tools like STAR, HISAT2, or Salmon. Quantification involves estimating the abundance of RNA molecules (genes or transcripts) in terms of read counts or normalized expression values.

Gene expression analysis involves comparing the expression levels of genes or transcripts across different conditions or samples. Normalization methods, such as Reads Per Kilobase Million (RPKM) or Transcripts Per Million (TPM), are used to account for differences in library sizes and transcript lengths. Differential expression analysis identifies genes or transcripts that are significantly upregulated or downregulated between different conditions. This is crucial for understanding how gene expression changes in response to specific treatments, environments, or experimental factors. Statistical tests like edgeR, DESeq2, or limma are used to determine differential expression. These tests account for variability and provide adjusted p-values to control for false positives (e.g., using the Benjamini-Hochberg correction). Once differentially expressed genes are identified, functional annotation involves assigning biological functions to these genes. Gene Ontology (GO) analysis and pathway enrichment analysis identifies overrepresented functions and pathways among differentially expressed genes. These analyses provide insights into the biological processes that are affected by changes in gene expression. Visualization tools, such as heatmaps, scatter plots, and volcano plots, help interpret the results by highlighting patterns and trends in the data. Integrating results with existing knowledge and databases helps researchers interpret the biological significance of differentially expressed genes. Metatranscriptomics data analysis requires robust computational skills, familiarity with relevant software tools, and a solid understanding of statistical methods. It is essential to choose appropriate parameters, apply quality control measures, and interpret results in the context of the biological questions being addressed (Vats 2021).

3.19 Metatranscriptomics: Unraveling Gene Expression Patterns and Functional Insights

Metatranscriptomics is pivotal in profiling gene expression and functionally characterizing microbial communities across diverse environments. It offers profound insights into gene activities, enabling a deep understanding of how microorganisms respond to varied conditions, interact, and contribute to ecosystem processes within these communities. Here are some applications of metatranscriptomics in gene expression profiling and functional characterization:

Environmental Microbiome Studies: Metatranscriptomics reveals the active genes and functions within microbial communities, providing insights into their roles in nutrient cycling, carbon fixation, and other ecosystem processes. Researchers can monitor how microbial communities respond to changing conditions such as temperature shifts, pollutant exposure, or nutrient availability. This helps predict ecosystem responses to environmental perturbations (Carvalhais et al. 2012). Metatranscriptomics serves as a crucial tool for unraveling the functional roles of microbes in the human gut, skin, oral cavity, and other body sites. It uncovers how the microbiome contributes to host health, metabolism, and immune system modulation. Plant-Microbe Interactions: Metatranscriptomics reveals gene expression patterns in plant-associated microbial communities. This sheds light on symbiotic relationships, nutrient uptake, and plant defense mechanisms (Zhang et al. 2021).

Biotechnological Applications: Metatranscriptomics identifies genes and pathways involved in pollutant degradation, aiding in the development of bioremediation strategies. By analyzing gene expression, researchers can identify microbial enzymes with potential industrial applications, such as biofuel production or pharmaceutical synthesis.

Disease Pathogenesis and Host Response: Metatranscriptomics can reveal how pathogens modulate their gene expression during infection, providing insights into virulence factors and potential drug targets. In host-microbiome interactions, metatranscriptomics helps understand how the host immune system responds to microbial colonization and infections. Metatranscriptomics allows functional characterization of microbes that are difficult to culture. This aids in understanding their roles in ecosystems and their potential applications. By analyzing gene expression profiles, researchers can identify cooperative or competitive interactions among microbial species in complex communities. Metatranscriptomics validates metagenomics predictions by revealing which genes are actively expressed, providing functional evidence for the presence of specific pathways or activities. Microbiome-Based Therapies: Metatranscriptomics contributes to the development of microbiome-based therapies by identifying specific microbial functions that can be targeted for therapeutic interventions (Baldrian and López-Mondéjar 2014). Metatranscriptomics bridges the gap between genetic potential and actual functional activities within microbial communities. By unveiling real-time gene expression patterns, metatranscriptomics enhances our comprehension of the intricate interactions and dynamics within these communities across various contexts (Huang et al. 2019).

3.20 Comparative Analysis of Metagenomics and Metatranscriptomics

Refer to Table 3.2 for a comparison of the similarities and differences between metagenomics and metatranscriptomics.

Table 3.2 Comparison between metagenomics and metatranscriptomics techniques

S. No	Metagenomics	Metatranscriptomics
Difference		
1	It is the field of study about the DNA sequences in a community of organisms	It is the field of study about the transcriptomes in a community of organisms
2	DNA	Transcribed DNA (type of RNA), primarily mRNA
3	DNA extraction method	RNA extraction method
4	To analyze mutations and specific genes in a community of microbes	To analyze transcribed DNA sequences and patterns of genes expression
Similarities		
5	It uses high-throughput techniques to analyze communities of organisms	It also uses high-throughput techniques to analyze communities of organisms
6	These techniques use microarray technology	It also use microarray technology
7	These techniques do not focus on one organism	It focuses on a group or community organisms
8	It studies mostly in relation to microbes	It studies mainly microbial communities

3.20.1 Similarities Between Metagenomics and Metatranscriptomics

Study of Microbial Communities: Both metagenomics and metatranscriptomics concentrate on investigating microbial communities in their native habitats, eliminating the necessity for cultivating individual microorganisms.

High-Throughput Sequencing: Both methods employ high-throughput sequencing technologies to produce extensive DNA or RNA sequence data from intricate microbial samples.

Insight into Microbial Diversity: Both techniques offer insights into the diversity of microorganisms existing in a specific environment, enabling the identification of novel species or strains.

Functional Potential: Metagenomics and metatranscriptomics both offer insights into the functional potential of microbial communities. Metagenomics predicts the presence of functional genes, while metatranscriptomics reveals which of these genes are actively expressed, providing a dynamic view of the community's functional activity.

Environmental and Clinical Applications: Both approaches have applications in environmental studies, human health, biotechnology, and more. They offer insights into ecosystem dynamics, microbial interactions, host-microbe relationships, and disease pathogenesis.

3.20.2 Differences Between Metagenomics and Metatranscriptomics

Nature of Data: Metagenomics produces DNA sequence data, offering insights into the genetic potential of the microbial community. Metatranscriptomics generates RNA sequence data, capturing the gene expression patterns and revealing which genes are actively being transcribed and translated (Jo et al. 2021).

Information Captured: Metagenomics reveals details about the genetic content, potential functions, and metabolic pathways of microorganisms within the community. In contrast, metatranscriptomics offers insights into the real-time gene expression patterns and functional activities of the microbial community at a specific moment.

Temporal Aspect: Metagenomics captures a static snapshot of a microbial community's genetic potential, while metatranscriptomics offers a dynamic view of gene expression patterns, unveiling how the community responds to environmental changes or perturbations.

Gene Expression Analysis: While metagenomics can predict functional genes, it does not provide information about which genes are actively being expressed. Metatranscriptomics directly analyzes gene expression, allowing for the identification of upregulated/downregulated genes and differential expression analysis (Li et al. 2019).

Analysis Complexity: Metatranscriptomics involves an additional layer of complexity compared to metagenomics, as it requires studying RNA molecules that are more transient and susceptible to degradation.

Experimental Design: Experimental design for metagenomics and metatranscriptomics may differ due to the need to preserve RNA integrity for transcriptomics and to capture gene expression responses.

Application Focus: Metagenomics is often used for understanding microbial community composition, functional potential, and biodiversity. Metatranscriptomics is particularly valuable for understanding gene expression patterns, host-microbiome interactions, and real-time responses of microbial communities (Jo et al. 2021).

3.20.3 Complementary Insights Provided by Each Approach

Metagenomics and metatranscriptomics are two complementary approaches that provide distinct insights into microbial communities and their functions. When used together, these approaches offer a more comprehensive understanding of microbial ecosystems (Jovel et al. 2022). Here are the complementary insights provided by each approach:

Metagenomics: Metagenomics reveals the genetic potential and diversity of microbial communities. It identifies the presence of functional genes, metabolic pathways, and potential interactions that the community is capable. Metagenomics can predict the functional capabilities of microorganisms based on the presence of specific genes and pathways. It offers a blueprint of what the microbial community could do under different conditions. Metagenomics can identify novel species, strains, or lineages that have not been previously described based on the genetic information present in the community (Qian et al. 2021). Metagenomics offers insights into the taxonomic composition of the community, aiding researchers in understanding the presence of organisms and their proportions within the ecosystem.

Metatranscriptomics: Metatranscriptomics directly reveals which genes are actively expressed at a specific point in time. It provides insights into the functional activities that are happening in response to environmental conditions (Perbal 2015). Metatranscriptomics captures real-time dynamics by revealing how gene expression patterns change in response to various environmental cues, stressors, or treatments, providing valuable insights into the adaptive responses of microbial communities. Metatranscriptomics validates the functional relevance of genes predicted by metagenomics. It confirms which genes are actually being used by the community in a given context. By identifying highly expressed genes or pathways, metatranscriptomics helps identify key players responsible for specific functions within the community. Metatranscriptomics offers valuable insights into the intricate interactions between microbial communities and their hosts. It unveils details about host immune responses, microbial colonization dynamics, and various other host-microbe interactions, shedding light on the complex relationships within biological systems. Metatranscriptomics helps researchers understand the functional roles of specific microbial species or groups within the community and their contributions to ecosystem processes. When combined, metagenomics and metatranscriptomics offer a holistic view of microbial communities. Metagenomics provides the blueprint of potential functions and diversity, while metatranscriptomics provides real-time insights into which functions are being actively utilized and how the community responds to its environment. This integration of genomic potential and functional activity is essential for a comprehensive understanding of microbial ecosystems and their roles in various environments (Li et al. 2017).

3.21 Synergizing Metagenomic and Metatranscriptomic Data: Achieving a Holistic Understanding

Integrating metagenomic and metatranscriptomic data stands as a powerful strategy, providing a comprehensive insight into microbial communities, their functional dynamics, and adaptive responses to environmental changes. This combination allows researchers to bridge the gap between the potential functions predicted by metagenomics and the actual gene expression patterns revealed by metatranscriptomics. This holistic approach leads to deeper insights into ecosystem interactions, microbial behavior, and environmental adaptation (Li et al. 2017).

The integration process initiates by annotating metagenomic data, foreseeing the functional capabilities of the microbial community. This involves identifying genes, metabolic pathways, and functional modules. Subsequently, the same reference databases used for functional annotation are employed to map metatranscriptomic data, effectively pinpointing which genes are actively transcribed within the community. Quantifying the expression levels of genes based on the mapped metatranscriptomic data provides valuable information about the abundance of actively transcribed genes. Performing differential expression analysis enhances the understanding of how the community responds to changes in environmental conditions, pinpointing genes that are significantly upregulated or downregulated under specific circumstances (Zhang et al. 2021) (see Fig. 3.5).

Comparing the functional potential predicted by metagenomics with the actual gene expression patterns from metatranscriptomics validates functions that are genuinely utilized within the community. Constructing functional networks that connect genes with similar functions or those involved in common pathways offers a visual representation of interactions among genes and pathways. The temporal aspect captured by metatranscriptomics is leveraged to correlate gene expression patterns with fluctuations in environmental conditions over time. This temporal analysis uncovers dynamic responses and adaptive strategies. Moreover, when studying host-microbiome interactions, integrating host gene expression data with metatranscriptomic data provides insights into the reciprocal influence between the host and microbiome (Aguiar-Pulido et al. 2016). The integration approach enables the identification of pivotal microbial species or groups that significantly contribute

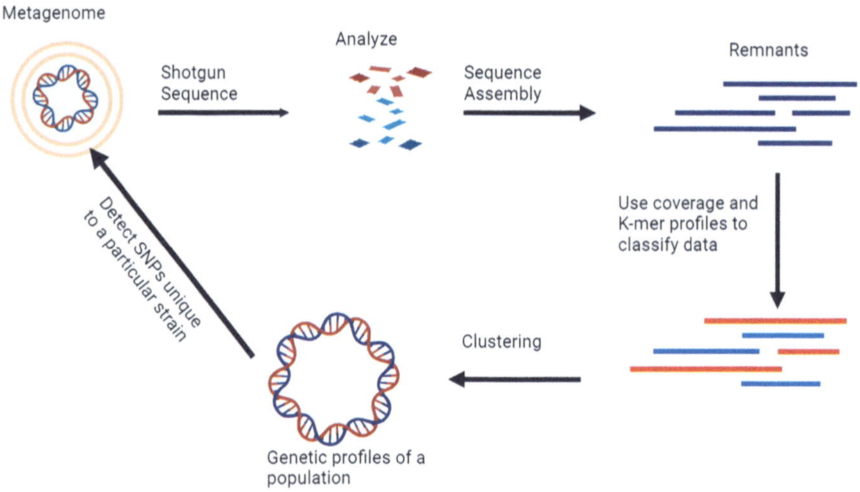

Fig. 3.5 Integration of metagenomics and metatranscriptomics data for a comprehensive understanding

to specific functions within the community. Linking abundant organisms with highly expressed genes in relevant pathways sheds light on key players responsible for critical functions. By effectively integrating metagenomic and metatranscriptomic data, researchers obtain a multidimensional perspective of microbial communities. This methodology not only validates predicted functions but also unveils the timing and utilization of these functions. The insights gained facilitate a comprehensive analysis of microbial behavior, functional adaptations, and ecosystem processes, making this integration a powerful tool in microbiome research (Jovel et al. 2022).

3.22 Challenges and Future Perspectives

In the domains of metagenomics and metatranscriptomics, numerous technical challenges and limitations affect the quality and depth of insights derived from Next-Generation Sequencing (NGS) technologies. The intricate nature of microbial communities in both fields poses challenges in capturing the complete spectrum of genetic or transcriptomic information. Moreover, variations in DNA and RNA extraction methods can introduce biases that affect the accurate representation of different community members or transcripts. From an NGS perspective, the relatively short read lengths can hinder the precise assembly of intricate genomes and the identification of full-length transcripts, particularly influencing the accurate depiction of gene structures and regulatory regions. Additionally, inherent error rates in NGS platforms can compromise sequence data accuracy, especially in regions with low coverage. Dealing with these challenges necessitates substantial computational resources and expertise, owing to the vast volume of data generated by Next-Generation Sequencing (NGS) technologies (Ojala et al. 2023). Significant strides have been made in addressing the challenges associated with NGS in metagenomics and metatranscriptomics. Long-read sequencing technologies, like PacBio and ONT, have emerged, offering the capability to improve genome assembly accuracy and uncover complete transcript structures. The continuous enhancement of high-throughput capabilities and multiplexing options has increased the cost-effectiveness of large-scale studies. On the data analysis front, algorithms for alignment, quantification, and assembly have evolved, bolstering the accuracy of analyses and facilitating biologically meaningful insights. Furthermore, the emergence of single-cell sequencing technologies has enabled the investigation of microbial diversity at the individual cell level, illuminating intra-community variability and shedding light on rare microbial taxa (Shakya et al. 2019). The convergence of these advancements paves the way for promising applications and future directions. The precise understanding of host-microbiome interactions can lead to tailored microbiome interventions for personalized medicine, agriculture, and environmental restoration. Microbiome-based therapeutics hold the potential to reshape healthcare by capitalizing on insights derived from NGS data. Looking ahead, multi-omics integration, encompassing metagenomics, metatranscriptomics, and beyond, is poised to provide a holistic view of microbial communities. Longitudinal studies leveraging NGS will uncover the dynamic evolution of microbial ecosystems over time, shedding light on responses to changing environments. The integration of functional assays and AI-driven analysis techniques will illuminate specific gene functions and enable predictive modeling. However, ethical considerations must also be at the forefront, navigating issues of data privacy, informed consent, and responsible data sharing as NGS technology continues to evolve (Terrón-Camero et al. 2022). While metagenomics and metatranscriptomics face technical challenges, ongoing advancements in NGS technologies and data analysis tools are overcoming these limitations. The field holds immense potential for applications in precision microbiome engineering, therapeutic development, and a deeper understanding of microbial communities and their functions. The future landscape of metagenomics and metatranscriptomics research will be shaped by the integration of multi-omics data and the ethical considerations surrounding data usage.

3.23 Conclusion

In this exploration of metagenomics and metatranscriptomics, we delved into the intricacies of understanding complex microbial communities through Next-Generation Sequencing (NGS) technologies. Metagenomics, which involves studying genetic material directly extracted from environmental samples, provides valuable insights into community composition, genetic diversity, and functional potential within microbial ecosystems. On the other hand, metatranscriptomics provides a dynamic snapshot of gene expression patterns, revealing how these communities actively respond to their environments. We delved into the processes of sample collection, DNA/RNA extraction, library preparation, sequencing platforms, and data analysis, demonstrating the exhaustive path from raw samples to valuable biological insights. Through the lens of data analysis, we unraveled the significance of quality control, read alignment, quantification, gene expression analysis, and functional annotation. Moreover, the applications of these disciplines are far-reaching, ranging from environmental studies to host-microbiome interactions and biotechnological advancements. Through the efficient integration of metagenomic and metatranscriptomic data, researchers can unlock complementary insights, offering a multidimensional perspective on microbial communities and their functions. The advent of Next-Generation Sequencing (NGS) technologies has revolutionized metagenomics and metatranscriptomics research. NGS's ability to generate vast amounts of DNA and RNA sequence data from complex microbial samples has enabled researchers to explore microbial communities at an unprecedented depth. This advancement has transcended the limitations of traditional culture-based methods, allowing for the study of unculturable microorganisms and providing insights into the intricate relationships within microbial ecosystems. NGS has brought precision and scalability to metagenomics by revealing genetic diversity, community structures, and potential functions present in environmental samples. Moreover, in metatranscriptomics, NGS provides a real-time snapshot of gene expression, shedding light on the active role microbes play in their environments. By bridging metagenomic potential with metatranscriptomic activity, NGS offers a holistic understanding of how microbial communities function, adapt, and respond to changing conditions.

NGS technologies stand as the cornerstone of modern metagenomics and metatranscriptomics research, propelling these fields into new frontiers of discovery. Their ability to offer detailed insights into the genetic composition, functional dynamics, and interactions of microbial communities is reshaping our comprehension of the microbial world and its implications across diverse domains, from ecology to human health.

References

Aguiar-Pulido V, Huang W, Suarez-Ulloa V, Cickovski T, Mathee K, Narasimhan G (2016) Metagenomics, metatranscriptomics, and metabolomics approaches for microbiome analysis. Evol Bioinforma 12:5–16

Ansorge WJ (2009) Next-generation DNA sequencing techniques. New Biotechnol 25:195–203

Araujo FA, Barh D, Silva A, Guimarães L, Ramos RTJ (2018) GO FEAT: a rapid web-based functional annotation tool for genomic and transcriptomic data. Sci Rep 8(1):1–4

Auger H, van Dijk EL, Jaszczyszyn Y, Thermes C (2014) Ten years of next-generation sequencing technology. Trends Genet 30(9):418–426

Ayling M, Clark MD, Leggett RM (2020) New approaches for metagenome assembly with short reads. Brief Bioinform 21:584–594

Baldrian P, López-Mondéjar R (2014) Microbial genomics, transcriptomics and proteomics: new discoveries in decomposition research using complementary methods. Appl Microbiol Biotechnol 98:1531–1537

Bashiardes S, Zilberman-Schapira G, Elinav E (2016) Use of metatranscriptomics in microbiome research. Bioinform Biol Insights 10:19–25

Bhargava P, Khan M, Verma A, Singh A, Vats S et al (2019a) Metagenomics as a tool to explore new insights from plant-microbe interface. In: Plant microbe interface, pp 271–289

Bhargava P, Vats S, Gupta N (2019b) Metagenomics as a tool to explore mycorrhizal fungal communities. In: Mycorrhizosphere and pedogenesis, pp 207–219

Carvalhais LC, Dennis PG, Tyson GW, Schenk PM (2012) Application of metatranscriptomics to soil environments. J Microbiol Methods 91(2):246–251

Chiu CY, Miller SA (2019) Clinical metagenomics. Microb Genomics 20:341–355

Diao Z, Han D, Zhang R, Li J (2022) Metagenomics next-generation sequencing tests take the stage in the diagnosis of lower respiratory tract infections. J Adv Res 38:201–212

Driscoll CB, Otten TG, Brown NM, Dreher TW (2017) Towards long-read metagenomics: complete assembly of three novel genomes from bacteria dependent on a diazotrophic cyanobacterium in a freshwater lake co-culture. Stand Genomic Sci 12:9

Fujii T, Uchiyama T, Matsuoka M, Myojin T, Sugimoto S, Nitta Y, Okabe F, Sugimoto A, Sekita-Hatakeyama Y, Morita K, Itami H, Hatakeyama K, Ohbayashi C (2020) Evaluation of DNA and RNA quality from archival formalin-fixed paraffin-embedded tissue for next-generation sequencing retrospective study in Japanese single institution. Pathol Int 70:602–611

Gann ER, Kang Y, Dyhrman ST, Gobler CJ, Wilhelm SW (2021) Metatranscriptome library preparation influences analyses of viral community activity during a brown tide bloom. Front Microbiol 12:664189

Gaulke CA, Schmeltzer ER, Dasenko M, Tyler BM, Thurber RV, Sharpton TJ (2021) Evaluation of the effects of library preparation procedure and sample characteristics on the accuracy of metagenomic profiles. mSystems 6(5):e00440-21

Godlewska U, Brzoza P, Kwiecień K, Kwitniewski M, Cichy J (2020) Metagenomic studies in inflammatory skin diseases. Curr Microbiol 77:3201–3212

Gökdemir FŞ, İşeri ÖD, Sharma A, Achar PN, Eyidogan F (2022) Metagenomics next generation sequencing (mNGS): an exciting tool for early and accurate diagnostic of fungal pathogens in plants. J Fungi 8:1195

Gorni C (2013) Metatranscriptomics in the NGS era. Transcriptomics 2:1

Huang W, Yin C, Wang G, Rosenblum J, Krishnan S, Dimitrova N, Fallon JT (2019) Optimizing a metatranscriptomic next-generation sequencing protocol for bronchoalveolar lavage diagnostics. J Mol Diagn 21(2):251–260

Huang J, Jiang E, Yang D et al (2020) Metagenomic next-generation sequencing versus traditional pathogen detection in the diagnosis of peripheral pulmonary infectious lesions. Infect Drug Resist 13:567–576

Jang JS, Berg B, Holicky E, Eckloff B, Mutawe M, Carrasquillo MM, Ertekin-Taner N, Cuninngham JM (2020) Comparative evaluation for the globin gene depletion methods for mRNA sequencing using the whole blood-derived total RNAs. BMC Genomics 21:890

Jo Y, Back C-G, Kim K-H, Chu H, Lee JH, Moh SH, Cho WK (2021) Comparative study of metagenomics and metatranscriptomics to reveal microbiomes in overwintering pepper fruits. Int J Mol Sci 22:6202

Jovel J, Nimaga A, Jordan T, O'Keefe S, Patterson J, Thiesen A, Hotte N, Bording-Jorgensen M, Subedi S, Hamilton J, Carpenter EJ, Lauga B, Elahi S, Madsen KL, Wong GK-S, Mason AL (2022) Metagenomics versus metatranscriptomics of the murine gut microbiome for assessing microbial metabolism during inflammation. Front Microbiol 13:829378

Kopylova E, Noe L, Touzet H (2012) SortMeRNA: fast and accurate filtering of ribosomal RNAs in metatranscriptomic data. Bioinformatics 28:3211–3217

Li F, Neves ALA, Ghoshal B, Guan LL (2017) Symposium review: mining metagenomic and metatranscriptomic data for clues about microbial metabolic functions in ruminants. J Dairy Sci 101:5605–5618

Li F, Hitch TCA, Chen Y, Creevey CJ, Guan LL (2019) Comparative metagenomic and metatranscriptomic analyses reveal the breed effect on the rumen microbiome and its associations with feed efficiency in beef cattle. Microbiome 7:6

Michealsamy A, Thangamani L, Manivel G, Kumar P, Sundar S, Piramanayagam S, Natarajan J (2021) Current research and applications of meta-omics stratagems in bioremediation: a bird's-eye view. J Appl Biotechnol Rep 8(2):109–115
Mukherjee A, Sudhakara Reddy M (2020) Metatranscriptomics: an approach for retrieving novel eukaryotic genes from polluted and related environments. 3 Biotech 10:71
Nakazato T, Ohta T, Bono H (2013) Experimental design-based functional mining and characterization of high-throughput sequencing data in the sequence read archive. PLoS One 8(10):e77910
Ojala T, Häkkinen A-E, Kankuri E, Kankainen M (2023) Current concepts, advances, and challenges in deciphering the human microbiota with metatranscriptomics. Trends Genet 39(9):686–701
Oniciuc EA, Likotrafiti E, Alvarez-Molina A, Prieto M, Santos JA, Alvarez-Ordóñez A (2018) The present and future of whole genome sequencing (WGS) and whole metagenome sequencing (WMS) for surveillance of antimicrobial resistant microorganisms and antimicrobial resistance genes across the food chain. Genes (Basel) 9:E268
Oulas A, Pavloudi C, Polymenakou P, Pavlopoulos GA, Papanikolaou N, Kotoulas G, Arvanitidis C, Iliopoulos I (2015) Metagenomics: tools and insights for analyzing next-generation sequencing data derived from biodiversity studies. Bioinform Biol Insights 9:75–88
Perbal L (2015) The case of the gene: post-genomics between modernity and postmodernity. EMBO Rep 16(7):777–781
Qian M, Zhan Y, Wu D, Ji L, Chen H, Cheng Y (2021) Clinical standardization of metagenomic next generation sequencing (mNGS) in the pathogen diagnosis. Clin Transl Dis 1:e12
Ramachandran PS, Wilson MR (2020) Metagenomics for neurological infections expanding our imagination. Nat Rev Neurol 16:547–556
Sanchez-Flores A, Abreu-Goodger C (2014) A practical guide to sequencing genomes and transcriptomes. Curr Top Med Chem 14(3):398–406
Sandhu SS, Aunna Pourang MD, Sivamani RK (2019) A review of next generation sequencing technologies used in the evaluation of the skin microbiome: what a time to be alive. Dermatol Online J 25(7):1
Sayers EW, Beck J, Bolton EE, Bourexis D, Brister JR, Canese K, Comeau DC, Funk K, Kim S, Klimke W, Marchler-Bauer A, Landrum M, Lathrop S, Zhiyong L, Madden TL, O'Leary N, Phan L, Rangwala SH, Schneider VA, Skripchenko Y, Wang J, Ye J, Trawick BW, Pruitt KD, Sherry ST (2021) Database resources of the National Center for Biotechnology Information. Nucleic Acids Res 49:D10–D17
Shakya M, Lo C-C, Chain PSG (2019) Advances and challenges in metatranscriptomic analysis. Front Genet 10:904
Suzuki T, Kawada J-I, Okuno Y, Hayano S, Horiba K, Torii Y, Takahashi Y, Umetsu S, Sogo T, Inui A, Ito Y (2017) Comprehensive detection of viruses in pediatric patients with acute liver failure using next-generation sequencing. J Clin Virol 96:67–72
Takeuchi S, Kawada J, Horiba K, Okuno Y, Okumura T, Suzuki T, Torii Y, Kawabe S, Wada S, Ikeyama T, Ito Y (2019) Metagenomic analysis using next-generation sequencing of pathogens in bronchoalveolar lavage fluid from pediatric patients with respiratory failure. Sci Rep 9:12909
Terrón-Camero LC, Gordillo-González F, Salas-Espejo E, Andrés-León E (2022) Comparison of metagenomics and metatranscriptomics tools: a guide to making the right choice. Genes 13:2280
Thomas T, Gilbert J, Meyer F (2012) Metagenomics—a guide from sampling to data analysis. Microb Inform Exp 2:3
Vats S (2021) Bio-informatics analysis of meta-transcriptomics sequencing. Trends Telemed E-Health 2(5):000549
Wang J, Han Y, Feng J (2019) Metagenomic next-generation sequencing for mixed pulmonary infection diagnosis. BMC Pulm Med 19(1):252
Yang IS, Kim S (2015) Analysis of whole transcriptome sequencing data: workflow and software. Genomics Inform 13(4):119–125
Zerbino DR (2010) Using the velvet de novo assembler for short-read sequencing technologies. Curr Protoc Bioinformatics 31(1):11–15

Zhang Y, Thompson KN, Branck T, Yan Y, Nguyen LH, Franzosa EA, Huttenhower C (2021) Metatranscriptomics for the human microbiome and microbial community functional profiling. Annu Rev Biomed Data Sci 4:279–311

Zhao N, Cao J, Jiayue X, Liu B, Liu B, Chen D, Xia B, Chen L, Zhang W, Zhang Y, Zhang X, Duan Z, Wang K, Xie F, Xiao K, Yan W, Xie L, Zhou H, Wang J (2021) Targeting RNA with next- and third-generation sequencing improves pathogen identification in clinical samples. Adv Sci 8:2102593

4 Bioinformatics Databases and Tools for Analysis of Multi-omics

Chung Anh Pham, Anh Dao Ngo, Nhat Le Bui, and Dinh-Toi Chu

4.1 Introduction

Multi-omics implies thinking about various sorts of "omics" data, including genomics (DNA), transcriptomics (RNA), proteomics (proteins), and metabolomics (metabolites) via several naturally genetic processes (Table 4.1) (Agamah et al. 2022). Previous studies mostly focused on the results of a single omics study; however those "omics" components are deeply related to each other. The complicated associations between proteins, metabolites, and other mRNA transcripts make it hard for scientists to analyze these complex common systems (Agamah et al. 2022; Maghsoudi et al. 2022). The rise of multi-omics developments has revolutionized our capacity to look at these brilliantly and pick up encounters into the workings of natural systems at uncommon significance. This approach allows scientists to investigate how different levels of nuclear course are associated and impact on each other, giving a more comprehensive understanding of cellular shapes than any single omic advancement. By integrating data from multiple stages, examiners can recognize plans and connections that would be troublesome or inconceivable to recognize with a single advancement (Huang et al. 2017).

Multi-omics plays a basic part in advancing our understanding of complex diseases such as cancer, neurological diseases, and metabolic disorders. It enables

C. A. Pham · A. D. Ngo
Center for Biomedicine and Community Health, International School, Vietnam National University, Hanoi, Vietnam

N. Le Bui · D.-T. Chu (✉)
Center for Biomedicine and Community Health, International School, Vietnam National University, Hanoi, Vietnam

Faculty of Applied Sciences, International School, Vietnam National University, Hanoi, Vietnam
e-mail: toicd@vnu.edu.vn

I. Mani, V. Singh (eds.), *Multi-Omics Analysis of the Human Microbiome*, https://doi.org/10.1007/978-981-97-1844-3_4

Table 4.1 The definition of omic terms

Approach	Definition
Genomics	Focuses on the organization, composition, history, distribution, and editing of information encoded in an organism's genome (Lederberg and McCray 2001)
Epigenetics	The investigation of nongenetic alterations like DNA methylation and histone modification that have the ability to influence how cells govern gene activity (Dupont et al. 2009)
Transcriptomics	Studying how the transcriptome, or the entire collection of RNA transcripts generated by the genome, is changed in response to regulatory mechanisms, splicing, illness, or other factors (Adams et al. 1991)
Proteomics	Determines the characteristics and patterns of protein expression in response to certain stimuli or after changes in the genome or transcriptome (Wilkins et al. 1996)

examiners to recognize potential steady targets, make personalized pharmaceutical approaches, and optimize helpful techniques. In order to better understand how innate varieties impact protein function, investigators are analyzing expression coupled with protein abundance and post-translational modifications (Vihinen 2021). Moreover, multi-omics makes a difference in recognizing novel biomarkers for early infection discovery, observing infection movement, and fitting treatment reactions (Reilly and Bornfeldt 2021). Many omics have tapped into scientists to investigate the intricate relationships between the human body and its inhabitant life forms in the setting of the human microbiome. According to some opinions, the changes in gut microbiota are associated with a variety of medical issues, including chronic gut diseases, weight issue, and even mental health status (Durack and Lynch 2019).

The human microbiome comprises trillions of microorganisms living in several parts of the body, affecting human well-being (Aggarwal et al. 2023). Previous studies have indicated that alterations inside the microbiome are associated with diverse disorders, including inflammatory bowel disease, weight, and mental problems (Vijay and Valdes 2022). To better understand these relationships, scientists have utilized multi-omics examination to look at the interaction between the microbiome and the genetic products. For example, Wu et al. utilized a multi-omics approach to look at the digestive system microbiome's portion in tweaking the safe system (Wu and Wu 2012). Another study of Lloyd-Price et al. combined shotgun metagenomics, RNA sequencing, and mass cytology to examine the association between living organisms and the stomach in patients with inflammatory bowel disease. They found that certain species of bacteria were associated with enhanced cell activation safety and unpleasant effects, providing useful insights into the causes of disease (Lloyd-Price et al. 2019). They found that certain species of bacteria were associated with enhanced cell activation safety and unpleasant effects, providing useful insights into the causes of disease.

Information about how are proteins, metabolites, and other mRNA transcripts related to cellular shapes and phenotypes is given by the competent multi-omics approach to comprehending complex characteristic systems (Hasin et al. 2017). As innovation progresses, we can expect more advanced multi-omics stages to develop, creating vast amounts of data that require advanced informatics for examination and interpretation. The collaboration between multi-omics and bioinformatics holds great promise for transformative discoveries in science and pharmaceuticals, ultimately leading to advancements in diagnostics, drugs, and personalized medications (O'Connor et al. 2023).

The exponential development of multi-omics information presents noteworthy challenges in terms of information examination, elucidation, and integration. Bioinformatics, a subfield of informatics centered on computational strategies for analyzing organic information, plays a significant part in tending to these challenges. Bioinformatics apparatuses empower analysts to prepare expansive datasets, recognize designs and connections, and coordinate information from distinctive sources. For illustration, machine learning calculations offer assistance in distinguishing gene-disease affiliations, anticipating protein structures and capacities, and cluster tests based on similitudes in quality expression profiles (Libbrecht and Noble 2015). In addition, advanced visualization devices energize communication among investigators and offer assistance in the disclosure of secured plans and designs in high-dimensional data (Zakaria 2021). In this chapter, the thought of multi-omics and its utilization in considering the human microbiome will all be presented. In addition, we also discuss how vital informatics tools—particularly bioinformatics—are used for disentangling and assessing multi-omic information.

4.2 Information Technology and Tools for Analysis of Multi-omics

Over the past few decades, science and technology have been significantly developed, leading to an increasing demand to measure and prudently study biomolecules. It not only studies single biomolecular systems such as genomics or proteomics to clarify their functions, but now multi-omics approaches can also provide an overview of the interactions between molecules biomolecules together, thereby could apply to many fields of science (Hasin et al. 2017; Weckwerth et al. 2020). There have been studies on the application of multi-omics analysis in agriculture to elucidate the genetic, metabolic, or environmental mechanisms of plants (Yang et al. 2021). As a result, scientists can research and develop plant varieties with good yield and quality, contributing to ensuring food security and increasing the economic value of crops. In medicine, information technology based on multi-omics analysis has provided equitable and accurate information to determine the conformity between the donor and the transplant recipient (Li et al. 2014). Smart devices with integrated health monitoring also give doctors and users information about health conditions, combined with omics analysis to produce detailed predictions and disease risks (Li et al. 2017). However, the development of multi-omics methods and tools has given people more insight into the pathogenesis and helps overcome the limitations of single omics (Karczewski and Snyder 2018; Mohammadi-Shemirani et al. 2023). The information collected by analyzing multi-omics data can be arranged into profiles, providing links between biomolecules in disease pathogenesis and helping to make accurate and economical disease diagnoses (Schussler-Fiorenza Rose et al. 2019). A number of recent studies have suggested that multi-omics acquisition and analysis technologies make a crucial contribution to the development of precision medicine and the recommendation of cancer treatments (Hasin et al. 2017; Gallo Cantafio et al. 2018; Turanli et al. 2018). An example is the integration of genomic and proteomic data, which demonstrated exceptional ability in the prognosis and accurate characterization of tumors (Zhu et al. 2017).

Scientists need to investigate and evolve multi-omics data analysis tools because these data appraise different aspects and often do not have a good interplay with other omics (Olivier et al. 2019). To solve practical problems, many analytical tools integrating various types of omics have been developed and applied in research (Table 4.2). Recently, Jarmund and colleagues developed a new R package

Table 4.2 Several tools applied in multi-omics analysis

Year	Name	Technology platform	Combined omics data types	References
2008	SIGMA^2	Java	Genomes, epigenomes, and transcriptomes	Chari et al. (2008)
2011	Paintomics	Perl and Python	Transcriptomes, metabolomes	Garcia-Alcalde et al. (2011), Hernandez-de-Diego et al. (2018)
2015	Omics Pipe	Python package	Genomes, transcriptomes	Fisch et al. (2015)
2018	Graphical interface to visualize multi-omics networks (Grimon)	R package	Phenome, genome, and transcriptome	Kanai et al. (2018)
2019	iOmicsPASS	C++	Genome, proteome, and transcriptome	Koh et al. (2019)
2019	Gaussian graphical models (GGMs)	R package	Genome, proteome, epigenome, and transcriptome	Altenbuchinger et al. (2020)
2022	ALASCA	R package	Proteome, metabolome, and transcriptome	Jarmund et al. (2022)

(ALASCA) to analyze data from various omics (Jarmund et al. 2022). The ability to adjust variables and analysis results for proteomics, metabolomics, and transcriptomics is the strength of this tool and has given positive results of this function in practice. In 2008, the SIGMA^2 tool was researched and developed, allowing researchers to analyze information about genomes, epigenomes, and transcriptomes for cancer diagnosis (Chari et al. 2008). Another tool for data analysis and data visualization in transcriptomics and metabolomics was developed in 2011 and is progressively being improved and equipped with more useful features for users (Garcia-Alcalde et al. 2011; Hernandez-de-Diego et al. 2018). The Omics Pipe, published in 2015, indicated that it is a tool with a positive future for development in the analysis of multi-omics data to better understand the basis of cancer and biological processes (Fisch et al. 2015). Masahiro et al. conducted and reported on Grimon to demonstrate the correlation between omics data in a specific and intelligible way (Kanai et al. 2018). With this tool, they elucidated the relationships between omics information networks such as phenome, genome, and transcriptome. In addition, a tool developed from C++ language that allows phenotypic diagnosis based on multi-omics such as genomics, proteomics, and transcriptomes was reported in 2019 (Koh et al. 2019). To clearly perform the correlation between data, Gaussian Graphical Models are useful tools for analyzing the interactions between biological variables and their association with many diseases (Altenbuchinger et al. 2020). The improvement of artificial intelligence and machine learning also contributes to the study of more tools to support multi-omics analysis in science (Nicora et al. 2020).

4.3 Bioinformatics for Analysis of Multi-omics

Thanks to technological advances and bioinformatics tools, multi-omics data can be understood more clearly, providing an objective view of pathological phenomena or characteristics of molecular biological systems (Long et al. 2019). Bioinformatics not only maintains researchers with insights into single omics but also serves diverse informational tools about biological system change (de Anda-Jauregui and Hernandez-Lemus 2020). The data input to perform multi-omics analysis can be the output of biomedical techniques such as DNA, RNA sequencing, next-generation sequencing (NGS), or tandem mass-spectrometry (Pettini et al. 2021). That huge amount of information will be stored in libraries, shared and used in medical research and other scientific fields (Jansen et al. 2019). Besides, several bioinformatics technologies and tools can also be used in multi-omics analysis (Huber et al. 2015; Cock et al. 2009). Machine learning models can classify and link data fields together to analyze multi-omics information accurately and efficiently about interactions between biomolecules in other physiological conditions together (Syed-Abdul et al. 2017). Precision medicine uses deep learning and machine learning analytics to unravel multi-omics information, discovering new biomarkers and understanding the pathogenesis of many diseases, including cancer (Reel et al. 2021; Chen et al. 2023). Bioinformatics analysis combined with omics data layers has shown positive results in determining the role of genes and biomolecules in cancer diseases (Garcia-Campos et al. 2015; Rocha et al. 2020). Several modular network inference technologies used in the analysis of multi-omics results have contributed to promoting the clinical and etiological predictions of cancer (Bonnet et al. 2015; Huang et al. 2017). The strong advancement of information technology in biomedical applications also supports the promotion of effective multi-omics analysis, such as Deep Network or Survival Analysis Learning with Multi-Omics Neural Networks (SALMON) (Tang et al. 2019; Huang et al. 2019). They provide a more comprehensive and reliable approach to patient survival, specific biomarkers, and drug response prediction (Sharifi-Noghabi et al. 2019). Numerous bioinformatics tools available for application in multi-staged and meta-dimensional multi-omics analysis have been evaluated for their effectiveness and strengths in each type of study (Sathyanarayanan et al. 2020).

The microbiota ecosystems are important to human health and development, and the instability of the microbiota in the body can lead to pathological conditions (Dominguez-Bello et al. 2019). Bioinformatics and multi-omics analysis help humanity deeper understand how changes in microbial species composition in the body can lead to inflammation and a number of negative health circumstances (Sarkar et al. 2023). A study on the application of multi-omics analysis of the gut microbiome reports that alterations in brain structure and the development of dementia are strongly associated with the composition and quantity of the gut microbiota (Liang et al. 2022). Studies in patients with inflammatory bowel disease and gastrointestinal inflammation using multi-omics measurements and analyses have displayed that there is a biological interaction between the host body and the microbiome; disturbances in these biological processes will lead to disease symptoms (Lloyd-Price et al. 2019). In the report by Robert H et al., a group of microorganisms overproducing proteases was detected through bioinformatics technologies and multi-omics analysis, and this suggested a possible cause for ulcerative colitis, as well as the treatment of this disease (Mills et al. 2022). A multi-omics meta-analysis provided evidence for the etiology and development of gingivitis and periodontitis with respect to plaque microbiota change (Huang et al. 2021). This study also highlights the role of oral hygiene and care for the preservation and positive functioning of the oral microbiota, thus limiting the risk of gingivitis and periodontitis.

4.4 Bioinformatics Databases for Analysis of Multi-omics

The biomedical field has seen a speedy improvement in the enhancement and application of multi-omics developments for a long time. The appearance of next-generation sequencing (NGS) and other high-throughput informative procedures has enabled examiners to simultaneously examine various sorts of natural particles, such as DNA, RNA, proteins, and metabolites, in a single run (Satam et al. 2023). This approach has driven to a change in our understanding of complex common systems and the human microbiome. In this chapter, we discuss the concept of multi-omics examination, its potential benefits, and the challenges related with it, particularly inside the setting of bioinformatics database enhancement (Ghosh and Das 2022).

Bioinformatics databases make a fundamental contribution to multi-omics testing, playing a role in coordinating and analyzing huge volumes of data performed through multivariate computational tools. These databases provide scientists with compelling utilities for comparing, dissociating, and describing data from experiments, allowing important interactions to be extracted from complex multi-omics data (Fig. 4.1) (Berg et al. 2020).

The National Center for Biotechnology Information (NCBI) is a valuable source for bioinformatics tools and services and is home to a number of databases pertinent to biotechnology and biomedicine. GenBank, a database of DNA sequences, and PubMed, a bibliographic database of biomedical literature, are both significant databases that belong to NCBI. Besides, NCBI supplies the Cancer Genome Anatomy Project in collaboration with the National Cancer Institute. Additionally, NCBI also offers the Gene database, Online Mendelian Inheritance in Man, Molecular Modeling Database (3D protein structures), database of single-nucleotide polymorphisms, Reference Sequence Collection, a map of the human genome, and a taxonomy browser. Each species of organism receives a special identification (taxonomy ID number) from the NCBI (Li et al. 2021). The database engages analysts to examine complex exchanges between innate and characteristic components, developing an understanding of the essential disobedience that drives human prosperity and disease. Another important resource for multi-omics analysis is the Human Microbiome Project (HMP) Data Repository, developed by the National Institutes of Health (NIH). This comprehensive database stores a wealth of metadata, statistical data, recovery history, and clinical epidemiological information, connected to human genomic information, transcription, and metabolism from different body

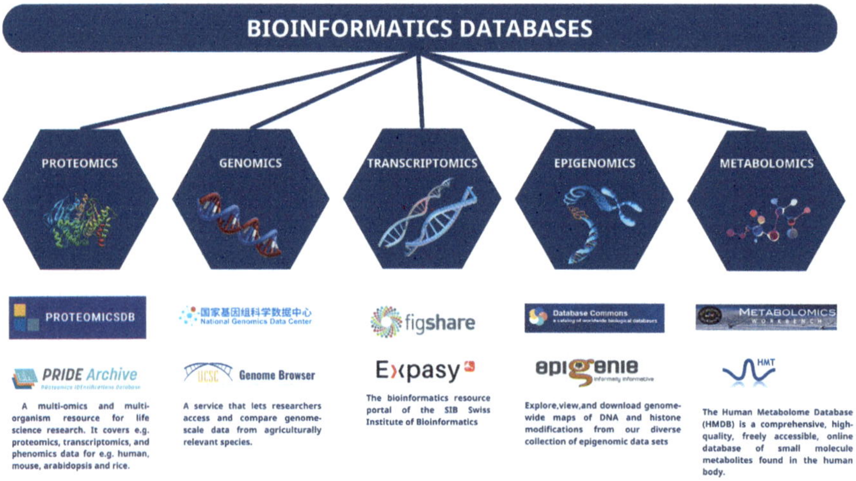

Fig. 4.1 Some of the most popular and reputable multi-omics bioinformatics databases

locations (Eicher et al. 2020). By leveraging this resource, analysts can excellently investigate the multifaceted human microbiome, which has genetic qualities and does not defend against disease. The HMP data warehouse provides investigators with a curious opportunity to examine the vibrant interaction between microbial communities and human well-being, paving the way for the evolution of new support approaches and personalized pharmaceutical approaches (Kashyap et al. 2017).

While there are many advantages to biomedical researchers being powered by bioinformatics databases, there are still significant challenges that need to be addressed. A major obstacle is data integration, as different datasets are generated by different technologies and platforms; thus, it requires harmony and synchronization before comparison and analysis (Subramanian et al. 2020). Furthermore, there is an urgent need for improved computational methods and software tools capable of handling huge volumes of data and the complexity of multi-omics data. Addressing these challenges will be critical in harnessing the full power of bioinformatics databases in multi-omics analysis and unlocking the secrets of the human microbiome (Labory et al. 2020).

In summary, bioinformatics databases play an important part in multi-omics investigation, empowering analysts to unleash important experiences from huge and complex datasets. In summary, bioinformatics databases play an important part in multi-omics investigation; these resources and tools have empowered analysts to draw important insights from huge and complex datasets. While challenges remain, continued movements in computational methodologies and program rebellious will lock in analysts to totally utilize the potential of bioinformatics databases in multi-omics examination, in the long run driving to breakthroughs in human medical conditions (Chen et al. 2023).

4.5 Conclusion, Limitations, and Future Directions

In conclusion, multi-omics is a multiple approach of several biological fields, including genomics, proteomics, transcriptomics, and metabolomics. Due to the huge amount of high-throughput multi-omic data and its complicated format, promising analyzing methods and tools are needed for integrating and interpreting data. With the help of information technology, including bioinformatics, machine learning, and even AI, the results from multi-omic research will provide a comprehensive flow in cells. There is emerging evidence that the human microbiome's composition is related to a variety of diseases such as metabolic diseases, gastrointestinal diseases, and even mental health issues. Multi-omic analysis is expected to decipher their underlying mechanisms. We presented a brief description of several popular bioinformatic tools and databases, as well as their methodology and limitations in this chapter. Despite recent advances in bioinformatics, challenges in data integration and interpretation still remain, which raises the need for more efficient bioinformatic tools in the future.

References

Adams MD et al (1991) Complementary DNA sequencing: expressed sequence tags and human genome project. Science 252(5013):1651–1656

Agamah FE et al (2022) Computational approaches for network-based integrative multi-omics analysis. Front Mol Biosci 9:967205

Aggarwal N et al (2023) Microbiome and human health: current understanding, engineering, and enabling technologies. Chem Rev 123(1):31–72

Altenbuchinger M et al (2020) Gaussian and mixed graphical models as (multi-)omics data analysis tools. Biochim Biophys Acta Gene Regul Mech 1863(6):194418

Berg G et al (2020) Microbiome definition re-visited: old concepts and new challenges. Microbiome 8(1):103

Bonnet E, Calzone L, Michoel T (2015) Integrative multi-omics module network inference with Lemon-Tree. PLoS Comput Biol 11(2):e1003983

Chari R et al (2008) SIGMA2: a system for the integrative genomic multi-dimensional analysis of cancer genomes, epigenomes, and transcriptomes. BMC Bioinformatics 9:422

Chen C et al (2023) Applications of multi-omics analysis in human diseases. MedComm (2020) 4(4):e315

Cock PJ et al (2009) Biopython: freely available Python tools for computational molecular biology and bioinformatics. Bioinformatics 25(11):1422–1423

de Anda-Jauregui G, Hernandez-Lemus E (2020) Computational oncology in the multi-omics era: state of the art. Front Oncol 10:423

Dominguez-Bello MG et al (2019) Role of the microbiome in human development. Gut 68(6):1108–1114

Dupont C, Armant DR, Brenner CA (2009) Epigenetics: definition, mechanisms and clinical perspective. Semin Reprod Med 27(5):351–357

Durack J, Lynch SV (2019) The gut microbiome: relationships with disease and opportunities for therapy. J Exp Med 216(1):20–40

Eicher T et al (2020) Metabolomics and multi-omics integration: a survey of computational methods and resources. Metabolites 10(5):202

Fisch KM et al (2015) Omics Pipe: a community-based framework for reproducible multi-omics data analysis. Bioinformatics 31(11):1724–1728

Gallo Cantafio ME et al (2018) From single level analysis to multi-omics integrative approaches: a powerful strategy towards the precision oncology. High Throughput 7(4):33

Garcia-Alcalde F et al (2011) Paintomics: a web based tool for the joint visualization of transcriptomics and metabolomics data. Bioinformatics 27(1):137–139

Garcia-Campos MA, Espinal-Enriquez J, Hernandez-Lemus E (2015) Pathway analysis: state of the art. Front Physiol 6:383

Ghosh TS, Das M (2022) Chapter two—Emerging tools for understanding the human microbiome. In: Das B, Singh V (eds) Progress in molecular biology and translational science. Academic Press, pp 29–51

Hasin Y, Seldin M, Lusis A (2017) Multi-omics approaches to disease. Genome Biol 18(1):83

Hernandez-de-Diego R et al (2018) PaintOmics 3: a web resource for the pathway analysis and visualization of multi-omics data. Nucleic Acids Res 46(W1):W503–W509

Huang S, Chaudhary K, Garmire LX (2017) More is better: recent progress in multi-omics data integration methods. Front Genet 8:84

Huang Z et al (2019) SALMON: survival analysis learning with multi-omics neural networks on breast cancer. Front Genet 10:166

Huang S et al (2021) Longitudinal multi-omics and microbiome meta-analysis identify an asymptomatic gingival state that links gingivitis, periodontitis, and aging. mBio 12(2):e03281-20

Huber W et al (2015) Orchestrating high-throughput genomic analysis with bioconductor. Nat Methods 12(2):115–121

Jansen P et al (2019) Research data stewardship for healthcare professionals. In: Fundamentals of clinical data science, pp 37–53
Jarmund AH, Madssen TS, Giskeødegård GF (2022) ALASCA: an R package for longitudinal and cross-sectional analysis of multivariate data by ASCA-based methods. Front Mol Biosci 9:962431
Kanai M, Maeda Y, Okada Y (2018) Grimon: graphical interface to visualize multi-omics networks. Bioinformatics 34(22):3934–3936
Karczewski KJ, Snyder MP (2018) Integrative omics for health and disease. Nat Rev Genet 19(5):299–310
Kashyap PC et al (2017) Microbiome at the frontier of personalized medicine. Mayo Clin Proc 92(12):1855–1864
Koh HWL et al (2019) iOmicsPASS: network-based integration of multiomics data for predictive subnetwork discovery. NPJ Syst Biol Appl 5:22
Labory J et al (2020) Multi-omics approaches to improve mitochondrial disease diagnosis: challenges, advances, and perspectives. Front Mol Biosci 7:590842
Lederberg J, McCray AT (2001) Ome SweetOmics—a genealogical treasury of words. Scientist 15(7):8
Li YR et al (2014) Making the genomic leap in HCT: application of second-generation sequencing to clinical advances in hematopoietic cell transplantation. Eur J Hum Genet 22(6):715–723
Li X et al (2017) Digital health: tracking physiomes and activity using wearable biosensors reveals useful health-related information. PLoS Biol 15(1):e2001402
Li B et al (2021) GPCards: an integrated database of genotype-phenotype correlations in human genetic diseases. Comput Struct Biotechnol J 19:1603–1611
Liang X et al (2022) Gut microbiome, cognitive function and brain structure: a multi-omics integration analysis. Transl Neurodegener 11(1):49
Libbrecht MW, Noble WS (2015) Machine learning applications in genetics and genomics. Nat Rev Genet 16(6):321–332
Lloyd-Price J et al (2019) Multi-omics of the gut microbial ecosystem in inflammatory bowel diseases. Nature 569(7758):655–662
Long Y et al (2019) Multiomics-based signaling pathway network alterations in human non-functional pituitary adenomas. Front Endocrinol (Lausanne) 10:835
Maghsoudi Z et al (2022) A comprehensive survey of the approaches for pathway analysis using multi-omics data integration. Brief Bioinform 23(6):bbac435. https://doi.org/10.1093/bib/bbac435
Mills RH et al (2022) Multi-omics analyses of the ulcerative colitis gut microbiome link Bacteroides vulgatus proteases with disease severity. Nat Microbiol 7(2):262–276
Mohammadi-Shemirani P, Sood T, Pare G (2023) From ‘omics to multi-omics technologies: the discovery of novel causal mediators. Curr Atheroscler Rep 25(2):55–65
Nicora G et al (2020) Integrated multi-omics analyses in oncology: a review of machine learning methods and tools. Front Oncol 10:1030
O’Connor LM et al (2023) Integrative multi-omics and systems bioinformatics in translational neuroscience: a data mining perspective. J Pharm Anal 13:836–850
Olivier M et al (2019) The need for multi-omics biomarker signatures in precision medicine. Int J Mol Sci 20(19):4781
Pettini F et al (2021) Multi-omics model applied to cancer genetics. Int J Mol Sci 22(11):5751
Reel PS et al (2021) Using machine learning approaches for multi-omics data analysis: a review. Biotechnol Adv 49:107739
Reilly MP, Bornfeldt KE (2021) Integrative multiomics approaches for discovery of new drug targets for cardiovascular disease. Circulation 143(25):2471–2474
Rocha D et al (2020) Pan-cancer molecular patterns and biological implications associated with a tumor-specific molecular signature. Cells 10(1):45
Sarkar P et al (2023) Multi-omics analysis demonstrates the critical role of non-ethanolic components of alcoholic beverages in the host microbiome and metabolome: a human- and animal-based study. Microorganisms 11(6):1501

Satam H et al (2023) Next-generation sequencing technology: current trends and advancements. Biology 12(7):997

Sathyanarayanan A et al (2020) A comparative study of multi-omics integration tools for cancer driver gene identification and tumour subtyping. Brief Bioinform 21(6):1920–1936

Schussler-Fiorenza Rose SM et al (2019) A longitudinal big data approach for precision health. Nat Med 25(5):792–804

Sharifi-Noghabi H et al (2019) MOLI: multi-omics late integration with deep neural networks for drug response prediction. Bioinformatics 35(14):i501–i509

Subramanian I et al (2020) Multi-omics data integration, interpretation, and its application. Bioinform Biol Insights 14:1177932219899051

Syed-Abdul S, Iqbal U, Jack Li YC (2017) Predictive analytics through machine learning in the clinical settings. Comput Methods Prog Biomed 144:A1–A2

Tang B et al (2019) Recent advances of deep learning in bioinformatics and computational biology. Front Genet 10:214

Turanli B et al (2018) A network-based cancer drug discovery: from integrated multi-omics approaches to precision medicine. Curr Pharm Des 24(32):3778–3790

Vihinen M (2021) Functional effects of protein variants. Biochimie 180:104–120

Vijay A, Valdes AM (2022) Role of the gut microbiome in chronic diseases: a narrative review. Eur J Clin Nutr 76(4):489–501

Weckwerth W et al (2020) PANOMICS meets germplasm. Plant Biotechnol J 18(7):1507–1525

Wilkins MR et al (1996) Progress with proteome projects: why all proteins expressed by a genome should be identified and how to do it. Biotechnol Genet Eng Rev 13:19–50

Wu HJ, Wu E (2012) The role of gut microbiota in immune homeostasis and autoimmunity. Gut Microbes 3(1):4–14

Yang Y et al (2021) Applications of multi-omics technologies for crop improvement. Front Plant Sci 12:563953

Zakaria MS (2021) Data visualization as a research support service in academic libraries: an investigation of world-class universities. J Acad Librariansh 47(5):102397

Zhu B et al (2017) Integrating clinical and multiple omics data for prognostic assessment across human cancers. Sci Rep 7(1):16954

Metagenomics in the Census of Microbial Diversity

5

Jahnvi Hora and Indra Mani

5.1 Introduction

Microbiomes exist in every ecological niche and constitute the community of microorganisms found in different environments like in or on humans, animals, or plants. This community, which includes bacteria, fungi, viruses, protozoa, etc., works towards maintaining the health and functioning of the ecosystem it inhabits. The human body is host to a wide variety of microorganisms, particularly those that reside in the gastrointestinal tract, skin, oral cavity, urogenital tract, respiratory tract, and other mucosal surfaces. According to estimates, the human body houses trillions of microbes (Davenport et al. 2017). A multitude of microorganisms residing in the human microbiome establish a mutualistic association with the host organism, leading to advantageous outcomes for both entities. Perturbations in the microbiome's configuration give rise to dysbiosis, which is correlated with numerous health disorders including gastrointestinal ailments, allergic reactions, autoimmune conditions, and obesity (Manos 2022; Gupta et al. 2022). However, the microbiota refers to the assemblage of microorganisms that inhabit a specific ecological niche such as the skin, gut, soil, or any other environment. The composition and diversity of microbiota may vary depending upon the specific habitat or environment (Scarpellini et al. 2015). Depending upon the location, the microbiota serves various functions, for example being involved in digestion and metabolism (gastrointestinal tract), nutrient cycling and organic matter decomposition (soil), defence against pathogens, and contributing to skin health (El-Sayed et al. 2021).

For numerous decades, technologies utilizing microorganisms' ability to grow under predetermined, specified conditions (culture-dependent methods) have been

J. Hora · I. Mani (✉)
Department of Microbiology, Gargi College, University of Delhi, New Delhi, India
e-mail: indra.mani@gargi.du.ac.in

I. Mani, V. Singh (eds.), *Multi-Omics Analysis of the Human Microbiome*,
https://doi.org/10.1007/978-981-97-1844-3_5

the gold standard in detecting, recognizing, and delineating the microbial populace within a specific area. Nevertheless, the emergence of culture-independent techniques, which have progressed alongside high-throughput sequencing technologies, has also engendered methods that offer significant comprehension of the multiplicity, composition, and configuration of microbial populations. Such methods rely on the use of conserved regions—marker genes that are universal. These genes allow primer binding for PCR. For bacterial and archaeal communities, a region of the 16S ribosomal RNA is sequenced (Lane et al. 1985; Yarza et al. 2014; Edgar 2018). 16S rRNA is a part of bacterial and archaeal ribosomes, which is universally present. Internal Transcribed Spacer (ITS) is performed for fungal species (Wagner et al. 2018). These regions are noncoding sequences between the small and large subunits of rRNA genes (Schoch et al. 2012). Sequencing of these regions allows species identification and taxonomic classification, microbiome studies, biodiversity and ecological analysis, community research, phylogenetic research, disease research, and biotechnological applications (Council et al. 2007). While these methods are largely used, they limit our understanding of the complex biodiversity. Metagenomics, on the other hand, uses sequencing strategies to sequence the entire genomic DNA without using any target-specific primers (Yang et al. 2018; Mani 2020a, b). An overview of the procedure followed for metagenomic analysis tells us that the first step in any metagenomic study is the sampling and processing of data (Thomas et al. 2012). There must be a maximum DNA yield of good quality. One major issue in sampling DNA from human sources for metagenomic studies is the presence of a high quantity of host DNA. For this, host DNA depletion is carried out by a number of methods (Shi et al. 2022). The next step is shotgun sequencing, which has shifted from Sanger sequencing to next-generation sequencing (NGS) techniques (Mani 2020a, b). Technologies like 454/Roche and Illumina are in great demand when it comes to metagenomics (Thomas et al. 2012). Following sequencing comes the assembly of the obtained sequences. This can either be reference-based or de novo, both of which require specific algorithms and software. Binning is performed next to sort DNA into different groups which correspond to closely related organisms. This is followed by genome annotation, for which again many software and algorithms are used (Thomas et al. 2012).

Metagenomics offers a variety of advantages such as a more comprehensive sampling due to the analysis of the entire genetic material present in the entire sample and not just specific cultured organisms with known genomes; the ability to identify viruses which do not have a conserved region of genetic material like bacteria and fungi; a more comprehensive assessment of biodiversity; discovery of novel genes and ability to perform a functional analysis of the community such as identifying antibiotic resistance or virulence genes (Browne et al. 2016; Truong et al. 2015; Whiteside et al. 2021). It also ensures ample resolution for different species as well as strains (Conlan et al. 2012; Byrd et al. 2018). An illustrative representation of the construction of unknown and known genomes using metagenomics data is shown in Fig. 5.1 (Yang et al. 2021). The advantages of using metagenomics make it a great tool for microbiome studies.

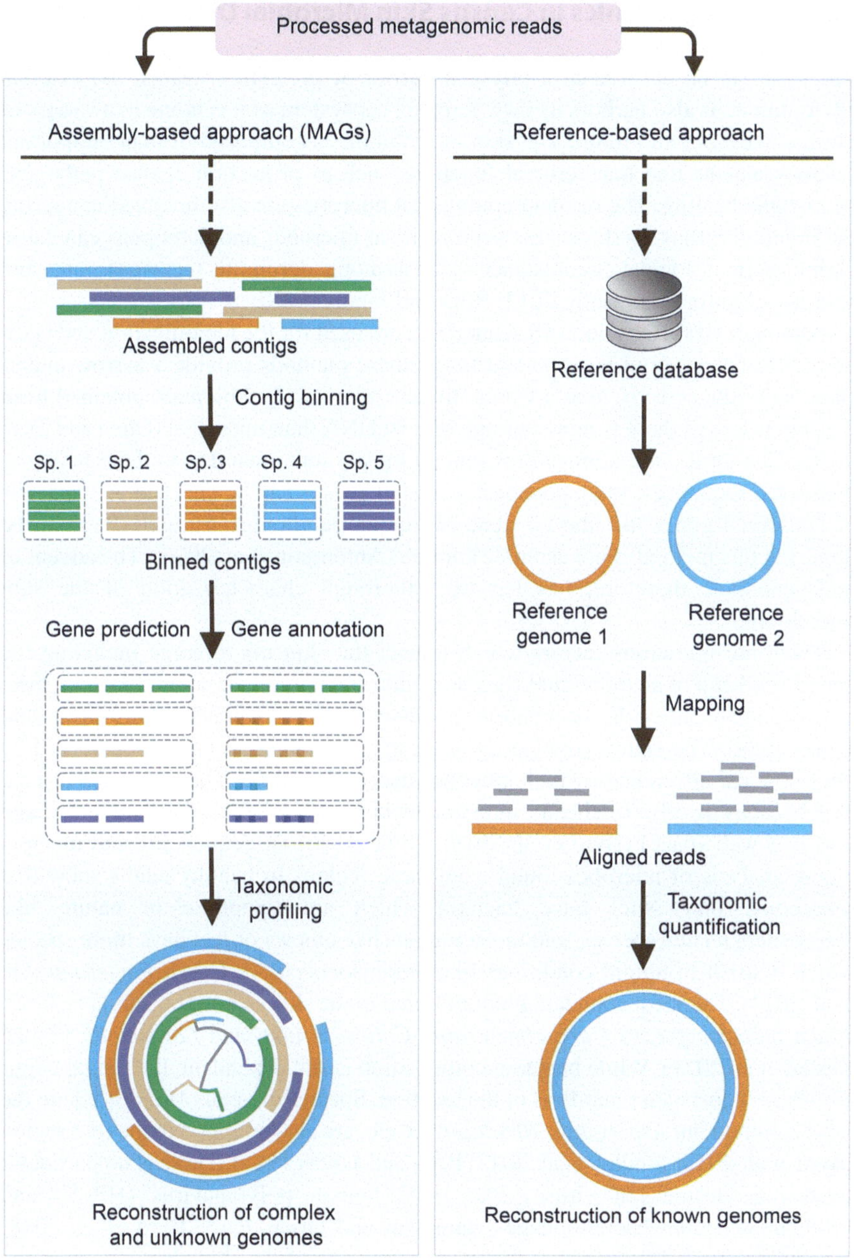

Fig. 5.1 A diagrammatic representation of the construction of unknown and known genomes using metagenomics data. (Adapted with permission from Yang et al. (2021))

5.2 Metagenomics in Census Skin Microbial Diversity

The skin can be considered a physical barrier of protection against the external environment. It also harbors its own complex ecosystem as it is home to a variety of viruses, bacteria, and fungi. The skin microbiome is composed of many important microorganisms that have crucial functions such as protection against pathogens and wound healing. The resident commensal microbes are also involved in altering the immunity; thus, dysbiosis between resident microbes and pathogens can cause various dermatological circumstances such as atopic dermatitis (eczema), acne, and psoriasis (Sanford and Gallo 2013; Roth and James 1988).

Although 16S rRNA and ITS sequencing are used for the identification and taxonomic classification of bacteria and fungi, these methods provide a narrow understanding of the ecosystem as a whole. In general, microbial biomass obtained from the skin is less as there is more amount of host DNA than microbial (Chen and Tsao 2013). The viral communities thus remain mostly unknown due to there being no conserved sequences. Phylogenetically related groups can be checked and analysis of conserved genes like the L1 gene of human papilloma viruses (HPV) can be done, but this method is not accurate enough (Antonsson et al. 2000). The advent of metagenomics, therefore, has led to a thorough characterization of the skin microbiome.

There are numerous factors which impact the skin microbiome including the genetic makeup of an individual, age, anatomic site or biogeography (like feet, face, arms), climate, lifestyle, occupation, hygiene, and use of medications (Roth and James 1988). The skin is divided into four major microenvironments governed by physiological differences, which influence the resident flora. These sites are termed as sebaceous or oily (forehead), moist (antecubital fossa), dry (volar forearm), and foot (toe web space) (Byrd et al. 2018; Oh et al. 2014). Sequencing and metagenomic analysis of microbes found from these regions in healthy adults show that sebaceous (oily) sites have bacteria which are lipophilic in nature like *Propionibacterium* species, and moist regions like elbows or feet have more species which flourish in humid conditions like *Staphylococcus* or *Corynebacterium* (Oh et al. 2014). The most abundant phylum found in the skin is Actinobacteria (~38%), which includes genera *Corynebacterium, Cutibacterium,* and *Lawsonella* (Saheb Kashaf et al. 2021). While bacterial composition is site dependent, the fungal populations are consistent regardless of the location. Species of genus *Malassezia* are the most common but *Aspergillus, Rhodotorula,* etc. are also common in the foot region (Byrd et al. 2018; Findley et al. 2013; Roth and James 1988). There is no particular conserved virome but viruses like HPV, human polyomavirus (HPyV) and *Propionibacterium* bacteriophage (Hannigan and Grice 2013; Byrd et al. 2018) have been found. The microbes that occupy any site of the skin also influences the metabolic potential of that part (Oh et al. 2014).

In certain conditions, commensal microbes may become pathogenic. The skin conditions due to this dysbiosis are associated with many different microbes, as confirmed by different metagenomic studies. Acne is an inflammatory teen skin condition caused due to *Propionibacterium acnes* (Leyden et al. 1975), which is a major colonizer of the healthy human microbiome (Fitz-Gibbon et al. 2013; Tomida et al. 2013). But the fact that only a few individuals have acne can be blamed on the genetic differences and the immunity of different individuals. Similarly, atopic dermatitis (AD) is associated with *Staphylococcus aureus* (Leyden et al. 1974). It is an inflammatory condition also known as eczema. Other *Staphylococcus* strains like *S. epidermis* have also been associated with AD (Byrd et al. 2017). Additional study showed that apart from *Staphylococcus* spp., increased *Gemella* spp. and decreased *Dermacoccus* spp. are seen (Chng et al. 2016). Other than these, microbes also affect chronic wounds such as diabetic foot ulcers (Prompers et al. 2006). Bacteria associated with diabetic foot ulcers are *Staphylococcus* spp., *Streptococcus* spp., and Proteobacteria spp. (Gardner et al. 2013). Fungi like *Cladosporium herbarum* and *Candida albicans* have also been identified (Kalan et al. 2016). Metagenomics has significantly advanced our understanding of the skin microbiome, uncovering its intricate composition and potential implications for human health. This powerful tool has illuminated the diversity and dynamic nature of microbial communities residing on our skin. Through metagenomic insights, we are better poised to explore the connections between the skin microbiome and dermatological conditions, as well as broader health outcomes. As research in this field continues to evolve, metagenomics is likely to play a pivotal role in personalized skincare and medical treatments, ultimately enhancing our knowledge of the skin microbiome's impact on well-being.

5.3 Metagenomics in Census Oral Microbial Diversity

The oral cavity is an environment that is highly abundant within the human body. The microbial community that resides within the oral cavity is known to inhabit various locations, such as the tongue, teeth, gingiva, lips, and hard and soft palate (Dewhirst et al. 2010). It is apparent why the microbes here are diverse as well as abundant. The mouth is in constant contact with the external environment by means of ingestion, breathing through mouth, talking, etc. The oral cavity is continuous with the digestive tract, the respiratory tract, and also the ears; thus, it is likely that the microbes in the mouth affect the microbial populations of these regions as well (Dewhirst et al. 2010). The microbes found in the oral cavity are not only dependent on the region where a person lives (geographical location and climate) (Li et al. (2014a, b), but also on other factors like pH (Sharma et al. 2018), diet (Sharma et al. 2018), whether a person smokes or not (Mason et al. 2014), etc.

The primary evidence of oral microbiota came long ago when Leeuwenhoek found bacteria in dental plaque (Gest 2004). The last decade has seen an increase in the number of studies to identify microbial species in the oral cavity. The oral microbiota is highly important because it is responsible for the maintenance of homeostasis and protection of the oral cavity from various pathogenic microbes. This can be attributed to the fact that the oral microbiome naturally occurs as a biofilm. Biofilms can be formed on the surfaces of soft or hard tissues in the oral cavity, and they are also important characteristics of oral infections. They develop within hours and this enforces the importance of oral hygiene. Streptococci are very potent biofilm developers that are the first to occur even after a few hours of brushing teeth (Xu and Gunsolley 2014). The microbes are generally necessary for immunological, metabolic, and physiological functions (Kilian et al. 2016). The need to study more about the oral microbial community arises from the fact that infections of the oral cavity are omnipresent. People of all age groups, races, and geographical locations are widely affected by dental caries (Selwitz et al. 2007) and periodontal disease (Pihlstrom et al. 2005; Horz and Conrads 2007). These microbes are now also getting associated with other problems like diabetes (Genco et al. 2005), cardiovascular disease (Joshipura et al. 1996; Beck and Offenbacher 2005), rheumatoid arthritis (Yeoh et al. 2013), inflammatory bowel disease (Ismail et al. 2012), bacterial pneumonia (Paju and Scannapieco 2007), and others.

Not all bacteria in the oral cavity can be cultivated. Thus, the culture-independent methods using 16S rRNA and metagenomic sequencing are widely being used to study the oral microbiota (Paster et al. 2001). Sample used for metagenomic studies of oral microbiome is supragingival dental plaque and for people with cavities, sample is obtained from the cavities as well (Belda-Ferre et al. 2011). Saliva can also be used as a sample for metagenomic sequencing of the oral microbiota (Hasan et al. 2014). The core microbiome of the oral cavity consists of the following phyla: Firmicutes (~33%), Proteobacteria (~27%), Actinobacteria (~14%), Bacteroidetes (~16%), Fusobacteria (6.7%), and, Spirochetes (<1%) (Palmer 2013; Zaura et al. 2009). Among various genera, *Actinomyces, Neisseria, Mannheimia, Corynebacterium, Streptococcus, Rothia, Bergeyella, Cardiobacterium*, etc. constitute the majority (Bik et al. 2010; Alcaraz et al. 2012). The commensal bacteria in the mouth are known to colonize specific locations (Jørn et al. 2005). Like *Prevotella* and *Flavobacterium* prefer the tongue, Firmicutes and Actinobacteria colonize dental plaques (Jørn et al. 2005). The core microbiome not only consists of bacteria but also includes other organisms such as fungi, viruses, and ultra-small bacteria that belong to the newly classified "Candidate Phyla Radiation" (CPR) group, which cannot be cultured; example is the TM7 group (Paster et al. 2001). *Candida* spp. is the major fungal taxa associated with biofilm formation in the oral cavity (Janus et al. 2017). But fungal species are in low proportions, overall. Different methods for identifying and profiling the mycobiome and virome of the oral cavity have been employed like MiCoP (Microbial Community Profiling method for detecting viral and fungal organisms in metagenomic samples) (LaPierre et al. 2019) and MetaPhlAn (Metagenomic Phylogenetic Analysis) (Truong et al. 2015). The presence of *Wickerhamomyces* (~42%), *Penicillium* (~30%), *Enterocytozoon* (~16%), and *Candida* (~11%) genera has been established by the above methods (LaPierre et al. 2019). Among viral species, *Streptococcus* phage is the most abundant (~75%), followed by *Actinomyces* phage (~13%) and Human herpesvirus 7 (~4%) (LaPierre et al. 2019). *Dialister, Basfia, Treponema, Oligotropha,* and *Parvibaculum* are some genera associated with dental caries (Alcaraz et al. 2012). Dental caries are mostly caused by acidogenic, biofilm producing bacteria like *Streptococcus* spp. (Peterson et al. 2011). Microbes associated with periodontal disease include the genera

Prevotella, Veillonella, Treponema, Leptotrichia, and *Porphyromonas* (Wang et al. 2013).

The Human Oral Microbiome Database (HOMD) (https://www.homd.org/), an extensively curated and specifically designated phylogeny-based database, is dedicated to the examination of the oral microbiome. Comprising 16S rRNA sequences of all characterized bacteria inhabiting the oral cavity and the aerodigestive tract, which encompasses the upper digestive and respiratory tracts such as the pharynx, sinuses, nasal passage, and esophagus (Chen et al. 2010). HOMD is called a "foundation for the oral microbiome and metagenome" (Chen et al. 2010). With a total of around 774 prokaryotic species, this database has information about 74% cultivable and 26% non-cultivable bacteria. Only 58% are officially named. Available at https://www.homd.org/ (Accessed 23 March 2024). Metagenomics has the potential to revolutionize dental and healthcare practices, as it enables us to study the microbial composition of the oral cavity with precision. As this field continues to evolve, metagenomics is poised to play a pivotal role in personalized dental care, contributing to a deeper understanding of how the oral microbiome influences health and disease.

5.4 Metagenomics in Census Lung Microbial Diversity

Lungs are vital organs which have a very important role in gas exchange. Traditionally, they have been regarded as a completely sterile environment, and any presence of microbes originating from the oral microbiota, as observed in culture-dependent methods, was attributed to sample contamination arising from exposure to the upper respiratory airways in the lower respiratory airways (Segal et al. 2014). However, the emergence of culture-independent techniques such as metagenomics has firmly established that the lungs are not devoid of microbial presence. This notion is easily comprehensible considering the constant interaction they have with the microorganism-rich upper respiratory tract (Whiteside et al. 2021; Yi et al. 2022).

The pulmonary microbiome plays a crucial role in the immune response of the airway mucosa. This is in accordance with studies conducted previously which show that children are less prone to developing allergies and asthma when they grow up in environments with diverse microbes (Ege et al. 2006, 2011), as is also seen during immune maturation of the gut by exposure to certain bacteria at an early age (Ivanov et al. 2008; Atarashi et al. 2011). The involvement of the lung microbiome in the regulation of airway inflammation is facilitated by the activation of Th17 cells. The quantity of bacterial load is directly proportional to the number of Th17 lymphocytes and pro-inflammatory cytokines that are activated (Jin et al. 2019; Cheng et al. 2014; Segal et al. 2016). It has been hypothesized that the gut and lung microbiome are in constant communication via a gut-lung axis, leading to the direct translocation of bacteria and indirect translocation of immune cells and cytokines (Halley et al. 2020; Beck et al. 2012; Helmink et al. 2019; Bachem et al. 2019).

It is a widely recognized fact that respiratory diseases are among the primary causes of illness and mortality worldwide. The presence of microorganisms, whether directly or indirectly, plays a significant role in the development of these conditions (Cox et al. 2019). Within the microbiome of a healthy lung, the phyla Firmicutes and Bacteroidetes are the most prevalent. The predominant genera from protected specimen brushes (PSB) are *Acinetobacter, Pseudomonas,* and *Ruminococcus* (~29%), while *Prevotella, Yersinia,* and *Pseudomonas* are from airway wall (~28%) (Charlson et al. 2011; Dickson et al. 2017). The dynamicity of the lung microbiome could be blamed mostly on microbial migration, which includes immigration and emigration but in case of diseases, it also includes the replication of existing microbes (Yi et al. 2022). In comparison to the upper respiratory tract, the lower tract has significantly lower biomass, which is maintained through clearance mechanisms such as coughing, mucociliary transport, and the innate immune system (Natalini et al. 2022). The existence of dysbiosis, caused by the constantly changing nature of the microbiome, leads to the manifestation of various disease states, such as cystic fibrosis, chronic obstructive pulmonary disease (COPD), idiopathic pulmonary fibrosis, asthma, etc. (Natalini et al. 2022). Chronic respiratory diseases are characterized by different microbes in disease states such as members of Proteobacteria like *Pseudomonas*, *Haemophilus*, and *Moraxella* (Hilty et al. 2010). Non-Proteobacteria taxa like *Fusobacterium*, *Porphyromonas*, etc. are linked with poor prognosis of severe asthma (Chung 2017). Increased fungi like *Aspergillus, Penicillium,* and *Cryptococcus* can also be associated with asthmatic conditions (MacAogáin et al. 2018). In acute respiratory issues such as diseases caused by viruses, metagenomics and metatranscriptomics studies have revealed the presence of different bacteria in severely ill patients, like *Burkholderia cepacia*, *Staphylococcus epidermidis*, and *Mycoplasma* spp. which were codetected with viruses (Huang et al. 2021).

Besides culturing to identify the microbes, studies for characterizing the lung microbiome have focused on sequencing of conserved regions and metagenomics. For bacteria, 16s rRNA gene can be sequenced, and the 18s rRNA gene or the ITS region for fungi. The other sequencing approach that can be used is metagenomics shotgun sequencing. As the biomass is very low in case of lungs, the host DNA is usually a challenge in metagenomics-based studies and can be usually depleted, while microbial DNA is enriched (Sulaiman et al. 2021; MacAogáin et al. 2020 Millares et al. 2015; Cameron et al. 2017). This technique tells us about the composition as well as the gene content which focuses on the functional potential of the environment. It can also identify viruses which do not share any conserved genes like fungi and bacteria. The technique majorly focuses on the DNA content, and thus, cannot distinguish between live and dead organisms. This is a matter of concern in the case of the microbes in the lung as the balance between the microbes, via immigration and emigration, determines disease conditions (Piva et al. 2020; Rogers et al. 2010). To obtain samples of the lung microbiome, commonly employed approaches include the utilization of sputum and bronchial alveolar fluid (BALF). While sputum collection is noninvasive, it can easily be contaminated, so the method is more useful in the case of conditions like cystic fibrosis when the microbial burden is high (Lu et al. 2020). On the other hand, BALF is hardly influenced by contamination but it is invasive (Charlson et al. 2012). Metagenomics has the potential to transform the way we approach lung-related conditions and may lead to more precise diagnostics and personalized treatment strategies. As research in this field continues to progress, metagenomics is poised to play a pivotal role in enhancing our understanding of the lung microbiome's influence on pulmonary health, ultimately leading to improved patient care and outcomes.

5.5 Metagenomics in Census Gut Microbial Diversity

The gastrointestinal system, commonly referred to as the gut, encompasses the stomach, intestines, and colon. Its primary functions include digestion, the absorption of nutrients, and the elimination of waste products. The interdependence between the human gut and its microbiome is increasingly acknowledged and regarded as significant. This acknowledgment stems from the realization that the microbial composition of the gut largely determines an individual's overall well-being. Recent attention has been directed towards the gut microbiome due to the Human Microbiome Project (HMP) (Huttenhower et al. 2012, Blottière et al. 2013) as well as the MetaHIT (Metagenomics of the Human Intestinal tract) (Qin et al. 2010). The healthy gut microbiota is a multifaceted and dynamic ecosystem that encompasses not only bacteria but also protozoans, fungi, and viruses (Weinstock 2012). Majority of these microorganisms are commensal and in a symbiotic relationship with the body. They play very important roles in immune modulation (Viaud et al. 2013; Maynard et al. 2012), safety against pathogens (Fukuda et al. 2011; Endt et al. 2010), management of nutrient and drug metabolism (Cani 2013; Tremaroli and Bäckhed 2012), as well as maintenance of the structure and function of the gastrointestinal tract (Lutgendorff et al. 2008; Cario et al. 2007). It is believed that individuals that are healthy have a more diverse gut microbial population (Chatelier et al. 2013).

Although culture-dependent methods as well as 16S rRNA sequencing have been successful strategies for the census of microbial diversity in case of the gastrointestinal tract, populations such as viruses, fungi, and protozoa were still not well characterized (Wang et al. 2015). For investigations pertaining to the gut microbiota using metagenomic methods, the chosen specimen is feces. Based on previous metagenomic studies, it has been observed that a significant proportion (99%) of the microorganisms discovered in the gut are of bacterial origin. Over 160 bacterial species are present (Qin et al., 2010). The natural selection of microorganisms within the intestinal ecosystem is crucial for the maintenance of homeostasis, taking into account the influence of both the host and other microbial communities present in the surrounding environment. This is responsible for maintaining the numbers of microbes, some in high abundance and others lower (Dethlefsen et al. 2008). The proper characterization of the gut metagenome (from the feces) has shown that three groups of bacteria are most predominant; these are *Prevotella, Ruminococcus,* and *Bactericides* spp. (Arumugam et al. 2011). The most common phyla in the gut are Firmicutes and Bacteroidetes making up almost 90% of the population. And the most abundant genera are *Bacteroides, Clostridium, Lactobacillus,*

Faecalibacterium, Ruminococcus, and *Bifidobacterium* while the least abundant ones are *Coprobacillus* and *Gordonibacter* (Rinninella et al. 2019). *Methanobrevibacter smithii* is an archaebacteria commonly found in the gut (Gill et al. 2006). The protozoan population in the gut includes some commensal commonly occurring species such as *Entamoeba* spp. (~46%), *Blastocystis* spp. (~61%), and *Enteromonas* (~11%) as shown by metagenomic studies (Lokmer et al. 2019; Parfrey et al. 2014). The protozoa which exist commensally or even symbiotically increase the gut microbial biodiversity as well as modulate the bacterial population (Dubik et al. 2022). The resident mycobiome of the gut has been shown to contain the following genera: *Candida* (Moran et al. 2014; Hallen-Adams and Suhr 2016), *Malassezia* (Suhr et al. 2016), and *Cladosporium* (Cohen et al. 1969). Maintenance of a healthy gut microbiome is also heavily dependent on the virome, more specifically, the phagome (Manrique et al. 2016). The phages identified according to metagenomic studies include an abundance of the order Caudovirales (almost 90%) and the families within it *Podoviridae*, *Siphoviridae*, and *Myoviridae.* Other than those, *Microviridae* family is also present in the gut (Manrique et al. 2016; Hoyles et al. 2014; Shkoporov et al. 2018; Aggarwala et al. 2017; Sutton and Hill 2019).

The precise constitution of the microbial populace in the gastrointestinal tract is contingent upon the location, which in turn relies on the physiological characteristics as well as the metabolites that are accessible in that particular region (Macpherson and McCoy 2013). In the small intestine, fast growing bacteria and facultative anaerobes are present, but their proliferation is ordinarily restricted due to the prevalence of low pH, antimicrobial substances, and oxygen (Donaldson et al. 2015). Thus, microbes belonging to Lactobacillaceae and Enterobacteriaceae can grow (Donaldson et al. 2015). The colon has low concentration of antimicrobials and is abundant with a diverse population of microbes from the families *Bacteroidaceae, Ruminococcaceae, Prevotellaceae, Lachnospiraceae*, and *Rikenellaceae* (Donaldson et al. 2015). The microbial population of the gut is influenced by many factors. These include age (Yatsunenko et al. 2012; Biagi et al. 2013), geography (Suzuki and Worobey 2014), diet (Claesson et al. 2012), and drugs (Vrieze et al. 2014).

Dysbiosis of the gut microbiota refers to an imbalance between commensal microbes and pathogenic species. This phenomenon has been linked to numerous pathological conditions such as irritable bowel syndrome (IBS) (Carroll et al. 2012), colorectal cancer (Ahn et al. 2013), inflammatory bowel disease (IBD) (Manichanh 2006), diarrhea (Pop et al. 2014), and even psychiatric or neurodevelopmental problems (Hsiao et al. 2013). Different bacterial populations are involved in different diseases in humans. Patients with colorectal cancer experience an elevation in *Bacteroides fragilis* and *Fusobacterium* levels, while experiencing a decline in *Faecalibacterium* levels (Wang et al. 2011, 2013). A similar pattern can be observed in individuals with IBD, where *Ruminococcus* spp. levels increase alongside a decrease in other bacterial populations (Sokol et al. 2008; Png et al. 2010; Machiels et al. 2013). Conversely, patients with IBS exhibit an increase in *E. coli* levels and a decrease in *Clostridium leptum* (Duboc et al. 2012). Finally, psychiatric issues can also be linked with an imbalance in the microbiota. People with depression have an increase in genera *Eggerthella, Gelria, Holdemania, Papaprevotella,* etc. in the gut and a decrease in *Dialister* and *Prevotella* (Kelly et al. 2016).

Functional metagenomics is not only beneficial for the census of microbial population of the gut but also for identifying antimicrobial resistance genes (Hu et al. 2013), detecting alterations in microbial functions (Wei et al. 2013; Qin et al. 2012), determining host-microbe interactions (Lakhdari et al. 2010; Wang et al. 2015; Woyke 2019). Along with that, using metagenomics, the entire enzymatic profile of the gut microbiome can be characterized (Tanes et al. 2021). However, despite its efficacy in discerning microbial populations, metagenomics is not particularly useful in characterizing the functionality of the microbiome. For a more thorough and in-depth understanding of the gut, metatranscriptomics (Poretsky et al. 2009), metaproteomics (VerBerkmoes et al. 2008), and metabolomics (Nicholson et al. 2005) are preferred.

5.6 Metagenomics in Census Urinary Tract Microbial Diversity

The urinary system includes the kidneys, ureters, bladder, and urethra, with its primary function being the filtration of waste substances from the bloodstream and the subsequent expulsion of these substances in the form of urine. Until recently, the urinary system was widely believed to be an environment devoid of any microbial presence, and the presence of microbes was indicative of a potential infection. This was majorly due to the fact that urine is considered inhospitable to microbes keeping in mind that it has a high urea content, low pH, and other substances which are antimicrobial (Ipe et al. 2016). However, the emergence of advanced sequencing technologies and metagenomic analysis has demonstrated that the urinary tract contains a diverse and dynamic microbiota (Imirzalioglu et al. 2008; Zasloff 2007). The urinary microbiome, also known as the "urobiome," refers to the comprehensive microbial community inhabiting the urinary tract and remains an area of ongoing research. These microorganisms play pivotal roles in upholding the urinary well-being of individuals, and an imbalance in their composition can result in various disorders such as neurogenic bladder dysfunction, interstitial cystitis, and urinary tract infections (UTIs) (Whiteside et al. 2015).

The conditions for urine culture limit the microorganisms that can be detected as only fast growing bacteria like *E. coli* and *Enterococcus faecalis* can grow while fastidious, slow growing bacteria or those with specific temperature or oxygen requirements like *Corynebacterium*, *Lactobacillus*, and *Ureaplasma* evade the screening (Wolfe and Brubaker 2015; Brubaker and Wolfe 2015; Wolfe et al. 2012). For this purpose, expanded quantitative urine culture (EQUC) was developed for a better and more meticulous understanding of the urobiome (Hilt et al. 2014). Molecular techniques have allowed a more thorough characterization of the urobiome. 16S rRNA sequencing and metagenomic analysis have revealed the existence of taxonomic groups like Firmicutes (65%), Bacteroidetes (18%), Actinobacteria (12%), Fusobacteria (3%), and Proteobacteria (2%) (Siddiqui et al. 2011) and the major genera are *Lactobacillus, Staphylococcus, Streptococcus, Corynebacterium,* and *Prevotella* (Brubaker and Wolfe 2017). However, variability is observed among different people. The variations are mostly observed between different genders and age groups. Women harbor a more diverse mix of bacteria than males (Lewis et al. 2013). This could be due to the different compositions of the urine which may favor certain microbes over others (Ipe et al. 2016). The composition of the urobiome may also be influenced by age, as a result of alterations in diet, personal hygiene practices (Whiteside et al. 2015), changes in hormones (Karstens et al. 2016), etc. Both age and gender are also associated with susceptibility to certain diseases/infections. Bacteria make up a major proportion of the urobiome in normal, healthy individuals. However, viruses can also be found in healthy, asymptomatic or immunocompetent people. Most of these are opportunistic and cause infections only in individuals with a weakened immune system. Different viruses have been identified in the urine; these include JC and BK polyomaviruses, herpesvirus, papillomavirus, and more (Moustafa et al. 2018; Paduch 2007). The virome is also composed of bacteriophages infecting *E. coli, Enterococcus,* or *Staphylococcus* (Santiago-Rodríguez et al. 2015; Garretto et al. 2018; Moustafa et al. 2018). Bacteriophages present in the urinary tract continue to be a subject that is currently being examined, albeit one of the potential reasons for their presence could possibly be attributed to their capacity to safeguard the urinary tract from uropathogenic bacteria (Garretto et al. 2019). The mycobiome of the urinary tract is not that well characterized either. Lack of culture-dependent screening of fungi puts more attention on the metagenomic analysis. There are hardly any commensal fungi in the urinary tract; however pathogenic fungi are present in some conditions. *Candida* spp. is the most common pathogen among *Aspergillus, Histoplasma, Mucoraceae,* and *Cryptococcus* (Enoch et al. 2006; Sanglard 2016) as shown by culture-based methods.

The urobiome has similar genera as other locations in the body but it has low biomass and lesser diversity in comparison to other sites like the gut (Pearce et al. 2014). But, knowledge about the microbiome gives the opportunity of gaining a better understanding of the diseases associated, gynecological issues, etc. (Brubaker and Wolfe 2015). UTIs are a prevalent occurrence among a significant number of individuals and are predominantly associated with the female population (Flores-Mireles et al. 2015). The most commonly affecting bacteria are *E. coli*, followed by other *Enterobacteriaceae* (Foxman 2010) but, through the utilization of 16S rRNA and metagenomic sequencing studies, it has been demonstrated that *Klebsiella, Lactobacillus, Citrobacter, Pseudomonas, Gardnerella*, etc. are also present (Moustafa et al. 2018). Some of the fungal species which most commonly affect humans in UTIs are *Candida albicans, C. orthopsilosis, Clavispora lusitaniae, Meyerozyma guilliermondii,* and *Malassezia globosa* (Kauffman 2014; Moustafa et al. 2018). In conclusion, metagenomics has emerged as a powerful tool in the study of the urinary microbiome, offering invaluable insights into a previously underexplored aspect of human health. By enabling us to comprehensively catalog and analyze the diverse microbial communities residing within the urinary tract, metagenomics has not only expanded our understanding of the role of the urinary microbiome but also opened new avenues for research and clinical applications.

5.7 Metagenomics in Census Brain Microbial Diversity

The brain, which serves as a vital organ in the human body, is characterized by its high degree of complexity. It serves as the control center for the body, responsible for processing sensory information, controlling bodily functions, and enabling cognitive functions such as thinking, learning, and memory. There is barely any evidence of there being a brain microbiome and, up until recently, no neurological conditions were associated with any microorganism but a pioneer and influential study provided evidence against the sterility of the brain (Branton et al. 2013). The researchers found different bacteria and bacteriophage sequences upon total RNA high-throughput sequencing from the autopsy-derived brain injury of HIV/AIDS patients. 16S rRNA sequencing was also performed. Somewhat similar populations were found in control and HIV samples and α-proteobacteria is the predominant phylum found in all samples. In addition to this, it is crucial to take into account the reality that healthy human blood harbors microorganisms, as evidenced by metagenomic sequencing (Païssé et al. 2016). Furthermore, all human tissues have a certain quantity of blood, encompassing the white matter within the brain (Hase et al. 2019).

The utilization of metagenomics for the purpose of characterizing the microbial heterogeneity in the brain under diseased conditions is of considerable significance owing to numerous factors. Most neuroinvasive pathogens are rare and less studied, which makes the diagnosis difficult. Via metagenomic sequencing, a single test can help diagnose and characterize the cause of the infection (Ramachandran and Wilson 2020). This also includes viral infections with potential to cause neurological disorders like Zika virus, Ebola virus, Chikungunya, etc.; metagenomics can easily identify viral pathogens as well (Tyler 2009; Wilson and Tyler 2016; Wilson et al. 2017a; Howlett et al. 2018).

The quality of the sample for metagenomic sequencing is something that should be considered along with the timing for obtaining it. Cerebrospinal fluid (CSF) is usually collected for metagenomic analysis. Patients with chronic infections (like meningoencephalitis) may have symptoms for over a month and thus there is a larger amount of time to obtain the CSF sample which may give results (Wilson et al. 2014, 2018; Beck et al. 2019). Similarly, in patients with acute infections (like West Nile virus), the CSF is best collected within a few days of illness; otherwise the virus will not be present in the CNS (DeBiasi and Tyler 2004; Wilson et al. 2017b). A person undergoing an antibiotic treatment may give a negative result in metagenomic sequencing of the CSF (Wilson et al. 2019; Brink et al. 2015; Zhang et al. 2019). Also, in case of compartmentalized infections (like brain abscess) or low abundance of pathogens, CSF metagenomic analysis can give false results (Wilson et al. 2019). So, biopsy samples from the brain can serve as an alternative for metagenomic analysis, but it is invasive (Wilson et al. 2017a; Salzberg et al. 2016; Morfopoulou et al. 2016; Naccache et al. 2015).

According to a study (Branton et al. 2016), bacteria in the normal microbiota of the brain include Proteobacteria (~60%) and Actinobacteria (~40%) (including *Propionibacteriaceae*). The same microbes were found in the brains of patients with multiple sclerosis. Another study assessing the metagenome of brain abscesses indicates how diverse the microbiota here is (Lin et al. 2019). The most widespread genera are *Mycobacterium* (~19%), then *Streptococcus* (~16%). Fungal populations included *Debaryomyces* (~41%), *Aspergillus* (~24%), *Malassezia* (~19%), and *Mitosporidiym* (12%). A virome was also found; this consisted of *Herpesvirus* (~48%), *Moumouvirus* (~32%), and *Granulovirus* (~8%). Similarly, a study conducted on elderly people and patients diagnosed with Alzheimer's disease (Alonso et al. 2018) demonstrated that Proteobacteria was the major phylum followed by Actinobacteria and Firmicutes in AD patients. In the control group (elderly and younger people without AD), the proportion of *Micrococcaceae*, *Pseudomonadaceae*, and *Xanthomonaceae* was higher and pathogenic *Staphylococcaceae* was more prominent in AD patients, suggesting that it may be involved in worsening clinical symptoms. As for the mycobiome, genera like *Alternaria* and *Malassezia* were more in patients with AD, and *Candida* and *Aspergillus* were more in elderly people as compared to AD patients and younger people. *Propionibacterium acnes* can also be found in patients with AD (Emery et al. 2017).

There is always a possibility that the sequences detected by sequencing methods (16S rRNA or metagenomics) in human tissues are due to contamination during extraction of genetic material (Lusk 2014; Salter et al. 2014; Mangul et al. 2018). Considering that the microbes would have to cross the blood-brain barrier to enter the brain, the order of magnitude of the microbes would be far lesser than that of the gut. So, chances that the sequences detected by metagenomics are just a result of contamination are persistent.

5.8 Conclusion and Future Perspectives

Identifying the normal microflora of the human microbiome is of immense importance. This helps in not only associating these microorganisms with specific functions but also helps in understanding how dysbiosis affects the ecological niche and leads to diseases. The advent of metagenomics has provided a better understanding of the constituents of the microbial communities, the roles of the members of the communities, and the way they adapt to different ecological niches. Metagenomics has the ability to globally analyze the microbiome of a particular biogeographical location, from bacteria and fungi to viruses and protozoans. While it is a powerful and unbiased technique to identify microbes, it fails to distinguish the viability of microbes. In such cases, culture-dependent techniques can be considered better (Byrd et al. 2018). Functional characterization via metagenomics is based on the gene content and not the actual activity. Techniques with a combination of culture-dependent and independent technologies may prove to be optimum. Finally, multi-omics technologies exist and can be utilized for better indication of functional profiles of a microbiome. Metatranscriptomics (mRNA/transcript sequencing), metaproteomics (quantification of proteins), and metabolomics (analysis of metabolites) help in better understanding of the functional activities of the microbiome.

Acknowledgments The authors are thankful to the principal of Gargi College for providing the infrastructural support.

Conflict of Interest None.

References

Aggarwala V, Liang G, Bushman FD (2017) Viral communities of the human gut: metagenomic analysis of composition and dynamics. Mob DNA 8(1):12. https://doi.org/10.1186/s13100-017-0095-y

Ahn J, Sinha R, Pei Z et al (2013) Human gut microbiome and risk for colorectal cancer. J Natl Cancer Inst 105(24):1907–1911. https://doi.org/10.1093/jnci/djt300

Alcaraz LD, Belda-Ferre P, Cabrera-Rubio R et al (2012) Identifying a healthy oral microbiome through metagenomics. Clin Microbiol Infect 18:54–57. https://doi.org/10.1111/j.1469-0691.2012.03857.x

Alonso R, Pisa D, Fernández-Fernández AM, Carrasco L (2018) Infection of fungi and bacteria in brain tissue from elderly persons and patients with Alzheimer's disease. Front Aging Neurosci 10:159. https://doi.org/10.3389/fnagi.2018.00159

Antonsson A, Forslund O, Ekberg H, Sterner G, Hansson B (2000) The ubiquity and impressive genomic diversity of human skin papillomaviruses suggest a commensalic nature of these viruses. J Virol 74(24):11636–11641. https://doi.org/10.1128/jvi.74.24.11636-11641.2000

Arumugam M, Raes J, Pelletier E et al (2011) Enterotypes of the human gut microbiome. Nature 473(7346):174–180. https://doi.org/10.1038/nature09944

Atarashi K, Tanoue T, Shima T et al (2011) Induction of colonic regulatory T cells by indigenous Clostridium species. Science 331:337. https://doi.org/10.1126/science.1198469

Bachem A, Makhlouf C, Binger KJ et al (2019) Microbiota-derived short-chain fatty acids promote the memory potential of antigen-activated CD8+ T cells. Immunity 51(2):285–297.e5. https://doi.org/10.1016/j.immuni.2019.06.002

Beck JD, Offenbacher S (2005) Systemic effects of periodontitis: epidemiology of periodontal disease and cardiovascular disease. J Periodontol 76(11-s):2089–2100. https://doi.org/10.1902/jop.2005.76.11-s.2089

Beck JM, Young VB, Huffnagle GB (2012) The microbiome of the lung. Transl Res 160(4):258–266. https://doi.org/10.1016/j.trsl.2012.02.005

Beck E, Ramachandran P, Khan LM et al (2019) Clinicopathology conference: 41-year-old woman with chronic relapsing meningitis. Ann Neurol 85(2):161–169. https://doi.org/10.1002/ana.25400

Belda-Ferre P, Alcaraz LD, Cabrera-Rubio R et al (2011) The oral metagenome in health and disease. ISME J 6(1):46–56. https://doi.org/10.1038/ismej.2011.85

Biagi E, Candela M, Turroni S, Garagnani P, Franceschi C, Brigidi P (2013) Ageing and gut microbes: perspectives for health maintenance and longevity. Pharmacol Res 69(1):11–20. https://doi.org/10.1016/j.phrs.2012.10.005

Bik EM, Long CD, Armitage GC et al (2010) Bacterial diversity in the oral cavity of 10 healthy individuals. ISME J 4(8):962–974. https://doi.org/10.1038/ismej.2010.30

Blottière HM, De Vos WM, Ehrlich SD, Doré J (2013) Human intestinal metagenomics: state of the art and future. Curr Opin Microbiol 16(3):232–239. https://doi.org/10.1016/j.mib.2013.06.006

Branton WG, Ellestad KK, Maingat F et al (2013) Brain microbial populations in HIV/AIDS: α-proteobacteria predominate independent of host immune status. PLoS One 8(1):e54673. https://doi.org/10.1371/journal.pone.0054673

Branton WG, Lu J, Surette MG et al (2016) Brain microbiota disruption within inflammatory demyelinating lesions in multiple sclerosis. Sci Rep 6(1):37344. https://doi.org/10.1038/srep37344

Brink M, Welinder-Olsson C, Hagberg L (2015) Time window for positive cerebrospinal fluid broad-range bacterial PCR and *Streptococcus pneumoniae* immunochromatographic test in acute bacterial meningitis. Infect Dis 47(12):869–877. https://doi.org/10.3109/23744235.2015.1078907

Browne HP, Forster SC, Anonye BO et al (2016) Culturing of 'unculturable' human microbiota reveals novel taxa and extensive sporulation. Nature 533(7604):543–546. https://doi.org/10.1038/nature17645

Brubaker L, Wolfe AJ (2015) The new world of the urinary microbiota in women. Am J Obstet Gynecol 213(5):644–649. https://doi.org/10.1016/j.ajog.2015.05.032

Brubaker L, Wolfe AJ (2017) The female urinary microbiota, urinary health and common urinary disorders. Ann Transl Med 5:34. https://doi.org/10.21037/atm.2016.11.62

Byrd AL, Deming C, Cassidy SKB et al (2017) *Staphylococcus aureus* and *Staphylococcus epidermidis* strain diversity underlying pediatric atopic dermatitis. Sci Transl Med 9(397):eaal4651. https://doi.org/10.1126/scitranslmed.aal4651

Byrd AL, Belkaid Y, Segre JA (2018) The human skin microbiome. Nat Rev Microbiol 16(3):143–155. https://doi.org/10.1038/nrmicro.2017.157

Cameron S, Lewis K, Huws S et al (2017) A pilot study using metagenomic sequencing of the sputum microbiome suggests potential bacterial biomarkers for lung cancer. PLoS One 12(5):e0177062. https://doi.org/10.1371/journal.pone.0177062

Cani PD (2013) The gut microbiota manages host metabolism. Nat Rev Endocrinol 10(2):74–76. https://doi.org/10.1038/nrendo.2013.240

Cario E, Gerken G, Podolsky DK (2007) Toll-like receptor 2 controls mucosal inflammation by regulating epithelial barrier function. Gastroenterology 132(4):1359–1374. https://doi.org/10.1053/j.gastro.2007.02.056

Carroll IM, Ringel-Kulka T, Siddle JP, Ringel Y (2012) Alterations in composition and diversity of the intestinal microbiota in patients with diarrhea-predominant irritable bowel syndrome. Neurogastroenterol Motil 24(6):521–e248. https://doi.org/10.1111/j.1365-2982.2012.01891.x

Charlson ES, Bittinger K, Haas AR et al (2011) Topographical continuity of bacterial populations in the healthy human respiratory tract. Am J Respir Crit Care Med 184(8):957–963. https://doi.org/10.1164/rccm.201104-0655oc

Charlson ES, Bittinger K, Chen J et al (2012) Assessing bacterial populations in the lung by replicate analysis of samples from the upper and lower respiratory tracts. PLoS One 7(9):e42786. https://doi.org/10.1371/journal.pone.0042786

Chatelier EL, Nielsen T, Qin J et al (2013) Richness of human gut microbiome correlates with metabolic markers. Nature 500(7464):541–546. https://doi.org/10.1038/nature12506

Chen YE, Tsao H (2013) The skin microbiome: current perspectives and future challenges. J Am Acad Dermatol 69(1):143–155.e3. https://doi.org/10.1016/j.jaad.2013.01.016

Chen T, Yu W, Izard J, Baranova OV, Lakshmanan A, Dewhirst FE (2010) The Human Oral Microbiome Database: a web accessible resource for investigating oral microbe taxonomic and genomic information. Database 2010:baq013. https://doi.org/10.1093/database/baq013

Cheng M, Qian L, Shen G et al (2014) Microbiota modulate tumoral immune surveillance in lung through a γδT17 immune cell-dependent mechanism. Cancer Res 74(15):4030–4041. https://doi.org/10.1158/0008-5472.can-13-2462

Chng KR, Tay AAO, Li C et al (2016) Whole metagenome profiling reveals skin microbiome-dependent susceptibility to atopic dermatitis flare. Nat Microbiol 1(9):16106. https://doi.org/10.1038/nmicrobiol.2016.106

Chung KF (2017) Airway microbial dysbiosis in asthmatic patients: a target for prevention and treatment? J Allergy Clin Immunol 139(4):1071–1081. https://doi.org/10.1016/j.jaci.2017.02.004

Claesson MJ, Jeffery IB, Conde S et al (2012) Gut microbiota composition correlates with diet and health in the elderly. Nature 488(7410):178–184. https://doi.org/10.1038/nature11319

Cohen R, Roth FJ, Delgado E, Ahearn DG, Kalser MH (1969) Fungal flora of the normal human small and large intestine. N Engl J Med 280(12):638–641. https://doi.org/10.1056/nejm196903202801204

Conlan S, Mijares LA, Becker J et al (2012) Staphylococcus epidermidis pan-genome sequence analysis reveals diversity of skin commensal and hospital infection-associated isolates. Genome Biol 13(7):R64. https://doi.org/10.1186/gb-2012-13-7-r64

Cox MJ, Ege MJ, Von Mutius E (2019) The lung microbiome. European Respiratory Society, Lausanne

Davenport E, Sanders JG, Song SJ, Amato KR, Clark AG, Knight R (2017) The human microbiome in evolution. BMC Biol 15(1):127. https://doi.org/10.1186/s12915-017-0454-7

DeBiasi RL, Tyler KL (2004) Molecular methods for diagnosis of viral encephalitis. Clin Microbiol Rev 17(4):903–925. https://doi.org/10.1128/cmr.17.4.903-925.2004

Dethlefsen L, Huse SM, Sogin ML, Relman DA (2008) The pervasive effects of an antibiotic on the human gut microbiota, as revealed by deep 16S RRNA sequencing. PLoS Biol 6(11):e280. https://doi.org/10.1371/journal.pbio.0060280

Dewhirst FE, Chen T, Izard J et al (2010) The human oral microbiome. J Bacteriol 192(19):5002–5017. https://doi.org/10.1128/jb.00542-10

Dickson RP, Erb-Downward JR, Freeman CM et al (2017) Bacterial topography of the healthy human lower respiratory tract. MBio 8(1). https://doi.org/10.1128/mbio.02287-16

Donaldson GP, Lee SM, Mazmanian SK (2015) Gut biogeography of the bacterial microbiota. Nat Rev Microbiol 14(1):20–32. https://doi.org/10.1038/nrmicro3552

Dubik M, Pilecki B, Møeller JB (2022) Commensal intestinal protozoa—underestimated members of the gut microbial community. Biology 11(12):1742. https://doi.org/10.3390/biology11121742

Duboc H, Rainteau D, Rajca S et al (2012) Increase in fecal primary bile acids and dysbiosis in patients with diarrhea-predominant irritable bowel syndrome. Neurogastroenterol Motil 24(6):513–e247. https://doi.org/10.1111/j.1365-2982.2012.01893.x

Edgar RC (2018) Updating the 97% identity threshold for 16S ribosomal RNA OTUs. Bioinformatics 34(14):2371–2375. https://doi.org/10.1093/bioinformatics/bty113

Ege M, Bieli C, Frei R et al (2006) Prenatal farm exposure is related to the expression of receptors of the innate immunity and to atopic sensitization in school-age children. J Allergy Clin Immunol 117(4):817–823. https://doi.org/10.1016/j.jaci.2005.12.1307

Ege M, Mayer M, Normand A et al (2011) Exposure to environmental microorganisms and childhood asthma. N Engl J Med 364(8):701–709. https://doi.org/10.1056/nejmoa1007302

El-Sayed A, Aleya L, Kamel MS (2021) The link among microbiota, epigenetics, and disease development. Environ Sci Pollut Res 28(23):28926–28964. https://doi.org/10.1007/s11356-021-13862-1

Emery DC, Shoemark DK, Batstone TE et al (2017) 16S RRNA next generation sequencing analysis shows bacteria in Alzheimer's post-mortem brain. Front Aging Neurosci 9:195. https://doi.org/10.3389/fnagi.2017.00195

Endt K, Stecher B, Chaffron S et al (2010) The microbiota mediates pathogen clearance from the gut lumen after non-typhoidal Salmonella diarrhea. PLoS Pathog 6(9):e1001097. https://doi.org/10.1371/journal.ppat.1001097

Enoch D, Ludlam H, Brown NM (2006) Invasive fungal infections: a review of epidemiology and management options. J Med Microbiol 55(7):809–818. https://doi.org/10.1099/jmm.0.46548-0

Findley K, Oh J, Yang JH et al (2013) Topographic diversity of fungal and bacterial communities in human skin. Nature 498(7454):367–370. https://doi.org/10.1038/nature12171

Fitz-Gibbon S, Tomida S, Chiu BH et al (2013) Propionibacterium acnes strain populations in the human skin microbiome associated with acne. J Invest Dermatol 133(9):2152–2160. https://doi.org/10.1038/jid.2013.21

Flores-Mireles AL, Walker JN, Caparon MG, Hultgren SJ (2015) Urinary tract infections: epidemiology, mechanisms of infection and treatment options. Nat Rev Microbiol 13(5):269–284. https://doi.org/10.1038/nrmicro3432

Foxman B. (2010) The epidemiology of urinary tract infection. Nature Reviews Urology. 7(12): 653–660. https://doi.org/10.1038/nrurol.2010.190.

Fukuda S, Toh Y, Hase K et al (2011) Bifidobacteria can protect from enteropathogenic infection through production of acetate. Nature 469(7331):543–547. https://doi.org/10.1038/nature09646

Gardner SE, Hillis SL, Heilmann KP, Segre JA, Grice EA (2013) The neuropathic diabetic foot ulcer microbiome is associated with clinical factors. Diabetes 62(3):923–930. https://doi.org/10.2337/db12-0771

Garretto A, Thomas-White K, Wolfe AJ, Putonti C (2018) Detecting viral genomes in the female urinary microbiome. J Gen Virol 99(8):1141–1146. https://doi.org/10.1099/jgv.0.001097

Garretto A, Miller-Ensminger T, Wolfe AJ, Putonti C (2019) Bacteriophages of the lower urinary tract. Nat Rev Urol 16(7):422–432. https://doi.org/10.1038/s41585-019-0192-4

Genco RJ, Grossi SG, Ho A, Nishimura F, Murayama Y (2005) A proposed model linking inflammation to obesity, diabetes, and periodontal infections. J Periodontol 76(11-s):2075–2084. https://doi.org/10.1902/jop.2005.76.11-s.2075

Gest H (2004) The discovery of microorganisms by Robert Hooke and Antoni van Leeuwenhoek, Fellows of The Royal Society. Notes Rec 58(2):187–201. https://doi.org/10.1098/rsnr.2004.0055

Gill SR, Pop M, DeBoy RT et al (2006) Metagenomic analysis of the human distal gut microbiome. Science 312(5778):1355–1359. https://doi.org/10.1126/science.1124234

Gupta A, Singh V, Mani I (2022) Dysbiosis of human microbiome and infectious diseases. Prog Mol Biol Transl Sci 192(1):33–51. https://doi.org/10.1016/bs.pmbts.2022.06.016

Hallen-Adams HE, Suhr MJ (2016) Fungi in the healthy human gastrointestinal tract. Virulence 8(3):352–358. https://doi.org/10.1080/21505594.2016.1247140

Halley A, Leonetti A, Gregori A et al (2020) The role of the microbiome in cancer and therapy efficacy: focus on lung cancer. Anticancer Res 40(9):4807–4818. https://doi.org/10.21873/anticanres.14484

Hannigan GD, Grice EA (2013) Microbial ecology of the skin in the era of metagenomics and molecular microbiology. Cold Spring Harb Perspect Med 3(12):a015362. https://doi.org/10.1101/cshperspect.a015362

Hasan NA, Young B, Minard-Smith A et al (2014) Microbial community profiling of human saliva using shotgun metagenomic sequencing. PLoS One 9(5):e97699. https://doi.org/10.1371/journal.pone.0097699

Hase Y, Ding R, Harrison GL et al (2019) White matter capillaries in vascular and neurodegenerative dementias. Acta Neuropathol Commun 7(1):16. https://doi.org/10.1186/s40478-019-0666-x

Helmink BA, Khan MAW, Hermann A, Gopalakrishnan V, Wargo JA (2019) The microbiome, cancer, and cancer therapy. Nat Med 25(3):377–388. https://doi.org/10.1038/s41591-019-0377-7

Hilt EE, McKinley K, Pearce MM et al (2014) Urine is not sterile: use of enhanced urine culture techniques to detect resident bacterial flora in the adult female bladder. J Clin Microbiol 52(3):871–876. https://doi.org/10.1128/jcm.02876-13

Hilty M, Burke CM, Pedro H et al (2010) Disordered microbial communities in asthmatic airways. PLoS One 5(1):e8578. https://doi.org/10.1371/journal.pone.0008578

Homd (2005) HOMD:: Human Oral Microbiome Database. Copyright (C) 2005 by HOMD. https://www.homd.org/

Horz HP, Conrads G (2007) Diagnosis and anti-infective therapy of periodontitis. Expert Rev Anti-Infect Ther 5(4):703–715. https://doi.org/10.1586/14787210.5.4.703

Howlett P, Walder A, Lisk DR et al (2018) Case series of severe neurologic sequelae of Ebola virus disease during epidemic, Sierra Leone. Emerg Infect Dis 24(8):1412–1421. https://doi.org/10.3201/eid2408.171367

Hoyles L, McCartney AL, Neve H et al (2014) Characterization of virus-like particles associated with the human faecal and caecal microbiota. Res Microbiol 165(10):803–812. https://doi.org/10.1016/j.resmic.2014.10.006

Hsiao EY, McBride SW, Hsien S et al (2013) Microbiota modulate behavioral and physiological abnormalities associated with neurodevelopmental disorders. Cell 155(7):1451–1463. https://doi.org/10.1016/j.cell.2013.11.024

Hu Y, Yang X, Qin J et al (2013) Metagenome-wide analysis of antibiotic resistance genes in a large cohort of human gut microbiota. Nat Commun 4(1):2151. https://doi.org/10.1038/ncomms3151

Huang Z, Wang Y, Shi Z et al (2021) Characterization of respiratory microbial dysbiosis in hospitalized COVID-19 patients. Cell Discov 7(1):23. https://doi.org/10.1038/s41421-021-00257-2

Huttenhower C, Gevers D, Knight R et al (2012) Structure, function and diversity of the healthy human microbiome. Nature 486(7402):207–214. https://doi.org/10.1038/nature11234

Imirzalioglu C, Hain T, Chakraborty T, Domann E (2008) Hidden pathogens uncovered: metagenomic analysis of urinary tract infections. Andrologia 40(2):66–71. https://doi.org/10.1111/j.1439-0272.2007.00830.x

Ipe DS, Horton E, Ulett GC (2016) The basics of bacteriuria: strategies of microbes for persistence in urine. Front Cell Infect Microbiol 6:14. https://doi.org/10.3389/fcimb.2016.00014

Ismail Y, Mahendran V, Octavia S et al (2012) Investigation of the enteric pathogenic potential of oral Campylobacter concisus strains isolated from patients with inflammatory bowel disease. PLoS One 7(5):e38217. https://doi.org/10.1371/journal.pone.0038217

Ivanov II, De Frutos R, Manel N et al (2008) Specific microbiota direct the differentiation of IL-17-producing T-helper cells in the mucosa of the small intestine. Cell Host Microbe 4(4):337–349. https://doi.org/10.1016/j.chom.2008.09.009

Janus MM, Crielaard W, Volgenant CMC, Van Der Veen MH, Brandt BW, Krom BP (2017) *Candida albicans* alters the bacterial microbiome of early *in vitro* oral biofilms. J Oral Microbiol 9(1):1270613. https://doi.org/10.1080/20002297.2016.1270613

Jin C, Lagoudas GK, Zhao C et al (2019) Commensal microbiota promote lung cancer development via γδ T cells. Cell 176(5):998–1013.e16. https://doi.org/10.1016/j.cell.2018.12.040

Jørn A, Paster BJ, Stokes LN, Olsen I, Dewhirst FE (2005) Defining the normal bacterial flora of the oral cavity. J Clin Microbiol 43(11):5721–5732. https://doi.org/10.1128/jcm.43.11.5721-5732.2005

Joshipura K, Rimm EB, Douglass CW, Trichopoulos D, Ascherio A, Willett WC (1996) Poor oral health and coronary heart disease. J Dent Res 75(9):1631–1636. https://doi.org/10.1177/00220345960750090301

Kalan L, Loesche MA, Hodkinson BP et al (2016) Redefining the chronic-wound microbiome: fungal communities are prevalent, dynamic, and associated with delayed healing. MBio 7(5). https://doi.org/10.1128/mbio.01058-16

Karstens L, Asquith M, Davin S et al (2016) Does the urinary microbiome play a role in urgency urinary incontinence and its severity? Front Cell Infect Microbiol 6:78. https://doi.org/10.3389/fcimb.2016.00078

Kauffman CA (2014) Diagnosis and management of fungal urinary tract infection. Infect Dis Clin N Am 28(1):61–74. https://doi.org/10.1016/j.idc.2013.09.004

Kelly JR, Borre Y, O'Brien C et al (2016) Transferring the blues: depression-associated gut microbiota induces neurobehavioural changes in the rat. J Psychiatr Res 82:109–118. https://doi.org/10.1016/j.jpsychires.2016.07.019

Kilian M, Chapple I, Hannig M et al (2016) The oral microbiome—an update for oral healthcare professionals. Br Dent J 221(10):657–666. https://doi.org/10.1038/sj.bdj.2016.865

Lakhdari O, Cultrone A, Tap J et al (2010) Functional metagenomics: a high throughput screening method to decipher microbiota-driven NF-KB modulation in the human gut. PLoS One 5(9):e13092. https://doi.org/10.1371/journal.pone.0013092

Lane D, Pace B, Olsen GJ, Stahl DA, Sogin ML, Pace NR (1985) Rapid determination of 16S ribosomal RNA sequences for phylogenetic analyses. Proc Natl Acad Sci U S A 82(20):6955–6959. https://doi.org/10.1073/pnas.82.20.6955

LaPierre N, Mangul S, Alser M et al (2019) MiCoP: microbial community profiling method for detecting viral and fungal organisms in metagenomic samples. BMC Genomics 20(S5):423. https://doi.org/10.1186/s12864-019-5699-9

Lewis D, Brown RL, Williams J et al (2013) The human urinary microbiome; bacterial DNA in voided urine of asymptomatic adults. Front Cell Infect Microbiol 3:41. https://doi.org/10.3389/fcimb.2013.00041

Leyden JJ, Marples RR, Kligman AM (1974) Staphylococcus aureus in the lesions of atopic dermatitis. Br J Dermatol 90(5):525. https://doi.org/10.1111/j.1365-2133.1974.tb06447.x

Leyden JJ, McGinley KJ, Mills OH, Kligman AM (1975) Propionibacterium levels in patients with and without acne vulgaris. J Invest Dermatol 65(4):382–384. https://doi.org/10.1111/1523-1747.ep12607634

Li J, Jia H, Cai X et al (2014a) An integrated catalog of reference genes in the human gut microbiome. Nat Biotechnol 32(8):834–841. https://doi.org/10.1038/nbt.2942

Li J, Quinque D, Horz HP et al (2014b) Comparative analysis of the human saliva microbiome from different climate zones: Alaska, Germany, and Africa. BMC Microbiol 14(1):316. https://doi.org/10.1186/s12866-014-0316-1

Lin JH, Wu Z, Gong L et al (2019) Complex microbiome in brain abscess revealed by whole-genome culture-independent and culture-based sequencing. J Clin Med 8(3):351. https://doi.org/10.3390/jcm8030351

Lokmer A, Cian A, Froment A et al (2019) Use of shotgun metagenomics for the identification of protozoa in the gut microbiota of healthy individuals from worldwide populations with various industrialization levels. PLoS One 14(2):e0211139. https://doi.org/10.1371/journal.pone.0211139

Lu J, Carmody LA, Opron K et al (2020) Parallel analysis of cystic fibrosis sputum and saliva reveals overlapping communities and an opportunity for sample decontamination. MSystems 5(4):e00296-20. https://doi.org/10.1128/msystems.00296-20

Lusk RW (2014) Diverse and widespread contamination evident in the unmapped depths of high throughput sequencing data. PLoS One 9(10):e110808. https://doi.org/10.1371/journal.pone.0110808

Lutgendorff F, Akkermans LMA, Söderholm JD (2008) The role of microbiota and probiotics in stress-induced gastrointestinal damage. Curr Mol Med 8(4):282–298. https://doi.org/10.2174/156652408784533779

MacAogáin M, Chandrasekaran R, Lim AYH et al (2018) Immunological corollary of the pulmonary mycobiome in bronchiectasis: the CAMEB study. Eur Respir J 52(1):1800766. https://doi.org/10.1183/13993003.00766-2018

MacAogáin M, Lau KJX, Cai Z et al (2020) Metagenomics reveals a core macrolide resistome related to microbiota in chronic respiratory disease. Am J Respir Crit Care Med 202(3):433–447. https://doi.org/10.1164/rccm.201911-2202oc

Machiels K, Joossens M, Sabino J et al (2013) A decrease of the butyrate-producing species *Roseburia hominis* and *Faecalibacterium prausnitzii* defines dysbiosis in patients with ulcerative colitis. Gut 63(8):1275–1283. https://doi.org/10.1136/gutjnl-2013-304833

Macpherson AJ, McCoy KD (2013) Stratification and compartmentalisation of immunoglobulin responses to commensal intestinal microbes. Semin Immunol 25(5):358–363. https://doi.org/10.1016/j.smim.2013.09.004

Mangul S, Yang HT, Strauli N et al (2018) ROP: dumpster diving in RNA-sequencing to find the source of 1 trillion reads across diverse adult human tissues. Genome Biol 19(1):36. https://doi.org/10.1186/s13059-018-1403-7

Mani I (2020a) Metagenomics approach for bioremediation: challenges and perspectives. In: Pandey VC, Singh V (eds) Bioremediation of pollutants. Elsevier, pp 275–285., ISBN 9780128190258. https://doi.org/10.1016/B978-0-12-819025-8.00012-0

Mani I (2020b) Current status and challenges of DNA sequencing. In: Singh V (ed) Advances in synthetic biology. Springer, Singapore. https://doi.org/10.1007/978-981-15-0081-7_5

Manichanh C (2006) Reduced diversity of faecal microbiota in Crohn's disease revealed by a metagenomic approach. Gut 55(2):205–211. https://doi.org/10.1136/gut.2005.073817

Manos J (2022) The human microbiome in disease and pathology. APMIS 130(12):690–705. https://doi.org/10.1111/apm.13225

Manrique P, Bolduc B, Walk ST, Van Der Oost J, De Vos WM, Young M (2016) Healthy human gut phageome. Proc Natl Acad Sci U S A 113(37):10400–10405. https://doi.org/10.1073/pnas.1601060113

Mason MR, Preshaw PM, Nagaraja HN, Dabdoub SM, Rahman A, Kumar P (2014) The subgingival microbiome of clinically healthy current and never smokers. ISME J 9(1):268–272. https://doi.org/10.1038/ismej.2014.114

Maynard CL, Elson CO, Hatton RD, Weaver CT (2012) Reciprocal interactions of the intestinal microbiota and immune system. Nature 489(7415):231–241. https://doi.org/10.1038/nature11551

Millares L, Pérez-Brocal V, Ferrari RG et al (2015) Functional metagenomics of the bronchial microbiome in COPD. PLoS One 10(12):e0144448. https://doi.org/10.1371/journal.pone.0144448

Moran GP, Coleman DC, Sullivan DJ (2014) An introduction to the medically important *Candida* species. In: Candida and candidiasis. ASM Press EBooks, pp 9–25. https://doi.org/10.1128/9781555817176.ch2

Morfopoulou S, Mee ET, Connaughton SM et al (2016) Deep sequencing reveals persistence of cell-associated mumps vaccine virus in chronic encephalitis. Acta Neuropathol 133(1):139–147. https://doi.org/10.1007/s00401-016-1629-y

Moustafa A, Li W, Singh H et al (2018) Microbial metagenome of urinary tract infection. Sci Rep 8(1):4333. https://doi.org/10.1038/s41598-018-22660-8

Naccache SN, Peggs KS, Mattes F et al (2015) Diagnosis of neuroinvasive astrovirus infection in an immunocompromised adult with encephalitis by unbiased next-generation sequencing. Clin Infect Dis 60(6):919–923. https://doi.org/10.1093/cid/ciu912

Natalini JG, Singh S, Segal LN (2022) The dynamic lung microbiome in health and disease. Nat Rev Microbiol 21(4):222–235. https://doi.org/10.1038/s41579-022-00821-x

Nicholson JK, Holmes E, Wilson ID (2005) Gut microorganisms, mammalian metabolism and personalized health care. Nat Rev Microbiol 3(5):431–438. https://doi.org/10.1038/nrmicro1152

Oh J, Byrd AL, Deming C, Conlan S, NISC Comparative Sequencing Program, Kong HH, Segre JA (2014) Biogeography and individuality shape function in the human skin metagenome. Nature 514(7520):59–64. https://doi.org/10.1038/nature13786

Paduch DA (2007) Viral lower urinary tract infections. Curr Urol Rep 8(4):324–335. https://doi.org/10.1007/s11934-007-0080-y

Païssé S, Valle C, Servant F et al (2016) Comprehensive description of blood microbiome from healthy donors assessed by 16S targeted metagenomic sequencing. Transfusion 56(5):1138–1147. https://doi.org/10.1111/trf.13477

Paju S, Scannapieco FA (2007) Oral biofilms, periodontitis, and pulmonary infections. Oral Dis 13(6):508–512. https://doi.org/10.1111/j.1601-0825.2007.01410a.x

Palmer R (2013) Composition and development of oral bacterial communities. Periodontology 2000 64(1):20–39. https://doi.org/10.1111/j.1600-0757.2012.00453.x

Parfrey LW, Walters WA, Lauber CL et al (2014) Communities of microbial eukaryotes in the mammalian gut within the context of environmental eukaryotic diversity. Front Microbiol 5:298. https://doi.org/10.3389/fmicb.2014.00298

Paster BJ, Boches SK, Galvin JL et al (2001) Bacterial diversity in human subgingival plaque. J Bacteriol 183(12):3770–3783. https://doi.org/10.1128/jb.183.12.3770-3783.2001

Pearce MM, Hilt EE, Rosenfeld A et al (2014) The female urinary microbiome: a comparison of women with and without urgency urinary incontinence. MBio 5(4):e01283. https://doi.org/10.1128/mbio.01283-14

Peterson SN, Snesrud E, Schork NJ, Bretz WA (2011) Dental caries pathogenicity: a genomic and metagenomic perspective. Int Dent J 61:11–22. https://doi.org/10.1111/j.1875-595x.2011.00025.x

Pihlstrom BL, Michalowicz BS, Johnson NW (2005) Periodontal diseases. Lancet 366(9499):1809–1820. https://doi.org/10.1016/s0140-6736(05)67728-8

Piva TC, Luft C, Antunes KH, Maróstica PJC, Pinto LA, Donadio MVF (2020) Extracellular DNA in sputum is associated with pulmonary function and hospitalization in patients with cystic fibrosis. Respir Med 172:106144. https://doi.org/10.1016/j.rmed.2020.106144
Png CW, Lindén SK, Gilshenan K et al (2010) Mucolytic bacteria with increased prevalence in IBD mucosa augment in vitro utilization of mucin by other bacteria. Am J Gastroenterol 105(11):2420–2428. https://doi.org/10.1038/ajg.2010.281
Pop M, Walker AW, Paulson JN et al (2014) Diarrhea in young children from low-income countries leads to large-scale alterations in intestinal microbiota composition. Genome Biol 15(6):R76. https://doi.org/10.1186/gb-2014-15-6-r76
Poretsky R, Hewson I, Sun S, Allen AE, Zehr JP, Moran MA (2009) Comparative day/night metatranscriptomic analysis of microbial communities in the North Pacific subtropical gyre. Environ Microbiol 11(6):1358–1375. https://doi.org/10.1111/j.1462-2920.2008.01863.x
Prompers L, Huijberts MSP, Apelqvist J et al (2006) High prevalence of ischaemia, infection and serious comorbidity in patients with diabetic foot disease in Europe. Baseline results from the Eurodiale study. Diabetologia 50(1):18–25. https://doi.org/10.1007/s00125-006-0491-1
Qin J, Li R, Raes J et al (2010) A human gut microbial gene catalogue established by metagenomic sequencing. Nature 464(7285):59–65. https://doi.org/10.1038/nature08821
Qin J, Li Y, Cai Z et al (2012) A metagenome-wide association study of gut microbiota in type 2 diabetes. Nature 490(7418):55–60. https://doi.org/10.1038/nature11450
Ramachandran P, Wilson MR (2020) Metagenomics for neurological infections—expanding our imagination. Nat Rev Neurol 16(10):547–556. https://doi.org/10.1038/s41582-020-0374-y
Rinninella E, Raoul P, Cintoni M et al (2019) What is the healthy gut microbiota composition? A changing ecosystem across age, environment, diet, and diseases. Microorganisms 7(1):14. https://doi.org/10.3390/microorganisms7010014
Rogers GB, Marsh P, Stressmann AF et al (2010) The exclusion of dead bacterial cells is essential for accurate molecular analysis of clinical samples. Clin Microbiol Infect 16(11):1656–1658. https://doi.org/10.1111/j.1469-0691.2010.03189.x
Roth RR, James W (1988) Microbial ecology of the skin. Ann Rev Microbiol 42(1):441–464. https://doi.org/10.1146/annurev.mi.42.100188.002301
Saheb Kashaf S, Proctor DM, Deming C et al (2021) Integrating cultivation and metagenomics for a multi-kingdom view of skin microbiome diversity and functions. Nat Microbiol 7(1):169–179. https://doi.org/10.1038/s41564-021-01011-w
Salter SJ, Cox MJ, Turek E et al (2014) Reagent and laboratory contamination can critically impact sequence-based microbiome analyses. BMC Biol 12(1):87. https://doi.org/10.1186/s12915-014-0087-z
Salzberg SL, Breitwieser FP, Kumar A et al (2016) Next-generation sequencing in neuropathologic diagnosis of infections of the nervous system. Neuroimmunol nd Neuroinflamm 3(4):e251. https://doi.org/10.1212/nxi.0000000000000251
Sanford JA, Gallo RL (2013) Functions of the skin microbiota in health and disease. Semin Immunol 25(5):370–377. https://doi.org/10.1016/j.smim.2013.09.005
Sanglard D (2016) Emerging threats in antifungal-resistant fungal pathogens. Front Med 3:11. https://doi.org/10.3389/fmed.2016.00011
Santiago-Rodríguez TM, Ly M, Bonilla N, Pride DT (2015) The human urine virome in association with urinary tract infections. Front Microbiol 6:14. https://doi.org/10.3389/fmicb.2015.00014
Scarpellini E, Ianiro G, Attili F, Bassanelli C, De Santis A, Gasbarrini A (2015) The human gut microbiota and virome: potential therapeutic implications. Dig Liver Dis 47(12):1007–1012. https://doi.org/10.1016/j.dld.2015.07.008
Schoch CL, Seifert KA, Huhndorf SM et al (2012) Nuclear ribosomal internal transcribed spacer (ITS) region as a universal DNA barcode marker for *Fungi*. Proc Natl Acad Sci U S A 109(16):6241–6246. https://doi.org/10.1073/pnas.1117018109
Segal LN, Rom WN, Weiden MD (2014) Lung microbiome for clinicians. New discoveries about bugs in healthy and diseased lungs. Ann Am Thorac Soc 11(1):108–116. https://doi.org/10.1513/annalsats.201310-339fr

Segal LN, Clemente JC, Tsay JJ et al (2016) Enrichment of the lung microbiome with oral taxa is associated with lung inflammation of a Th17 phenotype. Nat Microbiol 1(5):16031. https://doi.org/10.1038/nmicrobiol.2016.31

Selwitz RH, Ismaïl AI, Pitts N (2007) Dental caries. Lancet 369(9555):51–59. https://doi.org/10.1016/s0140-6736(07)60031-2

Sharma N, Bhatia S, Sodhi AS, Batra N (2018) Oral microbiome and health. AIMS Microbiol 4(1):42–66. https://doi.org/10.3934/microbiol.2018.1.42

Shi Y, Wang G, Lau HCH, Yu J (2022) Metagenomic sequencing for microbial DNA in human samples: emerging technological advances. Int J Mol Sci 23(4):2181. https://doi.org/10.3390/ijms23042181

Shkoporov AN, Ryan FJ, Draper LA et al (2018) Reproducible protocols for metagenomic analysis of human faecal phageomes. Microbiome 6(1):68. https://doi.org/10.1186/s40168-018-0446-z

Siddiqui H, Nederbragt AJ, Lagesen K, Jeansson S, Jakobsen KS (2011) Assessing diversity of the female urine microbiota by high throughput sequencing of 16S rDNA amplicons. BMC Microbiol 11(1):244. https://doi.org/10.1186/1471-2180-11-244

Sokol H, Pigneur B, Watterlot L et al (2008) *Faecalibacterium prausnitzii* is an anti-inflammatory commensal bacterium identified by gut microbiota analysis of Crohn disease patients. Proc Natl Acad Sci U S A 105(43):16731–16736. https://doi.org/10.1073/pnas.0804812105

Suhr MJ, Banjara N, Hallen-Adams HE (2016) Sequence-based methods for detecting and evaluating the human gut mycobiome. Lett Appl Microbiol 62(3):209–215. https://doi.org/10.1111/lam.12539

Sulaiman I, Wu B, Li Y et al (2021) Functional lower airways genomic profiling of the microbiome to capture active microbial metabolism. Eur Respir J 58(1):2003434. https://doi.org/10.1183/13993003.03434-2020

Sutton TDS, Hill C (2019) Gut bacteriophage: current understanding and challenges. Front Endocrinol 10:784. https://doi.org/10.3389/fendo.2019.00784

Suzuki T, Worobey M (2014) Geographical variation of human gut microbial composition. Biol Lett 10(2):20131037. https://doi.org/10.1098/rsbl.2013.1037

Tanes C, Bittinger K, Gao Y et al (2021) Role of dietary fiber in the recovery of the human gut microbiome and its metabolome. Cell Host Microbe 29(3):394–407.e5. https://doi.org/10.1016/j.chom.2020.12.012

Council NR. et al. (2007) The New Science of Metagenomics: Revealing the Secrets of Our Microbial Planet. National Academies Press. https://www.ncbi.nlm.nih.gov/books/NBK54006/pdf/Bookshelf_NBK54006.pdf.

Thomas T, Gilbert J, Meyer P (2012) Metagenomics—a guide from sampling to data analysis. Microb Inform Exp 2(1):3. https://doi.org/10.1186/2042-5783-2-3

Tomida S, Nguyen L, Chiu BH et al (2013) Pan-genome and comparative genome analyses of propionibacterium acnes reveal its genomic diversity in the healthy and diseased human skin microbiome. MBio 4(3):e00003-13. https://doi.org/10.1128/mbio.00003-13

Tremaroli V, Bäckhed F (2012) Functional interactions between the gut microbiota and host metabolism. Nature 489(7415):242–249. https://doi.org/10.1038/nature11552

Truong DT, Franzosa EA, Tickle TL et al (2015) MetaPhlAn2 for enhanced metagenomic taxonomic profiling. Nat Methods 12(10):902–903. https://doi.org/10.1038/nmeth.3589

Tyler KL (2009) Emerging viral infections of the central nervous system. Arch Neurol 66(8):939. https://doi.org/10.1001/archneurol.2009.153

VerBerkmoes NC, Russell AL, Shah M et al (2008) Shotgun metaproteomics of the human distal gut microbiota. ISME J 3(2):179–189. https://doi.org/10.1038/ismej.2008.108

Viaud S, Saccheri F, Mignot G et al (2013) The intestinal microbiota modulates the anticancer immune effects of cyclophosphamide. Science 342(6161):971–976. https://doi.org/10.1126/science.1240537

Vrieze A, Out C, Fuentes S et al (2014) Impact of oral vancomycin on gut microbiota, bile acid metabolism, and insulin sensitivity. J Hepatol 60(4):824–831. https://doi.org/10.1016/j.jhep.2013.11.034

Wagner K, Springer B, Pires VP, Keller PM (2018) Molecular detection of fungal pathogens in clinical specimens by 18S rDNA high-throughput screening in comparison to ITS PCR and culture. Sci Rep 8(1):6964. https://doi.org/10.1038/s41598-018-25129-w

Wang T, Cai G, Qiu Y et al (2011) Structural segregation of gut microbiota between colorectal cancer patients and healthy volunteers. ISME J 6(2):320–329. https://doi.org/10.1038/ismej.2011.109

Wang J, Qi J, Zhao H et al (2013) Metagenomic sequencing reveals microbiota and its functional potential associated with periodontal disease. Sci Rep 3(1). https://doi.org/10.1038/srep01843

Wang W, Xu S, Ren Z, Tao L, Jiang J, Zheng S (2015) Application of metagenomics in the human gut microbiome. World J Gastroenterol 21(3):803. https://doi.org/10.3748/wjg.v21.i3.803

Wei X, Yan X, Zou D et al (2013) Abnormal fecal microbiota community and functions in patients with hepatitis B liver cirrhosis as revealed by a metagenomic approach. BMC Gastroenterol 13(1). https://doi.org/10.1186/1471-230x-13-175

Weinstock GM (2012) Genomic approaches to studying the human microbiota. Nature 489(7415):250–256. https://doi.org/10.1038/nature11553

Whiteside SA, Razvi H, Davé S, Reid G, Burton JP (2015) The microbiome of the urinary tract—a role beyond infection. Nat Rev Urol 12(2):81–90. https://doi.org/10.1038/nrurol.2014.361

Whiteside SA, McGinniss JE, Collman RG (2021) The lung microbiome: progress and promise. J Clin Invest 131(15):e150473. https://doi.org/10.1172/jci150473

Wilson MR, Tyler KL (2016) Emerging diagnostic and therapeutic tools for central nervous system infections. JAMA Neurol 73(12):1389. https://doi.org/10.1001/jamaneurol.2016.3617

Wilson MR, Naccache SN, Samayoa E et al (2014) Actionable diagnosis of neuroleptospirosis by next-generation sequencing. N Engl J Med 370(25):2408–2417. https://doi.org/10.1056/nejmoa1401268

Wilson MR, Suan D, Duggins A et al (2017a) A novel cause of chronic viral meningoencephalitis: Cache Valley virus. Ann Neurol 82(1):105–114. https://doi.org/10.1002/ana.24982

Wilson MR, Zimmermann L, Crawford E et al (2017b) Acute West Nile virus meningoencephalitis diagnosed via metagenomic deep sequencing of cerebrospinal fluid in a renal transplant patient. Am J Transplant 17(3):803–808. https://doi.org/10.1111/ajt.14058

Wilson MR, O'Donovan B, Gelfand JM et al (2018) Chronic meningitis investigated via metagenomic next-generation sequencing. JAMA Neurol 75(8):947. https://doi.org/10.1001/jamaneurol.2018.0463

Wilson MR, Sample H, Zorn KC et al (2019) Clinical metagenomic sequencing for diagnosis of meningitis and encephalitis. N Engl J Med 380(24):2327–2340. https://doi.org/10.1056/nejmoa1803396

Wolfe AJ, Brubaker L (2015) "Sterile urine" and the presence of bacteria. Eur Urol 68(2):173–174. https://doi.org/10.1016/j.eururo.2015.02.041

Wolfe AJ, Nelson DE, Shibata N et al (2012) Evidence of uncultivated bacteria in the adult female bladder. J Clin Microbiol 50(4):1376–1383. https://doi.org/10.1128/jcm.05852-11

Woyke T (2019) Beyond the census of human gut dwellers. Nat Rev Microbiol 17(7):401. https://doi.org/10.1038/s41579-019-0220-7

Wu N, Yang X, Zhang R et al (2013) Dysbiosis signature of fecal microbiota in colorectal cancer patients. Microb Ecol 66(2):462–470. https://doi.org/10.1007/s00248-013-0245-9

Xu P, Gunsolley JC (2014) Application of metagenomics in understanding oral health and disease. Virulence 5(3):424–432. https://doi.org/10.4161/viru.28532

Yang S, Gao X, Meng J et al (2018) Metagenomic analysis of bacteria, fungi, bacteriophages, and helminths in the gut of giant pandas. Front Microbiol 9:1717. https://doi.org/10.3389/fmicb.2018.01717

Yang C, Chowdhury D, Zhang Z, Cheung WK, Lu A, Bian Z, Zhang L (2021) A review of computational tools for generating metagenome-assembled genomes from metagenomic sequencing data. Comput Struct Biotechnol J 19:6301–6314. https://doi.org/10.1016/j.csbj.2021.11.028

Yarza P, Yilmaz P, Pruesse E et al (2014) Uniting the classification of cultured and uncultured bacteria and archaea using 16S rRNA gene sequences. Nat Rev Microbiol 12(9):635–645. https://doi.org/10.1038/nrmicro3330

Yatsunenko T, Rey FE, Manary MJ et al (2012) Human gut microbiome viewed across age and geography. Nature 486(7402):222–227. https://doi.org/10.1038/nature11053

Yeoh N, Burton JP, Suppiah P, Reid G, Stebbings S (2013) The role of the microbiome in rheumatic diseases. Curr Rheumatol Rep 15(3):314. https://doi.org/10.1007/s11926-012-0314-y

Yi X, Gao J, Wang Z (2022) The human lung microbiome—a hidden link between microbes and human health and diseases. iMeta 1(3). https://doi.org/10.1002/imt2.33

Zasloff M (2007) Antimicrobial peptides, innate immunity, and the normally sterile urinary tract. J Am Soc Nephrol 18(11):2810–2816. https://doi.org/10.1681/asn.2007050611

Zaura E, Keijser BJF, Huse SM, Crielaard W (2009) Defining the healthy "core microbiome" of oral microbial communities. BMC Microbiol 9(1):259. https://doi.org/10.1186/1471-2180-9-259

Zhang XX, Guo LY, Liu LL et al (2019) The diagnostic value of metagenomic next-generation sequencing for identifying Streptococcus pneumoniae in paediatric bacterial meningitis. BMC Infect Dis 19(1):495. https://doi.org/10.1186/s12879-019-4132-y

Metatranscriptomics, Metaproteomics, and Metabolomics Approaches for Microbiome Characterization

6

Anshu Mathuria, Kritika Jain, Ashok Saini, Chaitenya Verma, and Indra Mani

6.1 Introduction

Microbes are probably developed ~3.5 billion years ago and now present almost everywhere around us in the biosphere comprising ~60% of the earth's biomass (Whitman et al. 1998; Pace 1997). Broadly, we can categorize them into two categories—prokaryotes and eukaryotes on the presence or absence of nuclear membrane. Prokaryotes do not have any nuclear membrane and other membrane-bound organelles while eukaryotes have proper nuclear membrane around their genetic material. Prokaryotes include viruses, eubacteria, and archaebacterial whereas algae, fungi, protists, and protozoa belong to eukaryotic group.

Microbes are present everywhere in the environment and can be isolated from various sources such as soil, water, air and even from the exterior and interior surfaces of plants and animals. Thus, microorganisms play both beneficial and harmful roles in the environment, agriculture, and animal or human health. In an estimate, 1 g of soil contains ~10^{8-9} bacteria, 10^{5-8} actinomycetes, 10^{5-6} fungi, 10^{3-6} microalgae, and 10^{3-5} protozoa (Dindal 1990). Similarly, in a human body

Anshu Mathuria and Kritika Jain equally contributed as first author.

A. Mathuria · I. Mani (✉)
Department of Microbiology, Gargi College, University of Delhi, New Delhi, India
e-mail: indra.mani@gargi.du.ac.in

K. Jain · A. Saini (✉)
Department of Microbiology, Institute of Home Economics, University of Delhi, New Delhi, India
e-mail: ashok.saini@ihe.du.ac.in

C. Verma
Department of Pathology, Ohio State University, Columbus, OH, USA

I. Mani, V. Singh (eds.), *Multi-Omics Analysis of the Human Microbiome*,
https://doi.org/10.1007/978-981-97-1844-3_6

~10–100 trillion microbial cells belonging to diverse groups are found, mainly in the gut (Turnbaugh et al. 2007). Thus, we need to study the diversity of the microbes found in various environments and further characterize them to study their roles.

The diversity of microbes was first studied by Antonie van Leewenhoek in the 1680s in oral and fecal samples. He observed prominent differences between the microorganisms isolated from these two sites in healthy and diseased individuals (Dobell 1920; van Leeuwenhoek 1684).

Until recently, the characterization of microbes found in the environment and other sites such as the human gut is studied mainly using two approaches: culture-dependent and culture-independent methods. In the culture-based approach, microorganisms are isolated from a sample and then grown on various culture mediums in lab to further characterize them using various biochemical and molecular techniques (Kellenberger 2001). Culture-based methods are very simple, easy to use, and cost effective. But sometimes we need to use different kinds of media to grow specific microorganisms (Atlas 2010; Basu et al. 2015; Bonnet et al. 2020; Mani 2023).

The major advantages of this approach are that it is very useful to study single isolates, which can be grown at high cell density so that their physiological and metabolic potentials such as enzyme production and pigment production can be studied. But the culture-based method faces several limitations. It has been shown by various studies that 5–20% of microbes could not be cultured in lab on agar-based media (Kellenberger 2001) due to which majority of the important viable but non-culturable microorganisms are neglected and could not be studied (Laiz et al. 2003). Furthermore, one medium especially nutrient rich media is not suitable for the overall biodiversity characterization (Dziurzynski et al. 2020). Another major limitations of culture-based methods are long incubation period, chances of contamination, and only provide phenotypic data (Pangallo et al. 2015; Chaudhary et al. 2019). Additionally, meta-omics-aided cultivation strategies are promising for the isolation of yet uncultured microorganisms (Gutleben et al. 2018).

Since there are limitations in culture-based approaches various culture-independent methods have been developed for species identification and characterization. Initially, the culture-independent approach concept was introduced by Norman Pace and colleagues in 1980, utilizing rRNA gene amplification using PCR (Hugenholtz et al. 1998). Later on, many other culture-independent methods have been developed to characterize the microbes without culturing them. The initial culture-independent methodologies relied on low-throughput sequencing of the bacterial 16S ribosomal rRNA gene, and the popularity and efficacy of 16S-based surveys expanded substantially as sequencing methods improved (Schmidt et al. 1991; Tringe and Hugenholtz 2008). High-throughput metaomics studies analyze entire microbial communities without culturing, providing comprehensive insights into their composition, function, and interactions in various environments like the gut or soil (Segata et al. 2013).

Furthermore, advancement in the field of high-throughput DNA sequencing and bioinformatics led to the emergence of "omics" studies in which millions of samples can be identified in a very short duration (Wiwanitkit 2013; Mani 2020a). The studies are high-throughput in which either whole genome (genomics), total RNA content (transcriptomics), or total protein content (proteomics) of a microbial cell/community is analyzed. Identifying a collection of microbial species, genes, variations, pathways, or metabolic activities that represent the microbial community occupying an uncultured sample is usually the goal of a meta"-omic" investigation. Various useful *omic* techniques have been developed for microbiome characterization—metagenomics, metatranscriptomics, metaproteomics, and metabolomics (Tyson et al. 2004; Wilmes and Bond 2006; Turnbaugh and Gordon 2008; Aguiar-Pulido et al. 2016; Mani 2020b).

Metagenomics is the most widely used approach to study the genomes of microbial communities present in the sample. Many metagenomics studies have been done to characterize microbiota (Barberan et al. 2011; Freilich et al. 2010; Kuczynski et al. 2010; Faust et al. 2012; Qin et al. 2012). The genome of a microbial community organism is transcribed into various RNA. All the genes are not expressed at the same time or space in a microbial community. So, metatranscriptomics mainly deals with the study of the expression and functional profile of the genes present in the microbial community under a specific condition (Moran 2009; Poretsky et al. 2005; Carvalhais et al. 2012; Gilbert et al. 2008). Proteins are regarded as the core components of biological systems, act as building blocks, and coordinate biological processes. Thus, in metaproteomics approach we mainly focus on the identification and quantification of proteins isolated from any environmental, human, or animal samples containing different types of microorganisms (Wilmes and Bond 2006). In this approach, the metaproteomic data is compared and analyzed to identify the species composition or functional aspect of a microbial community (Gouveia et al. 2020; Hardouin et al. 2021; Mesuere et al. 2018).

Metabolomics is another approach in which identification and quantification of all the metabolites present in the sample of a microbial community is done (Fiehn 2002). Metabolomic studies directly indicate the health status of a particular environment or its alterations due to variation in the activities of microbial communities present (Bernini et al. 2009; Krumsiek et al. 2015).

The microbiota consists of a wide range of bacteria, archaea, fungus, and viruses, which colonize and establish intricate associations with their animal hosts. This amalgamation of microbes and their combined genetic material is known as the microbiome and plays a crucial role in different functions and biological events within the organisms they inhabit, including nutrition, immunological responses, energy metabolism, neurotransmission, and so on (Tremaroli and Backhed 2012). The microbiome composition and its function is susceptible to numerous abiotic, biotic, and anthropogenic factors, with dysbiosis potentially leading to adverse consequences for the host (Zhao et al. 2018; Evariste et al. 2019; Gupta et al. 2022). The human body houses dynamic microorganism populations, especially in the gastrointestinal tract (Thursby and Juge 2017), impacting health through host-microbiota interactions (Wang et al. 2020). The human-associated microbiota's genetic composition exceeds the human genome by over 100 times (Lloyd-Price et al. 2017; Li et al. 2014). Genome sequencing advancements have accelerated gut microbiota studies. Metagenomics is the most widely used approach to analyze various aspects of genomes of microorganisms present in a microbial community such as the gut. Many metagenomics studies have been done to characterize microbiota (Barberan et al. 2011; Freilich et al. 2010; Kuczynski et al. 2010; Faust et al. 2012; Qin et al. 2012). Thus, it reveals gut microbiota's taxonomic composition, functional analysis, like identifying activated bacterial metabolic pathways, remains limited (Starr et al. 2018). The genome of a microbial community organism is transcribed into various RNA. All the genes are not expressed at the same time or space in a microbial community. So, metatranscriptomics mainly deals with the study of the expression and functional profile of the genes expressed by the microbial community members under a specific condition (Moran 2009; Poretsky et al. 2005; Carvalhais et al. 2012; Gilbert et al. 2008).

Among the methods available for investigating microbiomes, metaproteomics stands out as a rapidly emerging approach. Metaproteomics can be described as the comprehensive identification and quantification of proteins found in microbial communities, enabling direct observation of the protein complement of microorganisms isolated from any environmental, human, or animal samples (Kleiner 2019; Wilmes and Bond 2006). In this approach, the metaproteomic data is compared and analyzed to identify the species composition or functional aspect of a microbial community (Gouveia et al. 2020; Hardouin et al. 2021; Mesuere et al. 2018). It employs liquid chromatography-tandem mass spectrometry (LC-MS/MS) and has shown the ability to provide extensive operational insights into the ever-changing interactions between hosts and microbiota (Wang et al. 2020; Starr et al. 2018; Ezzeldin et al. 2019). However, the complicated and diverse architecture of the metaproteome of the gut presents a considerable obstacle for the exact identification of proteins and quantification, especially when contrasted to conventional proteomics that focuses on a single organism. The bulk of current metaproteomic research uses label-free quantification (LFQ) via the data-dependent acquisition (DDA) methodology, which remains the most common strategy for tackling the intricacies of the metaproteome of the gut in studies (Tanca et al. 2017; Zhang et al. 2017, 2018, 2020; Li et al. 2019). In recent times, data-independent acquisition (DIA) approaches have been developed, allowing for the systematic collection of fragmentation details for all precursor ions within predetermined isolation periods. In a standard DIA investigation, peptide levels are assessed by targeted extraction of quantitative indicators from a spectrum library that includes identified peptides (Gillet et al. 2012). The quality and richness of the spectral library have a significant impact on the analysis outcomes (Rosenberger et al. 2017).

Metabolomics is another approach that identifies and quantifies all the metabolites present in a sample of a microbial population, which provides impartial insights on how a cell's genome interacts with its environment (Rochfort 2005; Spratlin et al. 2009; Fiehn 2002). It complements established approaches by showing complex biochemical controls and correcting uncertainty due to environmental factors (Vemuri and Aristidou 2005). Metabolomic studies directly indicate the health status of a particular environment or its alterations due to variation in the activities of microbes present in a community (Bernini et al. 2009; Krumsiek et al. 2015).

Changes in metabolites alter phenotypic and cellular functioning, making metabolome analysis useful in a variety of domains (Villas-Boas et al. 2005; Spratlin et al. 2009). The primary challenge lies in identifying and quantifying the diverse array of metabolites, each with unique characteristics (Villas-Boas et al. 2005; Hollywood et al. 2006; Vemuri and Aristidou 2005). Metabolites are dynamic in nature, changing swiftly in reaction to stimuli, which complicates study. The lack of a close connection between genes and metabolites, as observed in microbes, hampers metabolomics research (Vemuri and Aristidou 2005; Forster et al. 2003).

6.2 Metatranscriptomics

As we know, only 1% of microorganisms can be studied in the lab using culture-based methods. Remaining ~99% of the microorganisms can be studied using some culture-independent methods such as metagenomics, which helps us to understand their genetic diverseness, composition, and ecological roles (Singh et al. 2009). Metagenomics mainly study the structural and functional aspects of the whole genome. In a microbial community, microorganisms express their genes at a certain time or under particular conditions only. Therefore, using metagenomics it is not possible to understand the functional aspects of all the genes present in a microbiome at a particular time.

To study the expression of the gene and its regulation at a specific time point, it is necessary to examine the transcriptome. Thus, another field known as metatranscriptomics provides a view of genes expressed by a specific sample at a specific time or under specified conditions by the means of detecting the entire mRNA (Aguiar-Pulido et al. 2016).

Furthermore, the distinctiveness of active genes within microbial communities can be studied using metatranscriptomics to quantify their expression levels and to determine whether these levels differ under variable conditions, such as pathological or physiological conditions of an organism (Aguiar-Pulido et al. 2016). Functional annotations of expressed genes will ultimately help to deduce which genes are expressed under well-defined conditions. As metatranscriptomics focuses on expressed genes, it allows the functional characterization of complete microbial communities (Moran 2009).

Any metatranscriptomic study starts with the designing of the experimental plan that must have various steps (Fig. 6.1). After finalizing the experimental design such as diseased *vs* control, the first step is collection of sample and extraction of total RNA. Then, total RNA (or sometimes enriched targeted RNA) either directly or after converted into cDNA is inserted into a vector producing a library. Further, the library is sequenced using suitable high-throughput sequencing technique. The raw data which contains millions of short read sequences is then analyzed using a bioinformatics approach to either assembled into a group of similar sequences to derive the taxonomic and functional profile of the microbial community by compared against reference databases (Ojala et al. 2023). A summary of recent studies on analyzing gene expression based on metatranscriptomic analysis is given in Table 6.1

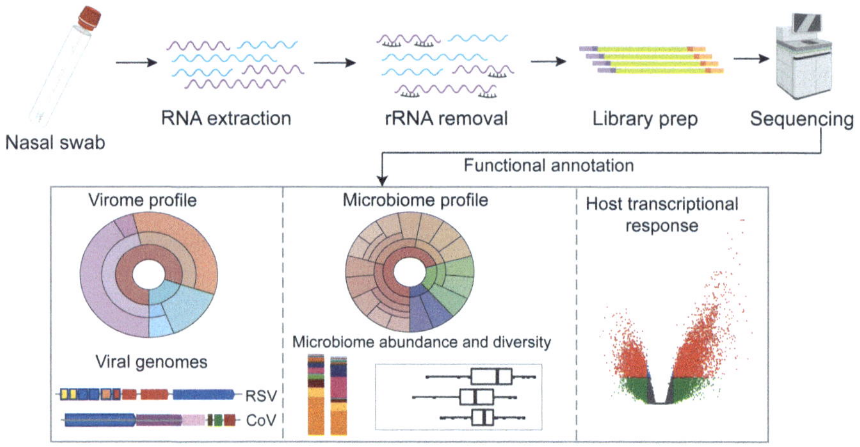

Fig. 6.1 Steps involved in the metatranscriptomic analysis include the study design, collection of samples, RNA isolation and enrichment followed by high-throughput sequencing. The raw data is then computationally analyzed for virome profile, microbiome profile and host transcriptional response (Adapted from Rajagopala et al. 2021)

In 2005, Rachel S. Poretsky, Nasreen Bano, and their colleagues embarked on pioneering research endeavors, seeking to pinpoint active genes within environmental samples. This marked the inception of metatranscriptomics (Poretsky et al. 2005).

Over the ensuing decade, metatranscriptomics has yielded numerous revelations, significantly enhancing our understanding of microbiomes across diverse environments. These revelations span from investigating burn wound infections in patients (Ojala et al. 2023) to exploring the intricacies of the human gut (Qin et al. 2010), soil ecosystems (Rondon et al. 2000), and vast oceanic expanses (Venter et al. 2004).

6.2.1 Significance of Metatranscriptomics in Microbial Studies

- It provides a comprehensive view of gene expression and helps to find out functional roles of different microbial species and their interactions within the ecosystem.
- It categorizes functional insights of expressed genes in response to various alterations in the environment.
- It reveals the previously unknown genes and metabolic pathways, aiding in discovering novel enzymes and metabolic pathways to assist development in medicine and biotechnology.
- It establishes time-resolved analysis providing specific information into the dynamics and adaptations of the microbial community.
- It allows researchers to study the gene expression of non-culturable organisms without the need for their isolation and cultivation.

Table 6.1 Recent studies on analyzing gene expression based on metatranscriptomic analysis

RNA extraction	Metatranscriptomic sequencing	Bioinformatics analysis	Taxonomic annotation	Functional annotation	Statistical analysis	Reference
– Spin column of RNeasy® Mini Kit (Qiagen, Germany) – QIAamp DNA mini kit (Qiagen, Germany)	NovaSeq 6000 (Illumina, San Diego, CA)	– Megahit – BLASTp – MIBiG database	AntiSMASH	BIG-SCAPE		El Samak et al. (2023)
Agilent RNA 6000 Pico Kit	Illumina HiSeq 2000 system	– SILVA database (v.132) using BLASTN – Bowtie2 – RSEM		– PERMANOVA – DESeq2 – BLASTx – eggNOG – Mapper – COG	CAT: DIAMOND blastx	Kozjek et al. (2023)
– Soil RNA isolation kit – DNase I	– ND-2000 – Ribo-Zero rRNA Removal Kit (Illumina) – HiSeq	– BLASTP	– NR database	– RSEM – BLASTP – KEGG	– GraphPad Prism 5 – Hiplot	Zhang et al. (2023)
Polyethersulfone filter	– Pyrosequencing	– Ribosomal Database Project database	– MEGAN	– COGs – KEGG – MG-RAST	– COG – COG category – KEGG pathway – Vegan (version, 2.0–10)	Chen et al. (2015)

(continued)

Table 6.1 (continued)

RNA extraction	Metatranscriptomic sequencing	Bioinformatics analysis	Taxonomic annotation	Functional annotation	Statistical analysis	Reference
– RiboPure-Bacteria kit – MessageAmp II-Bacteria kit	– Pyrosequencing with a Roche GS FLX sequencer	– BLASTX – Rfam database (version 9.1)	– Ribosomal Database roject-II (RDP) – Rfam database (version 9.1)	– COG – gCOGdb	– Bayesian statistical model	Gosalbes et al. (2011)
– Stool DNA isolation kit – Macaloid-based RNA isolation	– Dyenamic ET dye terminator cycle sequencing kit – MegaBace automated sequencer	– GenBank database – COG database	HITChip	BLAST	– BioNumerics (version 4.0) – UPGMA clustering	Booijink et al. (2010)

- It gives an unbiased approach, enabling the detection of both well-known and previously unidentified microbial members and their activities.
- It helps in overcoming the limitations of metagenomics, which does not directly reveal which genes are actively expressed and so complements metagenomics by providing information about the functional activities of the community, bridging the gap between potential capabilities and actual functions.

6.2.2 Application of Metatranscriptomics in Microbial Characterization

Since its initial introduction in the early 2000s, a significant surge has been observed in the metatranscriptomics projects—focused majorly on the sequencing of RNAs from microbiome. This technique has found many applications, from characterizing the evolution of viruses and bacteria in wild life (Ortiz Baez 2023) to revealing the important roles of the gut microbiome (Zhang et al. 2023). It has also contributed in revealing plastic-degrading potential in soil communities, in understanding concepts related to functional eukaryotic microbial biogeography in the ocean (Hoerstmann et al. 2023), improving productive agriculture and promoting a sustainable environment (Sharuddin et al. 2022), monitoring gene expression thereby describing the connection among viruses and their hosts (Moniruzzaman et al. 2017), and detecting regulatory antisense RNA (Bao et al. 2015), among other applications.

As one of the fundamental objectives of metatranscriptomics is to achieve a comparative quantification of gene expression, both assembly based and read-based workflows offer several types of abundance metric and/or gene coverage (such as number of reads per kb per million reads sequenced or raw count per gene). This numeric data (abundance values) when employed with other tools in addition can subsequently enable a comparison of relative gene expression during time-course experiments or under different growth conditions. The ultimate aim is to gain insights into the genes and pathways that may be crucial for a specific phenotype under study.

Regardless of certain limitations associated with metatranscriptome analysis, ongoing advancements in algorithms as well as various tools have helped in the systematic analyses of metatranscriptomic data. The evolution of tools and algorithms enhances with our understanding of various challenges related to these datasets and highlights the potential of the next generation of metatranscriptomic tools in stimulating our comprehension of the biologically active components of microbiomes and the related pathways employed.

6.2.2.1 Analyzing Gene Expression

By studying gene expression patterns, we can delve deeper into the functional pursuits of the microbes, such as metabolic pathways, nutrient breakdown and consumptions, and responses to environmental stimuli. This information helps in understanding the roles of specific microbial groups and their contribution to ecosystem processes.

Aquatic Environment

- A research study by El Samak et al. (2023) stated the use of combined metatranscriptomic and metagenomic analysis of three different samples of the Egyptian Red Sea sponge *Theonella* sp. microbiome to explore and understand its biosynthetic activity. Their analysis suggested a significant role of the sponge microbiome in the production of bioactive compounds. Full-length biosynthetic gene clusters (BGCs) of cytotoxic theonellamide and misakinolide were identified and came across active transcription. Most of the identified BGCs did not show any similarity with any known cluster under the MIBiG database, hence classifying them as novel.
- Another transcriptomic examination of the reef-building coral *Orbicella faveolata* showing a comprehensive holobiont response to White Plague Disease (Coral Disease) was demonstrated by Daniels et al. (2015). They implemented a lab-based approach to isolate the coral holobiont and further resolved its taxa-specific and compartment-specific responses to coral disease. Their ultimate aim was to interpret the holobiont function or, in other words, functional gene patterns along the holobiont members; therefore they employed a metatranscriptomic framework.

Terrestrial Environment

- The mechanism of carbon sequestration in agricultural soils through microbial gene expression in straw residue amendments was demonstrated by Kozjek et al. (2023). Their findings indicated that arable soils with less organic carbon may experience a greater dynamic microbial community with such amendments. DESeq2 tool in metatranscriptomic workflow (Love et al. 2014) was used to perform a differential gene expression by filtering reads/counts.
- In other research study on initial characterization of the nidovirus-like metatranscriptome of the newly proposed family Abyssoviridae (Bukhari et al. 2018). Their study elucidate the valuable role of metatransciptome in the phylogenetic analysis, in particular to revealing a sister group to Coronavirinae, known under the proposed genus *Alphaletovirus*.
- A metatranscriptomic study by Geisen et al. (2015) presented the most comprehensive data on communities of active protists found in soil. This proved to be important in studying the ecological role played by protists in complex and ambiguous soil systems.

Human Microbiomes

- Zhang et al. (2023) observed the transcriptional activity of characterized microbiomes present in the gut of primates associated with three different diets and compared the results to their prior reported metagenomic profile. They analyzed the gut microbiome using various metatranscriptomic tools such as $BLAST_P$ and carried out annotations against the Kyoto Encyclopedia of Genes and Genomes

(KEGG) database. By gaining insights in microbe's gene expression, they concluded that microbes in the gut contribute to host dietary adaptation by aiding the host in digesting nutrients that are inaccessible in a given diet.

6.2.2.2 Microbial Interaction

By identifying the patterns of expression of genes associated with communication, signaling, and competition, we can uncover the complex network of interactions established by different microbial species. A summary of current studies on analyzing microbial interactions based on metatranscriptomic study is provided in Table 6.2.

Animal-Microbe Interaction

- Karla Fabiola Corral-Jara et al. (2022) performed a holistic metatranscriptomic analysis using nontargeted total RNA sequencing and Metatrans open-source pipeline. Their work revealed differences in mechanisms of enteric methanogenesis in cows and goats. It was observed that the process of methanogenesis was influenced by the presence of starch and lipids in their diet, and their comparable effects were further studied.
- Another study by Xue et al. (2022) was conducted where metagenomics, metatranscriptomics, and metabolomics were used to elucidate potential microbial traits underlying feed efficiency in dairy cows. Further, on comparing metagenomic and metatranscriptomic data it was understood that metatranscriptomics is a better methodology to elucidate the relationship between the functional activity of microbes present in rumens and host competence.

Plant-Microbe Interaction

- Guo et al. (2021) presented a metatranscriptomic-based analogy of the interactions of *Arabidopsis* with both endophytic and pathogenic *Fusarium* sp., revealing their importance in plant plasticity at the transcriptional level. Their data showed a common expression pattern for the majority of plant genes in response to both fungal types at each instance from 12 to 96 h post-inoculation (HPI).
- It has been shown that the Carbon Catabolite Repression (CCR) emerges as a crucial factor influencing diverse adaptive and physiological responses in a wide array of microbes usually found in association with hosts species, which are eukaryotic (Franzino et al. 2021).
- Law et al. (2022) elaborated on how metatranscriptomics can help in identifying variable shifts in mycorrhizal coordination between fungus and plant roots found in boreal forests. The host–microbe interactions highlighted in this study based on functional analysis elucidate how these intricate and sensitive symbiotic relationships can facilitate the carbon storage capacity of boreal soils under varying nutrient conditions.

6.2.2.3 Human Health and Diseases

Although symbiotic bacteria protect us from pathogens, evidently under unusual conditions they can subdue these defense responses and elicit pathological effects in

Table 6.2 Recent studies on analyzing microbial interaction based on metatranscriptomic analysis

RNA extraction	Metatranscriptomic sequencing	Bioinformatics analysis	Taxonomic annotation	Functional annotation	Statistical analysis	Reference
– Bead-beating – Column purification	– Illumina HiSeq 300 – Trimmomatic (version 0.39) – Meta-Velvet (version 1.2.01)	– BWA v0.7.1 – Megahit v1.1.2 – MetaGene v0.3.38 – CD-HIT – SparCC	– DIAMOND – RefSeq database – HUMAnN v2.0 – (SAMSA) – MG-RAST pipeline	– UBLAST	– DIAMOND against the KEGG database – KEGG Orthology (KO)	Xue et al. (2022)
– ZR soil/fecal RNA microprep kit	– NanoDrop 2000 – Illumina seq	– HISAT – StringTie – DESeq2 pipelines	– HISAT2 version 2.0.5	– agriGO v2 – SEA (SEACOMPARE) – PFAM	– Metascape – Cytoscape version 3.8.0	Guo et al. (2021)
– RNeasy Plus Mini Kit	– MetaTrans open-source pipeline – SortMeRNA tool	– SOAP2 – fastq-Join – UCLUST – (KEGG)	– Silva v138 – DESeq2 R package	– SOAP2 – CD-HIT v4.6	– KEGGs – vegan R library	Corral-Jara et al. (2022)
– RNeasy PowerSoil total RNA kit	– Illumina Novaseq 6000 platform	– Hammer software – PFam – Swiss-Prot	– DESeq2 R package (version 1.26) – GAIA tool (v2.02)	– KEGG – Gene Ontology (GO) – Eggnog	– KEGG	Tartaglia et al. (2023)

the host. The metatranscriptomic analysis of such microbes can be used as a marker of health status of humans as well as an efficient tool for the diagnosis, treatment, and prevention of these microbe-specific diseases.

- Jagare et al. (2023) recently performed a metatranscriptomic analysis of the blood microbiome of healthy people against patients suffering from IBD. Their study suggested that the microbes in blood originated from gut and oral microbiomes and were less likely to have originated from the skin microbiome. They also discuss the limitations of metagenomic analysis in demonstrating the live active or inactive microbiome, its physiological functions, and subsequent interactions with host organisms, and discussed the use of metatranscriptomes to overcome various limitations. Metatranscriptomic approaches can provide taxonomic composition as well as the active functioning of intricate microbial communities.
- Mullish et al. (2018) demonstrated the bile acid metabolism associated with microbes present in the gut and their interaction in health and disease. They specifically targeted the metatranscriptome by performing microbial profiling using 16S rRNA gene sequencing.
- Belstrøm et al. (2017) in their multi-omic analysis of saliva revealed microbiota in patients associated with diseases like periodontitis and dental caries. The objective was to perform characterization of the metagenome and metatranscriptome of salivary microbiota such that their presence would provide a status of the patient's oral health. Their results revealed that the major bacterial genus was *Streptococcus*, constituting ~25 and ~50% of total DNA and RNA reads, respectively.

6.2.2.4 Agriculture-Microbe Dynamics

Plant growth is highly influenced by the microbial communities living on and around them as they provide the required nutrients. Moreover, in the view of crop health and its productivity in association with agricultural methods, it is very crucial for specific micro-communities to be actively present. The diverse agriculture-microbe interaction as to how soil enriched with microbes produces high yielding, healthy crops can be thus studied by exploiting the technique of metatranscriptomics.

- Recently, Parizadeh et al. (2023) in their study concluded that gene expression of soil microbiome in association with agricultural ecosystems was seen to vary with time along with neonicotinoid seed treatments. They elaborated on how metatranscriptomic approach gave a quantitative analysis of the enhanced gene expression of the diverse soil microbiota when subjected to 2-year soybean/corn crop rotation in Quebec, Canada, for 2 years.
- Falk et al. (2023) established combined DNA/RNA amplicon sequencing and stated that metatranscriptomics revealed microbial-driven nutrient transformations and keystone taxa in agriculturally impacted sediments. In this study, DNA and RNA-based 16S amplicon sequencing was paired with RNAseq metatran-

scriptomics to improve interpretations of the core microbial taxa and nutrient cycling genes across sediments of three sites of differing agricultural impact in Southern Ontario, Canada.

6.2.2.5 Functional Profiling

It allows us to interlink specific genes to biological processes, such as carbon cycling, nitrogen fixation, degradation of organic matter, and production of secondary metabolites. This information helps in unraveling the functional potential of the microbial community and understanding their ecological roles.

- Marshall et al. (2023) described how nitrogen cycling that is carried out by microbes is significant in the sediment benthic flux denitrification of coastal regions. They suggested the potential of metatranscriptomics to resolve spatiotemporal comparisons in the microbial contribution to benthic nitrogen cycling. The results of their work indicated that archaeal *nirK* undergoes active transcription in response to changing environmental conditions, which may be a critical yet overlooked feature of nitrogen cycling of coastal sediments.
- Antonioli et al. (2023) studied the oil-contaminated marine sediment under controlled lab conditions to find out the functional response of microbial communities present in the sample using metatranscriptomic data obtained from 16S rRNA amplicon sequencing of both contaminated and uncontaminated samples of shallow marine sediment. They also studied the taxonomic structure, diverseness of species, and shifts over a time period corresponding to the local population. Hydrogen-degrading action with a unique transcriptional profile in the late stages of degradation of oil was observed by performing co-expression analysis linked with functional enrichment in contaminated and controlled samples.

6.2.3 Challenges in Metatranscriptomic Analysis

- Methods involving the metatranscriptome are sometimes too extensive. Experimental designing is crucial. Collection of samples can be destructive and if the material required for sequencing is not present, it can result in insufficient data analysis.
- Metatranscriptomics does not always secure or acquire the whole metatranscriptome due to the vast diversity and relative proportions of members of the microbial community.
- Short and half-life of RNA can introduce a number of drawbacks related to technology.
- Many transcripts within microbial communities are present in low abundance, which can lead to their underrepresentation in sequencing data.
- Analyzing metatranscriptomic data requires sophisticated bioinformatics tools to map, annotate, and quantify transcripts accurately, which can be difficult due to the lack of complete reference databases for many microbial species.

- Environmental factors can strongly influence gene expression, and capturing a snapshot of gene activity might not represent the full dynamic range of responses.
- Integrating metatranscriptomic data with other omics data (metagenomics, metaproteomics) to gain a thorough understanding of the structure and function of a community can be challenging.

6.3 Metaproteomics in Microbial Characterization

Metaproteomics is the full assessment of the entire protein content within an environmental microbiota at a specific moment (Wilmes et al. 2015). This approach provides a detailed understanding of the proteins present and their concentrations, enabling the identification of the components crucial for the microbial ecosystem's sustenance and functioning. Metaproteomics typically employs a shotgun-like approach to identify and quantify protein species. This process involves fragmenting peptides and subjecting them to liquid chromatography-mass spectrometry (LC-MS) (Wilmes et al. 2015; Zhang et al. 2017). The generated data provide information about amino acid sequences, protein abundance, and the occurrence of post-translational changes such as phosphorylation. To identify the specific protein species, sequence homology is contrasted with online reference repositories such as Pipasic, Unipept, and MetaProteomeAnalyzer. Metaproteomics holds appeal for researchers due to its ability to provide more relevant and accurate protein-related data (Xiong et al. 2015).

Metaproteomics and metatranscriptomics are effective methods for analyzing the expression of genes in communities of microbes. Metatranscriptomics has gained popularity, as it involves sequencing the total mRNA in a microbial community (Ottesen et al. 2014), a method called RNAseq. Metaproteomics intends to analyze the whole range of gene translation products, providing further insight into post-translational modifications and protein spatial organization, which go beyond what can be gleaned from transcriptome data alone. In practice, the corresponding amounts of proteins in the community of microbes can be determined using several methods, such as label-free quantification, isobaric chemical labeling procedures, and metabolic labeling (Mosier et al. 2014; Li et al. 2012). It is worth noting that, unlike DNA and mRNA, proteins lack an equivalent to the PCR for their amplification. As a result, quantifying protein abundances presents unique challenges. Achieving a similar dynamic range to what is typically observed in mRNA sequencing can be more demanding, making metaproteomics a field that requires innovative approaches and techniques to overcome these challenges. The association between mRNA distributions and the related protein abundances, while it can be plausible for some fundamental metabolic events in specific microbial systems, generally displays poor or no correlation in most of the biological systems that have been examined to date (Maier et al. 2009; Vogel and Marcotte 2012). This suggests that, in many cases, proteomics data is a more reliable indicator of the biological phenotype compared to mRNA data alone. Proteomics data provides insights into the production of protein and its degradation, which are critical aspects of understanding cellular processes. However, it is essential to remember that protein abundance does not correctly reflect a protein's activity or functional state. For a thorough understanding of metabolic processes and cellular functions, it is frequently helpful to encompass diverse -omics data types, including genomics (DNA), transcriptomics (RNA), and proteomics (proteins). This multi-omics approach enables researchers to acquire a more complete and nuanced view of how biological systems operate and respond to different conditions.

So far, there has been a limited adoption of metaproteomics in skin microbiome research, a notable contrast to the prevalent use of metaproteomics in gut microbiome studies, where multiple researchers combine metaproteomics with metagenomic sequencing (Heyer et al. 2017; Xiong et al. 2015; Xiao et al. 2017). As it stands, in skin microbiome research, amplicon sequencing is widely in use due to its convenience and affordability. However, advances in biotechnology, particularly whole metagenome shotgun sequencing (WMS), are of greater worth and versatility than amplicon sequencing.

6.3.1 Investigations into the Metaproteome of the Microbiota of Human Gut

Metaproteomic analysis of the gut microbiome involves studying bacteria from different sources. A typical workflow, as depicted in Fig. 6.2 (Pettersen et al. 2021), includes protein extraction, preparation, LC-MS/MS analysis, database search, and taxonomic/functional analysis. Billions of microorganisms, comprising numerous bacterial species, reside in the human intestines, rendering it the largest and most intricate microbial ecology in the human body (Lozupone et al. 2012). The microbiota of gut is crucial for sustaining human wellness and is connected to numerous disorders. For instance, microbial imbalances have been linked to illnesses such as obesity and Crohn's disease (Nell et al. 2010; Berer et al. 2011; Clemente et al. 2012). The knowledge of the microbiota of the human gut is limited due to challenges such as heterogeneity in bacterial species, a broad spectrum of protein quantities and a lack of matching metagenomes, and informatics issues. This has hindered metaproteomic studies, despite extensive metagenomic research. The following section provides a concise overview of the existing human gut metaproteomics studies, emphasizing how they contribute to our comprehension of this distinctive ecosystem and its impact on health and disease.

6.3.1.1 Metaproteome of Human Infant Gut

The conventional belief was that the human infant gut is essentially sterile at birth, but recent evidence has emerged challenging this theory. New discoveries suggest the existence of microorganisms in amniotic fluid, infant's meconium, and the placenta (Ardissone et al. 2014; Aagaard et al. 2014; Wang et al. 2013). After birth, microorganisms rapidly colonize, and the microbial makeup changes significantly over the next couple of years before a stable microbiota develops (Matamoros et al. 2013). Consequently, the initial microbial community composition of the newborn

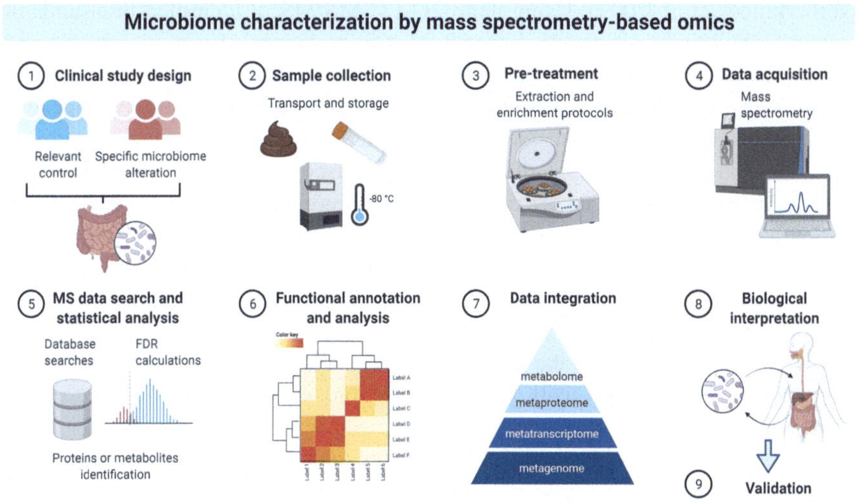

Fig. 6.2 Functional investigations of the human microbiome via liquid chromatography-mass spectrometry (LC-MS) involve a robust study design and controls. Careful assessment of sample transport, storage, and pretreatment is conducted to prevent unintended impacts. Mass spectra are acquired, and data is searched against a sample-specific database with statistical filtering. Bioinformatics and statistical analyses extract biological insights from metaproteome and metabolome datasets. Identified proteins and metabolites associated with specific phenotypes require further experimental validation. (Adapted with permission from Pettersen et al. 2021)

gut is simpler and more straightforward, posing less analytical problems, such as sampling dimensions, than the more complicated and diversified communities of microbes seen in the human adult gut.

Metaproteomics was applied to investigate changes in microbial protein composition in an infant's fecal sample (Klaassens et al. 2007), despite limited genome information. Although protein identification was restricted, the study emphasized the requirement for improved experimental and informatics techniques. Recent developments in sequencing technologies and analytics have considerably improved our knowledge of the newborn microbiome, i.e., remarkable advancements in DNA and protein sequencing technologies and substantial enhancements in the bioinformatics tools employed for data assembly, annotation, and analysis. A recent metaproteome examination of the baby microbiome in the gut revealed dynamic alterations in its functional signature (Young et al. 2015). The study found that functional diversity grew in the first 2–3 weeks, settled early, and remained stable. In addition, it identified various human proteins associated with the microbiome, indicating potential crosstalk between the infant's innate immunity and the microbiome.

6.3.1.2 Gut Proteome of Human Adult

Insight into the Microbiome of Human Gut of a Healthy Adult: The preliminary metaproteomic research of the adult human microbiota of the gut involved an appropriate healthy female monozygotic twin couple from a Swedish twin cohort

(Verberkmoes et al. 2009). Using nano-2D-LC-MS/MS, it identified thousands of proteins, providing an initial look into the gut microbiome's functional profile and its interactions with the host. Notably, around 30% of the detected proteins were of human source, spanning different functions such as innate immunity, digestive enzymes, and cell-to-cell adhesion. While the majority of proteins found in humans were identical between people, differences were discovered in fewer prevalent proteins due to the stochastic nature of sampling. Categorizing proteins with Cluster of Orthologous Groups (COGs) provided a comprehensive view of gut microbiota processes (Tatusov et al. 2000). This analysis revealed variations in COG abundances between the metaproteome and metagenome (Verberkmoes et al. 2009). The metaproteome focused on proteins involved in translation, carbohydrate metabolism, and energy production, whereas the metagenome was abundant in proteins involved in inorganic ion breakdown, cell wall construction, cell division, and secondary metabolite synthesis.

Metaproteomics is utilized to characterize the microbiome of the gut in colorectal cancer (CRC) patients and healthy individuals. Fecal samples were collected and analyzed using high-resolution LC-MS to identify and quantify proteins. The workflow included differential analysis, visualized through a volcano plot highlighting proteins with significant changes between CRC patients and healthy volunteers as shown in Fig. 6.3 (Long et al. 2020). The results offer insights into the gut microbiome composition in CRC, emphasizing the importance of a robust experimental design and advanced analytical techniques (Long et al. 2020).

Microbial functional difference between the healthy and sick states: The initial comparison involves the microbiota of the gut in healthy individuals and those with Crohn's disease (CD), revealing significant insights (Erickson et al. 2012). The study focused on five twin pairs, including both concordant and discordant CD cases, using advanced protein sequencing technology. It identified key functional signatures associated with CD, such as modifications in the metabolism of bacteria, bacterial-host interactions, and host responses. Importantly, these functional fingerprints corresponded to illness status instead of host genetics, indicating prospective targets for potential diagnostic and therapeutic studies. To study the involvement of the gut microbiome in obesity, a study compared metagenomics and metaproteomics in "lean" and "obese" adolescents (Ferrer et al. 2013). Proteomic analysis revealed significant shifts in microbial communities, with Bacteroidetes dominating in the "lean" gut and relatively equal Firmicutes and Bacteroidetes in the "obese" gut. This aligns with findings that Bacteroidetes increase as obese individuals lose weight (Ley et al. 2006), emphasizing the importance of comparative metaproteomics in understanding obesity-related functional changes.

Changes and shifts in microbiome functioning over time: Two significant studies have explored longitudinal changes in adult gut metaproteomes. In the first, the fecal metaproteomes of three healthy individuals remained relatively stable over a year (Kolmeder et al. 2012), revealing a common functional core related to carbohydrate processing and bacterial adaptation. The second study tracked gut microbiota before and after antibiotic treatment (Ferrer et al. 2014), providing a model for

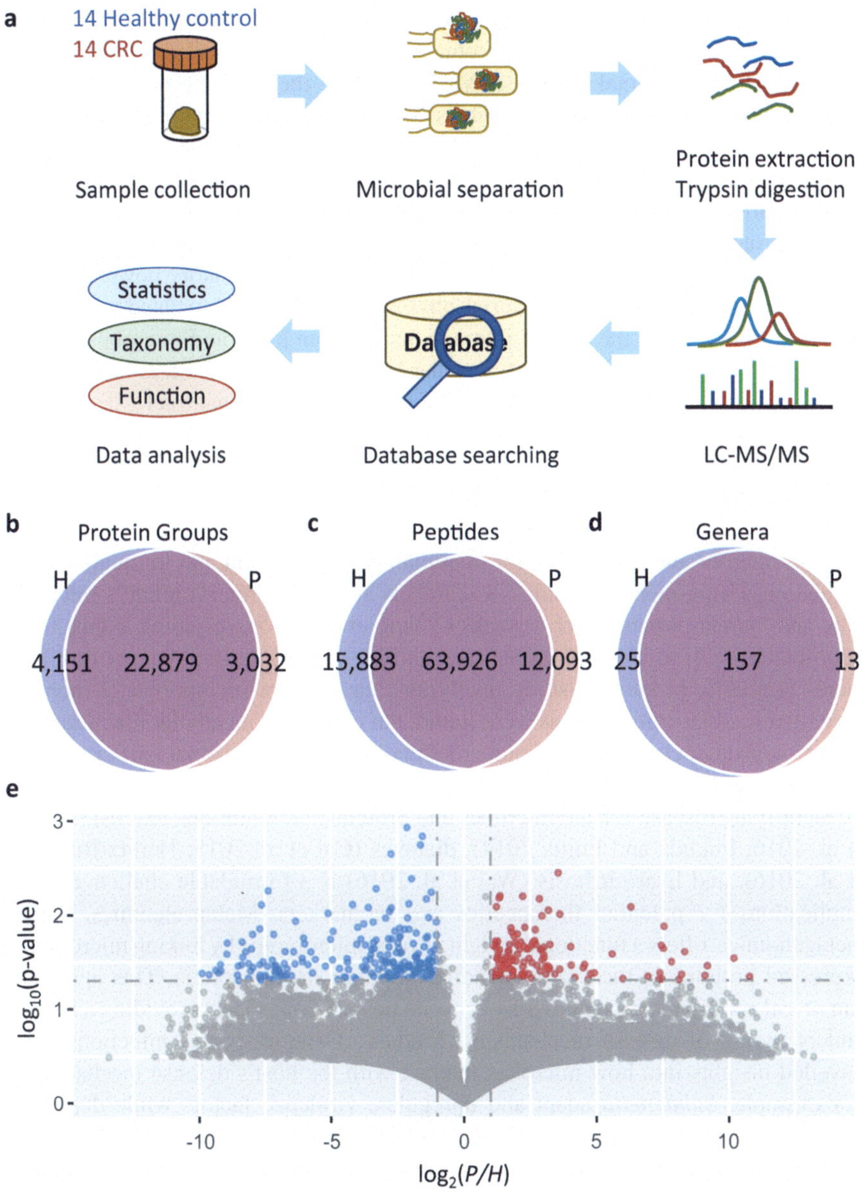

Fig. 6.3 The study aimed to characterize the gut microbiome in colorectal cancer (CRC) patients and healthy volunteers using metaproteomics. (**a**) Experimental design and workflow. Fecal samples were collected from CRC patients (P) and healthy volunteers (H). Protein groups (**b**), peptides (**c**), and genera (**d**) were quantified in both groups. A volcano plot (**e**) highlighted differential proteins, with red indicating those with a CRC/healthy fold change (FC) ≥ 2 and P value <0.05, while blue represented proteins with FC ≤ 0.5 and P value <0.05. Source data can be found in the provided Source Data file. (Adapted with permission from Long et al. 2020)

understanding the temporal reactions of gut microbiota to antibiotics, taking into account both the composition of microbial dynamics and metabolic control.

The Mucosal Luminal Interface (MLI): MLI on the intestine mucous surface provides a protective barrier and enables interactions among the host and the gut flora. Two investigations investigated MLI metaproteomes in mucosal lavage samples. The preliminary study analyzed 205 lavage samples from six colon areas of 38 healthy participants (Li et al. 2011); the results showed an enrichment of immune-related proteins and distinct differences in MLI metaproteomes between colon regions. The second study examined individuals with inflammatory bowel diseases (IBD) and identified specific bacterial phylotypes and IBD-associated proteins at the MLI, highlighting the role of interactions between hosts and microbes in IBD etiology (Presley et al. 2012). Analyzing the MLI metaproteome offers insights into biogeographic-specific metaproteomes along the gastrointestinal tract.

6.3.2 Application of Metaproteomics

The gut microbiome is essential for mammalian development and immune system regulation. Using metaproteomics, researchers studied a preterm infant's microbiome and human proteins over 3 weeks (Young et al. 2015), revealing a shift from microbial growth to metabolic functions and improved intestinal barrier development. Similarly, in another study involving four preterm infants over 3 months, varying microbial compositions were found, but consistent metabolic functions and host proteins associated with immune modulation and the intestinal barrier maintenance were detected (Xiong et al. 2017).

Identifying disease-related microbes and proteins for conditions like IBD (Tedjo et al. 2016; Fukuda and Fujita 2014), diabetes (Cai et al. 2015; Heintz-Buschart et al. 2016), and liver cirrhosis (Wei et al. 2016) is a formidable challenge, with implications for potential therapeutics and biomarkers. Metaproteomics, unlike metagenomics, offers a functional insight into the microbiome by linking microbiota-expressed proteins to their corresponding populations of bacteria. This approach can reveal communication networks between host and microbes, providing a deeper understanding of disease mechanisms. Studies of single-species infections have unveiled insights into how microbes interact with the host's defense mechanisms. For example, *Salmonella* adapt and upregulate virulence factors while depleting host metals. *Cryptococcus neoformans* and HIV use and alter the host kinase phosphorylation to survive. Picornaviruses redirect host proteins for viral replication (Liu et al. 2015). Comparing the fecal microbial communities of liver cirrhosis patients and their spouses reveals distinct microbial proteins with potential biomarker value.

Metaproteomics is commonly utilized in clinical immunoproteomics to design and diagnose vaccines. First introduced in 2001 (Jungblut 2001), immunoproteomics is a rapidly evolving field of research that allows for the detection of antigens produced by bacteria and characterization of the host's antibody reaction. Essentially, immunoproteomics approaches rely on the identification of immunoreactive antigens utilizing sera that include antibodies (Fulton and Twine 2013). It can also be used to generate new antibiotic medicines based on antimicrobial peptides.

6.4 Metabolomics in Microbial Characterization

Metabolomics involves the comprehensive study of the different metabolites inside an organism or biological system. This approach offers valuable qualitative and quantitative insights into the metabolic pathways and their alterations. It has become a valuable tool in microbial research, contributing to areas such as microbial identification, mutation breeding, functional gene analysis, metabolic engineering, and fermentation processes.

Recent advancements in gut microbiome characterization use high-resolutionLC-MS for large-scale profiling of proteins and metabolites. The workflow involves sample collection, medication, LC separation, and MS discovery. Data is also anatomized through bioinformatics, furnishing perceptivity into the functional and compositional aspects of the microbiome of the gut and its counteraccusations for mortal health (Fig. 6.2; Pettersen et al. 2021) illustrates the standard LC-MS grounded metaproteomic and metabolomics workflow (Pettersen et al. 2021).

The exploration of the metabolome, encompassing all the metabolites generated within an organism, reflects the intricate web of enzymatic pathways encoded in its genome. Furthermore, the comprehensive metabolite composition portrays the dynamic interplay between developmental processes and the ever-changing environment throughout the organism's life. Metabolomics has the promise to provide an enhanced portrayal of an organism's actual physiological condition due to its ability to track the overall influence of many factors on cells (Bundy et al. 2005; Garcia et al. 2008; Nicholson and Lindon 2008). Recent improvements in analytical techniques have resulted in the identification of many metabolites serving as biomarkers for many diseases, including mental disorders, coronary heart disease, cancer, and diabetes (Ohdoi et al. 2003; Li et al. 2008; Brindle et al. 2002; Morrow Jr 2010). Beyond these, metabolomics has also made significant contributions to health-related domains such as toxicology, in vitro fertilization, nutrition (nutrigenomics), and oncology, among others (Nicholson and Lindon 2008; Rochfort 2005; Seli et al. 2008; Spratlin et al. 2009). The growing enthusiasm for metabolomics and its expanding range of applications is clearly demonstrated by the substantial increase in publications in this field since 2003 (Rochfort 2005).

Metabolomics, the metabolic analogue to functional genomics, offers a broader perspective by emphasizing the importance of biochemical reaction intermediates in connecting multiple pathways inside a living cell. Metabolomics is an essential part of systems biology because of its critical significance (Villas-Boas et al. 2005; Bundy et al. 2005; Durot et al. 2009). In comparison, while microbial genomics was an early leader in sequencing technology, metabolomics in microorganisms entered the scene relatively later. Prokaryotes were at the forefront in terms of the number of completed sequenced organisms (Kyrpides 2009). Nonetheless, microbial metabolomics has made substantial progress and contributions. Prominent examples include initiatives like the Human Metabolome Project and the Consortium for Metabonomic Toxicology (COMET) (Lindon et al. 2005; Wishart et al. 2009).

6.4.1 Metabolomics Challenges

Metabolomics, a vital part of systems biology, examines an organism's complete metabolite profile under specific conditions. It uncovers how genes, proteins, and metabolites interact in cells and clarifies the impact of the environment (Rochfort 2005). Metabolomics is crucial in environmental research, bioenergy, gene discovery, and functional genomics across different organisms (Morrow Jr 2010; Hollywood et al. 2006; Rochfort 2005).

The fundamental issue in metabolomics is mostly due to technical restrictions, notably the capacity to detect and estimate the entire diversity of internal and external metabolites with a molecular weight below 1 kDa (Villas-Boas et al. 2005; Ohashi et al. 2008).

The key challenges include: (1) Diverse Metabolites, metabolites vary significantly in number and identity among various organisms, ranging from hundreds to potentially hundreds of thousands, making their comprehensive identification a challenge; (2) Chemical Variability, unlike genome, transcriptome, and proteome analyses, metabolites derived from metabolic reactions exhibit high variability in terms of properties and chemical structures (Hollywood et al. 2006); (3) Complex Metabolome, the metabolome can encompass a wide spectrum of substances, including volatile alcohols, hydrophilic carbohydrates, ketones, hydrophobic lipids, amino and non-amino organic acids, and complex natural compounds (e.g., antibiotics, non-ribosomal peptides, pigments, and cofactors) (Villas-Boas et al. 2005; Durot et al. 2009; Gill et al. 2006); (4) Simultaneous Identification, determining the whole array of metabolites in a specific physiological state is an extraordinarily difficult undertaking (Garcia et al. 2008; Ohashi et al. 2008); (5) Dynamic Nature, metabolites are always in flux, with concentrations and compositions quickly altering in response to external stimuli (Villas-Boas et al. 2005; Hollywood et al. 2006; Vemuri and Aristidou 2005); (6) Complex Correlations, establishing direct links between genes and metabolites is not always straightforward, especially when microbes have less metabolites than genes present (Vemuri and Aristidou 2005; Forster et al. 2003). Taken together, these elements add to the intricacy and challenge of carrying out research on metabolomics.

6.4.2 Application of Metabolomics

Given their importance to human health and the ecosystem, microorganisms are ideal for studying systems biology because of their manipulability. An enhanced understanding of microbial interactions and cellular processes is made possible by the integration of biological data into systems microbiology with microbial metabolomics.

Metabolomics in Human Ecosystem: The Human Microbiome Project (HMP), launched in 2007, changed our view of microbes in human biology. The microbiota—the microorganisms that live within and on humans—far outnumbers human cells (Gill et al. 2006; Peterson et al. 2009). The HMP focused on five body sites and integrated microbiota structure, gene content, gene expression, and metabolism (Turnbaugh et al. 2007). It offered insightful information about the microbiome's function in human health, with many previously unknown taxa (Ventura et al. 2009; Pennisi 2010; Lemon et al. 2010; Atarashi et al. 2011). As of the end of 2010, 500 out of 3000 relevant microbial genomes had been sequenced (Pennisi 2010). Metabolomics has a wide application in the study of gut microbiomes.

Metabolomics in Lactic Acid Bacteria LAB) (Research: Traditionally, LAB were classified based on structure and biochemical experiments. Metabolic profiling, which swiftly identifies strains based on unique extracellular metabolite peaks, has become the favored choice in metabolomics. Additionally, this approach can uncover changes in intestinal metabolites due to *Lactobacillus* activity and its influence on cytokine expression.

Metabolomics in Environmental Ecosystem: Microbial communities are integral to ecosystems, yet our comprehension was historically limited to cultured microorganisms (Tringe et al. 2005). DNA sequencing technology's introduction and innovative approaches have transformed environmental microbiology, enabling the study of entire communities, including uncultivated members (Torsvik and Ovreas 2002; Rusch et al. 2007; Tringe and Rubin 2005; Raes et al. 2007; Wilmes et al. 2010). To understand the communities and dynamics within microbial ecosystems, metabolomics is essential. To get over metagenomic analysis's constraint, especially in the case of uncultured community members, demands innovative approaches (Raes and Bork 2008; Schleper et al. 2005). Innovations like predictive metaproteomics, using fragmented genomic data to identify encoded proteins, offer a powerful approach for understanding microbial communities. Analyzing operons in diverse environmental samples reveals correlations with specific conditions, highlighting metabolic requirements. This method, rooted in genomic data, revolutionizes functional studies within microbial communities (Tringe et al. 2005). A different approach to metagenomic sequencing was provided by Sieracki and Stepanauskas to comprehend the metabolic capacities and ecological responsibilities of uncultured community members. They sequenced genes from individual bacterial cells within the community, enabling direct detection of metabolic genes. This technique illuminated important metabolic pathways in these ecosystems by identifying genes linked to vital biogeochemical processes in marine habitats, including photometabolic systems, nitrate usage, and nitrogen fixation (Stepanauskas and Sieracki 2007).

Metabolomics in Study of Intestinal Flora: Intestinal microorganisms are vital for the host's well-being, impacting health and disease. They prevent infections, provide energy, boost immunity, and influence metabolic traits. Microbial metabolomics technology reveals how gut microbes affect the host's health and disease by directly examining metabolic changes.

Metabolomics in Research of Pathogenic Bacteria: Metabolomics technology analyzes pathogenic bacteria, aiding disease diagnosis, mutation research, and secondary metabolite detection. Metabolomics in Food and Nutrition: Pathogens, toxins, and microbial by-products in food are central to food safety. Microbial metabolomics helps evaluate diet's impact on the body and reduces the influence of factors like age, lifestyle, and gender for accurate monitoring.

6.5 Conclusion and Future Perspectives

The dynamic nature of microbial populations can be understood through metatranscriptomics, as it takes the advantage of RNA transcripts rather than DNA, allowing conclusions to be drawn about how changes in the environment cause genetic change at the community scale. While metagenomics aids in generating a taxonomic profile of the sample, metatranscriptomics provides a functional profile by analyzing gene expression. Metaproteomics is a powerful and innovative approach for studying microbial communities, offering a deeper understanding of the functional aspects of these ecosystems. By analyzing the complete protein content within an environmental microbiota, metaproteomics provides valuable insights into the proteins present, their concentrations, and post-translational modifications, such as phosphorylation. This approach has found widespread use in gut microbiome research. Metaproteomics is particularly valuable because it bridges the gap between genomics and transcriptomics. Metaproteomics aids in vaccine development, diagnostics, and identifying disease-related microbes and proteins, offering potential therapeutic targets and biomarkers for various conditions like IBD, diabetes, and more. The limited adoption of metaproteomics in skin microbiome research highlights the need for its wider application in various microbial ecosystems. Metaproteomics continues to advance, offering deeper insights into microbiomes, microbial-host interactions, and their effects on the health of humans and disease. Metabolomics, the study of an organism's metabolites, is vital in microbial research, helping in microbial identification, gene analysis, and understanding complex interactions. It provides a comprehensive overview of an organism's physiological status, has applications in various fields, and overcomes challenges through advanced technologies. Metabolomics contributes to human microbiome, environmental research, and pathogenic bacteria studies, impacting health and disease. Metabolomics has evolved into a key method for enhancing our understanding of complicated biological systems.

Conflict of Interest None.

References

Aagaard K, Ma J, Antony KM, Ganu R et al (2014) The placenta harbors a unique microbiome. Sci Transl Med 6:237ra265

Aguiar-Pulido V, Huang W, Suarez-Ulloa V, Cickovski T, Mathee K, Narasimhan G (2016) Metagenomics, metatranscriptomics, and metabolomics approaches for microbiome analysis. Evol Bioinformatics Online 12(Suppl 1):5–16. https://doi.org/10.4137/EBO.S36436

Antonioli R, de Faria Poloni J, Escalona MAR, Dorn M (2023) Functional response of microbial communities in lab-controlled oil-contaminated marine sediment. Mol Omics 19(10):756–768

Ardissone AN, de la Cruz DM, Davis-Richardson AG, Rechcigl KT et al (2014) Meconium microbiome analysis identifies bacteria correlated with premature birth. PLoS One 9:e90784

Atarashi K, Tanoue T, Shima T, Imaoka A, Kuwahara T, Momose Y, Cheng G, Yamasaki S, Saito T, Ohba Y, Taniguchi T, Takeda K, Hori S, Ivanov II, Umesaki Y, Itoh K, Honda K (2011) Induction of colonic regulatory T cells by indigenous Clostridium species. Science 331(6015):337–341

Atlas R (2010) Handbook of microbiological media, 4th edn. CRC Press, Boca Raton, FL

Bao G, Wang M, Doak TG, Ye Y (2015) Strand-specific community RNA-seq reveals prevalent and dynamic antisense transcription in human gut microbiota. Front Microbiol 6:896. https://doi.org/10.3389/fmicb.2015.00896

Barberan A, Bates ST, Casamayor EO, Fierer N (2011) Using network analysis to explore cooccurrence patterns in soil microbial communities. ISME J 6:343–351

Basu S, Bose C, Ojha N, Das N, Das J, Pal M, Khurana S (2015) Evolution of bacterial and fungal growth media. Bioinformation 11:182–184. https://doi.org/10.6026/97320630011182

Belstrøm D, Constancias F, Liu Y, Yang L, Drautz-Moses DI, Schuster SC, Kohli GS, Jakobsen TH, Holmstrup P, Givskov M (2017) Metagenomic and metatranscriptomic analysis of saliva reveals disease-associated microbiota in patients with periodontitis and dental caries. NPJ Biofilms Microbiomes 3(1):23

Berer K, Mues M, Koutrolos M, Rasbi ZA et al (2011) Commensal microbiota and myelin autoantigen cooperate to trigger autoimmune demyelination. Nature 479:538–541

Bernini P, Bertini I, Luchinat C et al (2009) Individual human phenotypes in metabolic space and time. J Proteome Res 8(9):4264–4271

Bonnet M, Lagier JC, Raoult D, Khelaifia S (2020) Bacterial culture through selective and non-selective conditions: the evolution of culture media in clinical microbiology. New Microbes New Infect 34:100622

Booijink CC, Boekhorst J, Zoetendal EG, Smidt H, Kleerebezem M, de Vos WM (2010) Metatranscriptome analysis of the human fecal microbiota reveals subject-specific expression profiles, with genes encoding proteins involved in carbohydrate metabolism being dominantly expressed. Appl Environ Microbiol 76(16):5533–5540. https://doi.org/10.1128/AEM.00502-10

Brindle JT, Antti H, Holmes E, Tranter G, Nicholson JK, Bethell HW, Clarke S, Schofield PM, McKilligin E, Mosedale DE, Grainger DJ (2002) Rapid and noninvasive diagnosis of the presence and severity of coronary heart disease using 1H-NMR-based metabonomics. Nat Med 8(12):1439–1444

Bukhari K, Mulley G, Gulyaeva AA, Zhao L, Shu G, Jiang J, Neuman BW (2018) Description and initial characterization of metatranscriptomic nidovirus-like genomes from the proposed new family Abyssoviridae, and from a sister group to the Coronavirinae, the proposed genus Alphaletovirus. Virology 524:160–171

Bundy JG, Willey TL, Castell RS, Ellar DJ, Brindle KM (2005) Discrimination of pathogenic clinical isolates and laboratory strains of Bacillus cereus by NMR-based metabolomic profiling. FEMS Microbiol Lett 242(1):127–136

Cai L, Wu H, Li D, Zhou K, Zou F (2015) Type 2 diabetes biomarkers of human gut microbiota selected via iterative sure independent screening method. PLoS One 10(10):e0140827. https://doi.org/10.1371/journal.pone.0140827

Carvalhais LC, Dennis PG, Tyson GW, Schenk PM (2012) Application of metatranscriptomics to soil environments. J Microbiol Methods 91(2):246–251

Chaudhary DK, Khulan A, Kim J (2019) Development of a novel cultivation technique for uncultured soil bacteria. Sci Rep 9(1):6666. https://doi.org/10.1038/s41598-019-43182-x

Chen L, Hu M, Huang L et al (2015) Comparative metagenomic and metatranscriptomic analyses of microbial communities in acid mine drainage. ISME J 9:1579–1592. https://doi.org/10.1038/ismej.2014.245
Clemente JC, Ursell LK, Parfrey LW, Knight R (2012) The impact of the gut microbiota on human health: an integrative view. Cell 148:1258–1270
Corral-Jara KF, Ramayo-Caldas Y, Bernard L, Martin C, Tournayre J, Morgavi DP, Popova M (2022) An integrative metatranscriptomic analysis reveals differences in enteric methanogenesis mechanisms between cows and goats, 15 July 2022, PREPRINT (Version 1) available at Research Square. https://doi.org/10.21203/rs.3.rs-1828071/v1
Daniels CA, Baumgarten S, Yum LK, Michell CT, Bayer T, Arif C et al (2015) Metatranscriptome analysis of the reef-building coral Orbicella faveolata indicates holobiont response to coral disease. Front Mar Sci 2:62
Dindal DL (1990) Soil biology guide. Wiley, New York
Dobell C (1920) The discovery of intestinal protozoa in man. Proc R Soc Med 13:1–15
Durot M, Bourguignon PY, Schachter V (2009) Genome-scale models of bacterial metabolism: reconstruction and applications. FEMS Microbiol Rev 33(1):164–190
Dziurzynski M, Ciuchcinski K, Dyda M, Szych A, Drabik P, Laudy A, Dziewit L (2020) Assessment of bacterial contamination of air at the museum of King John III's palace at Wilanow (Warsaw, Poland): selection of an optimal growth medium for analyzing airborne bacteria diversity. Science 10:7128. https://doi.org/10.3390/app10207128
El Samak M, Zakeer S, Hanora A, Solyman SM (2023) Metagenomic and metatranscriptomic exploration of the Egyptian Red Sea sponge Theonella sp. associated microbial community. Mar Genomics 70:101032. https://doi.org/10.1016/j.margen.2023.101032
Erickson AR, Cantarel BL, Lamendella R, Darzi Y et al (2012) Integrated metagenomics/metaproteomics reveals human host-microbiota signatures of Crohn's disease. PLoS One 7:e49138
Evariste L et al (2019) Gut microbiota of aquatic organisms: a key endpoint for ecotoxicological studies. Environ Pollut 248:989–999
Ezzeldin S et al (2019) Current understanding of human metaproteome association and modulation. J Proteome Res 18:3539–3554. https://doi.org/10.1021/acs.jproteome.9b00301
Falk N, Droppo IG, Drouillard KG, Weisener C (2023) Combined DNA/RNA amplicon sequencing and metatranscriptomics reveals microbial-driven nutrient transformations and keystone taxa in agriculturally impacted sediments. Available at SSRN: https://ssrn.com/abstract=4441467 or https://doi.org/10.2139/ssrn.4441467
Faust K, Sathirapongsasuti JF, Izard J et al (2012) Microbial co-occurrence relationships in the human microbiome. PLoS Comput Biol 8(7):e1002606
Ferrer M, Ruiz A, Lanza F, Haange SB et al (2013) Microbiota from the distal guts of lean and obese adolescents exhibit partial functional redundancy besides clear differences in community structure. Environ Microbiol 15:211–226
Ferrer M, Martins dos Santos VA, Ott SJ, Moya A (2014) Gut microbiota disturbance during antibiotic therapy: a multi-omic approach. Gut Microbes 5:64–70
Fiehn O (2002) Metabolomics—the link between genotypes and phenotypes. Plant Mol Biol 48(1–2):155–171
Forster J, Famili I, Fu P, Palsson BO, Nielsen J (2003) Genome-scale reconstruction of the Saccharomyces cerevisiae metabolic network. Genome Res 13(2):244–253
Franzino T, Boubakri H, Cernava T, Abrouk D, Achouak W, Reverchon S, Nasser W, Haichar FZ (2021) Implications of carbon catabolite repression for plant-microbe interactions. Plant Commun 3:100272. https://doi.org/10.1016/j.xplc.2021.100272
Freilich S, Kreimer A, Meilijson I, Gophna U, Sharan R, Ruppin E (2010) The large-scale organization of the bacterial network of ecological co-occurrence interactions. Nucleic Acids Res 38(12):3857–3868
Fukuda K, Fujita Y (2014) Determination of the discriminant score of intestinal microbiota as a biomarker of disease activity in patients with ulcerative colitis. BMC Gastroenterol 14:49. https://doi.org/10.1186/1471-230X-14-49

Fulton KM, Twine SM (2013) Immunoproteomics: current technology and applications. Methods Mol Biol 1061:21–57. https://doi.org/10.1007/978-1-62703-589-7_2

Garcia DE, Baidoo EE, Benke PI, Pingitore F, Tang YJ, Villa S, Keasling JD (2008) Separation and mass spectrometry in microbial metabolomics. Curr Opin Microbiol 11(3):233–239

Geisen S, Tveit A, Clark I et al (2015) Metatranscriptomic census of active protists in soils. ISME J 9:2178–2190. https://doi.org/10.1038/ismej.2015.30

Gilbert JA, Field D, Huang Y et al (2008) Detection of large numbers of novel sequences in the metatranscriptomes of complex marine microbial communities. PLoS One 3(8):e3042

Gill SR, Pop M, Deboy RT, Eckburg PB, Turnbaugh PJ, Samuel BS, Gordon JI, Relman DA, Fraser-Liggett CM, Nelson KE (2006) Metagenomic analysis of the human distal gut microbiome. Science 312(5778):1355–1359

Gillet LC et al (2012) Targeted data extraction of the MS/MS spectra generated by data-independent acquisition: a new concept for consistent and accurate proteome analysis. Mol Cell Proteomics 11:O111.016717. https://doi.org/10.1074/mcp.O111.016717

Gosalbes MJ, Durbán A, Pignatelli M, Abellan JJ, Jiménez-Hernández N, Pérez-Cobas AE, Latorre A, Moya A (2011) Metatranscriptomic approach to analyze the functional human gut microbiota. PLoS One 6(3):e17447. https://doi.org/10.1371/journal.pone.0017447

Gouveia D, Grenga L, Pible O, Armengaud J (2020) Quick microbial molecular phenotyping by differential shotgun proteomics. Environ Microbiol 22:2996–3004

Guo L, Yu H, Wang B, Vescio K, DeIulio G, Yang H, Berg A, Zhang L, Steinberg C, Edel-Hermann V, Kistler H, Ma L-J (2021) Metatranscriptomic comparison of endophytic and pathogenic Fusarium-Arabidopsis interactions reveals plant transcriptional plasticity. Mol Plant-Microbe Interact 34:1071. https://doi.org/10.1094/MPMI-03-21-0063-R

Gupta A, Singh V, Mani I (2022) Dysbiosis of human microbiome and infectious diseases. Prog Mol Biol Transl Sci 192(1):33–51. https://doi.org/10.1016/bs.pmbts.2022.06.016

Gutleben J, Chaib De Mares M, van Elsas JD, Smidt H, Overmann J, Sipkema D (2018) The multi-omics promise in context: from sequence to microbial isolate. Rev Microbiol 44:212–229

Hardouin P, Chiron R, Marchandin H, Armengaud J, Grenga L (2021) Metaproteomics to decipher CF host–microbiota interactions: overview, challenges and future perspectives. Genes (Basel) 12:892

Heintz-Buschart A, May P, Laczny CC, Lebrun LA, Bellora C, Krishna A, Wampach L, Schneider JG, Hogan A, de Beaufort C, Wilmes P (2016) Integrated multi-omics of the human gut microbiome in a case study of familial type 1 diabetes. Nat Microbiol 2:16180. https://doi.org/10.1038/nmicrobiol.2016.180

Heyer R, Schallert K, Zoun R et al (2017) Challenges and perspectives of metaproteomic data analysis. J Biotechnol 261:24–36

Hoerstmann C, Aguiar-González BM, Barrillon S, Bastos CC, Grosso O, Pérez-Hernández MD et al (2023) Diazotroph connectomics in Gulf Stream eddies supports N2 fixation in the North Atlantic. https://doi.org/10.21203/rs.3.rs-3721371/v1

Hollywood K, Brison DR, Goodacre R (2006) Metabolomics: current technologies and future trends. Proteomics 6(17):4716–4723

Hugenholtz P, Goebel BM, Pace NR (1998) Impact of culture-independent studies on the emerging phylogenetic view of bacterial diversity. J Bacteriol 80:4765–4774

Jagare L, Rozenberga M, Silamikelis I, Ansone L, Elbere I, Briviba M, Megnis K, Konrade I, Birka I, Straume Z, Klovins J (2023) Metatranscriptome analysis of blood in healthy individuals and irritable bowel syndrome patients. J Med Microbiol 72(6):001719

Jungblut PR (2001) Proteome analysis of bacterial pathogens. Microbes Infect 3(10):831–840. https://doi.org/10.1016/S1286-4579(01)01441-1

Kellenberger E (2001) Exploring the unknown. The silent revolution of microbiology. EMBO Rep 2(1):5–7. https://doi.org/10.1093/embo-reports/kve014

Klaassens ES, De Vos WM, Vaughan EE (2007) Metaproteomics approach to study the functionality of the microbiota in the human infant gastrointestinal tract. Appl Environ Microbiol 73:1388–1392

Kleiner M (2019) Metaproteomics: much more than measuring gene expression in microbial communities. mSystems 4. https://doi.org/10.1128/mSystems.00115-19
Kolmeder CA, de Been M, Nikkila J, Ritamo I et al (2012) Comparative metaproteomics and diversity analysis of human intestinal microbiota testifies for its temporal stability and expression of core functions. PLoS One 7:e29913
Kozjek K, Mano L, Urich T, Ahrén D, Hedlund K (2023) Microbial gene activity in straw residue amendments reveals carbon sequestration mechanisms in agricultural soils. Soil Biol Biochem 179:108994. https://doi.org/10.1016/j.soilbio.2023.108994
Krumsiek J, Mittelstrass K, Do KT et al (2015) Gender-specific pathway differences in the human serum metabolome. Metabolomics 11(6):1815–1833
Kuczynski J, Liu Z, Lozupone C, McDonald D, Fierer N, Knight R (2010) Microbial community resemblance methods differ in their ability to detect biologically relevant patterns. Nat Methods 7:813–819
Kyrpides NC (2009) Fifteen years of microbial genomics: meeting the challenges and fulfilling the dream. Nat Biotechnol 27(7):627–632
Laiz L, Piñar G, Lubitz W, Saiz-Jimenez C (2003) Monitoring the colonization of monuments by bacteria: cultivation versus molecular methods. Microbiology 5:72. https://doi.org/10.1046/j.1462-2920.2003.00381.x
Law SR, Serrano AR, Daguerre Y, Sundh J, Schneider AN, Stangl ZR et al (2022) Metatranscriptomics captures dynamic shifts in mycorrhizal coordination in boreal forests. Proc Natl Acad Sci 119(26):e2118852119
Lemon KP, Klepac-Ceraj V, Schiffer HK, Brodie EL, Lynch SV, Kolter R (2010) Comparative analyses of the bacterial microbiota of the human nostril and oropharynx. MBio 1(3):e00129-10
Ley RE, Turnbaugh PJ, Klein S, Gordon JI (2006) Microbial ecology: human gut microbes associated with obesity. Nature 444:1022–1023
Li M, Wang B, Zhang M, Rantalainen M, Wang S, Zhou H, Zhang Y, Shen J, Pang X, Wei H, Chen Y, Lu H, Zuo J, Su M, Qiu Y, Jia W, Xiao C, Smith LM, Yang S, Holmes E, Tang H, Zhao G, Nicholson JK, Li L, Zhao L (2008) Symbiotic gut microbes modulate human metabolic phenotypes. Proc Natl Acad Sci U S A 105(6):2117–2122
Li X, LeBlanc J, Truong A, Vuthoori R et al (2011) A metaproteomic approach to study human-microbial ecosystems at the mucosal luminal interface. PLoS One 6:e26542
Li Z, Adams RM, Chourey K, Hurst GB et al (2012) Systematic comparison of label-free, metabolic labeling, and isobaric chemical labeling for quantitative proteomics on LTQ Orbitrap Velos. J Proteome Res 11:1582–1590
Li J et al (2014) An integrated catalog of reference genes in the human gut microbiome. Nat Biotechnol 32:834–841. https://doi.org/10.1038/nbt.2942
Li L et al (2019) An in vitro model maintaining taxon-specific functional activities of the gut microbiome. Nat Commun 10:4146. https://doi.org/10.1038/s41467-019-12087-8
Lindon JC, Keun HC, Ebbels TM, Pearce JM, Holmes E, Nicholson JK (2005) The Consortium for Metabonomic Toxicology (COMET): aims, activities and achievements. Pharmacogenomics 6(7):691–669
Liu Y, Zhang Q, Hu M, Yu K, Fu J, Zhou F, Liu X (2015) Proteomic analyses of intracellular Salmonella enterica serovar Typhimurium reveal extensive bacterial adaptations to infected host epithelial cells. Infect Immun 83(7):2897–2906. https://doi.org/10.1128/IAI.02882-14
Lloyd-Price J et al (2017) Erratum: strains, functions and dynamics in the expanded Human Microbiome Project. Nature 551:256–256. https://doi.org/10.1038/nature24485
Long S, Yang Y, Shen C et al (2020) Metaproteomics characterizes human gut microbiome function in colorectal cancer. NPJ Biofilms Microbiomes 6:14. https://doi.org/10.1038/s41522-020-0123-4
Love MI, Huber W, Anders S (2014) Moderated estimation of fold change and dispersion for RNA-seq data with DESeq2. Genome Biol 15(12):550. https://doi.org/10.1186/s13059-014-0550-8
Lozupone CA, Stombaugh JI, Gordon JI, Jansson JK, Knight R (2012) Diversity, stability and resilience of the human gut microbiota. Nature 489:220–230

Maier T, Guell M, Serrano L (2009) Correlation of mRNA and protein in complex biological samples. FEBS Lett 583:3966–3973
Mani I (2020a) Current status and challenges of DNA sequencing. In: Singh V (ed) Advances in synthetic biology. Springer, Singapore
Mani I (2020b) Metagenomics approach for bioremediation: challenges and perspectives. In: Pandey VC, Singh V (eds) Bioremediation of pollutants. Elsevier, pp 275–285., ISBN 9780128190258. https://doi.org/10.1016/B978-0-12-819025-8.00012-0
Mani I (2023) Cutting-edge tools to assess microbial diversity and their function in land remediation. In: Pandey VC (ed) Bio-inspired land remediation. Environmental contamination remediation and management. Springer, Cham. https://doi.org/10.1007/978-3-031-04931-6_5
Marshall AJ, Phillips L, Longmore A, Hayden HL, Tang C, Heidelberg KB, Mele P (2023) Using metatranscriptomics to better understand the role of microbial nitrogen cycling in coastal sediment benthic flux denitrification efficiency, vol 15. Environ Microbiol Rep, p 308
Matamoros S, Gras-Leguen C, Le Vacon F, Potel G, de La Cochetiere MF (2013) Development of intestinal microbiota in infants and its impact on health. Trends Microbiol 21:167–173
Mesuere B, Van der Jeugt F, Willems T, Naessens T, Devreese B, Martens L et al (2018) High-throughput metaproteomics data analysis with Unipept: a tutorial. J Proteome 171:11–22
Moniruzzaman M, Wurch LL, Alexander H, Dyhrman ST, Gobler CJ, Wilhelm SW (2017) Virus-host relationships of marine single-celled eukaryotes resolved from metatranscriptomics. Nat Commun 8:16054. https://doi.org/10.1038/ncomms16054
Moran MA (2009) Metatranscriptomics: eavesdropping on complex microbial communities. Microbiome 4(7):329–334
Morrow KJ Jr (2010) Mass spec central to metabolomics. Genet Eng Biotechnol News 30(7):1–3
Mosier AC, Li Z, Thomas BC, Hettich RL et al (2014) Elevated temperature alters proteomic responses of individual organisms within a biofilm community. ISME J 9:180
Mullish BH, Pechlivanis A, Barker GF, Thursz MR, Marchesi JR, McDonald JA (2018) Functional microbiomics: evaluation of gut microbiota-bile acid metabolism interactions in health and disease. Methods 149:49–58
Nell S, Suerbaum S, Josenhans C (2010) The impact of the microbiota on the pathogenesis of IBD: lessons from mouse infection models. Nat Rev Microbiol 8:564–577
Nicholson JK, Lindon JC (2008) Metabonomics. Nature 455(23):1054–1056
Ohashi Y, Hirayama A, Ishikawa T, Nakamura S, Shimizu K, Ueno Y, Tomita M, Soga T (2008) Depiction of metabolome changes in histidine-starved Escherichia coli by CE-TOF-MS. Mol BioSyst 4(2):135–147
Ohdoi C, Nyhan WL, Kuhara T (2003) Chemical diagnosis of Lesch-Nyhan syndrome using gas chromatography-mass spectrometry detection. J Chromatogr B Anal Technol Biomed Life Sci 792(1):123–130
Ojala T, Kankuri E, Kankainen M (2023) Understanding human health through metatranscriptomics. Trends Mol Med 29(5):376–389. https://doi.org/10.1016/j.molmed.2023.02.002
Ortiz Baez A (2023) Understanding virus and microbial evolution in wildlife through metatranscriptomics. Doctoral dissertation
Ottesen EA, Young CR, Gifford SM, Eppley JM et al (2014) Multispecies diel transcriptional oscillations in open ocean heterotrophic bacterial assemblages. Science 345:207–212
Pace NR (1997) A molecular view of microbial diversity and the biosphere. Science 276:734–740
Pangallo D, Bučková M, Kraková L, Puškárová A, Šaková N, Grivalský T, Chovanová K, Zemánková M (2015) Biodeterioration of epoxy resin: a microbial survey through culture-independent and culture-dependent approaches. Microbiology 17:462. https://doi.org/10.1111/1462-2920.12523
Parizadeh M, Mimee B, Kembel SW (2023) Soil microbial gene expression in an agricultural ecosystem varies with time and neonicotinoid seed treatments. Microbiology (Reading) 169(4):001318. https://doi.org/10.1099/mic.0.001318
Pennisi E (2010) Body's hardworking microbes get some overdue respect. Science 330(6011):1619
Peterson J, Garges S, Giovanni M, McInnes P, Wang L, Schloss JA, Bonazzi V, McEwen JE, Wetterstrand KA, Deal C, Baker CC, Di Francesco V, Howcroft TK, Karp RW, Lunsford RD,

Wellington CR, Belachew T, Wright M, Giblin C, David H, Mills M, Salomon R, Mullins C, Akolkar B, Begg L, Davis C, Grandison L, Humble M, Khalsa J, Little AR, Peavy H, Pontzer C, Portnoy M, Sayre MH, Starke-Reed P, Zakhari S, Read J, Watson B, Guyer M (2009) The NIH Human Microbiome Project. Genome Res 19(12):2317–2323

Pettersen VK, Antunes LCM, Dufour A, Arrieta MC (2021) Inferring early-life host and microbiome functions by mass spectrometry-based metaproteomics and metabolomics. Comput Struct Biotechnol J 20(20):274–286. https://doi.org/10.1016/j.csbj.2021.12.012

Poretsky RS, Bano N, Buchan A, LeCleir G, Kleikemper J, Pickering M, Pate WM, Moran MA, Hollibaugh JT (2005) Analysis of microbial gene transcripts in environmental samples. Appl Environ Microbiol 71(7):4121–4126. https://doi.org/10.1128/AEM.71.7.4121-4126.2005

Presley LL, Ye J, Li X, Leblanc J et al (2012) Host-microbe relationships in inflammatory bowel disease detected by bacterial and metaproteomic analysis of the mucosal-luminal interface. Inflamm Bowel Dis 18:409–417

Qin J, Li R, Raes J, Arumugam M, Burgdorf KS, Manichanh C et al (2010) A human gut microbial gene catalogue established by metagenomic sequencing. Nature 464:59–65. https://doi.org/10.1038/nature08821

Qin J, Li Y, Cai Z, Li S, Zhu J, Zhang F, Liang S, Zhang W, Guan Y, Shen D, Peng Y, Zhang D, Jie Z, Wu W, Qin Y, Xue W, Li J, Han L, Lu D, Wu P et al (2012) A metagenome-wide association study of gut microbiota in type 2 diabetes. Nature 490:55–60

Raes J, Bork P (2008) Molecular eco-systems biology: towards an understanding of community function. Nat Rev Microbiol 6(9):693–699

Raes J, Foerstner KU, Bork P (2007) Get the most out of your metagenome: computational analysis of environmental sequence data. Curr Opin Microbiol 10(5):490–498

Rajagopala SV, Bakhoum NG, Pakala SB, Shilts MH, Rosas-Salazar C, Mai A, Boone HH, McHenry R, Yooseph S, Halasa N, Das SR (2021) Metatranscriptomics to characterize respiratory virome, microbiome, and host response directly from clinical samples. Cell Rep Methods 1(6):100091. https://doi.org/10.1016/j.crmeth.2021.100091

Rochfort S (2005) Metabolomics reviewed: a new "omics" platform technology for systems biology and implications for natural products research. J Nat Prod 68(12):1813–1820

Rondon MR, August PR, Bettermann AD, Brady SF, Grossman TH, Liles MR et al (2000) Cloning the soil metagenome: a strategy for accessing the genetic and functional diversity of uncultured microorganisms. Appl Environ Microbiol 66:2541–2547. https://doi.org/10.1128/AEM.66.6.2541-2547.2000

Rosenberger G et al (2017) Statistical control of peptide and protein error rates in large-scale targeted data-independent acquisition analyses. Nat Methods 14:921–927. https://doi.org/10.1038/nmeth.4398

Rusch DB, Halpern AL, Sutton G, Heidelberg KB, Williamson S, Yooseph S, Wu D, Eisen JA, Hoffman JM, Remington K, Beeson K, Tran B, Smith H, Baden-Tillson H, Stewart C, Thorpe J, Freeman J, Andrews-Pfannkoch C, Venter JE, Li K, Kravitz S, Heidelberg JF, Utterback T, Rogers YH, Falcon LI, Souza V, Bonilla-Rosso G, Eguiarte LE, Karl DM, Sathyendranath S, Platt T, Bermingham E, Gallardo V, Tamayo-Castillo G, Ferrari MR, Strausberg RL, Nealson K, Friedman R, Frazier M, Venter JC (2007) The Sorcerer II Global Ocean Sampling expedition: Northwest Atlantic through eastern tropical Pacific. PLoS Biol 5(3):e77

Schleper C, Jurgens G, Jonuscheit M (2005) Genomic studies of uncultivated archaea. Nat Rev Microbiol 3(6):479–488

Schmidt TM, DeLong EF, Pace NR (1991) Analysis of a marine picoplankton community by 16S rRNA gene cloning and sequencing. J Bacteriol 173:4371–4378

Segata N, Boernigen D, Tickle TL, Morgan XC, Garrett WS, Huttenhower C (2013) Computational meta'omics for microbial community studies. Mol Syst Biol 9:666. https://doi.org/10.1038/msb.2013.22

Seli E, Botros L, Sakkas D, Burns DH (2008) Noninvasive metabolomic profiling of embryo culture media using proton nuclear magnetic resonance correlates with reproductive potential of embryos in women undergoing in vitro fertilization. Fertil Steril 90(6):2183–2189

Sharuddin SS, Ramli N, Yusoff MZM, Muhammad NAN, Ho LS, Maeda T (2022) Advancement of metatranscriptomics towards productive agriculture and sustainable environment: a review. Int J Mol Sci 23(7):3737. https://doi.org/10.3390/ijms23073737

Singh J, Behal A, Singla N, Joshi A, Birbian N, Singh S, Bali V, Batra N (2009) Metagenomics: concept, methodology, ecological inference and recent advances. Biotechnol J 4:480–494

Spratlin JL, Serkova NJ, Eckhardt SG (2009) Clinical applications of metabolomics in oncology: a review. Clin Cancer Res 15(2):431–440

Starr AE, Deeke SA, Li L, Zhang X, Daoud R, Ryan J, Ning Z, Cheng K, Nguyen LVH, Abou-Samra E, Lavallée-Adam M, Figeys D (2018) Proteomic and metaproteomic approaches to understand host-microbe interactions. Anal Chem 90(1):86–109. https://doi.org/10.1021/acs.analchem.7b04340

Stepanauskas R, Sieracki ME (2007) Matching phylogeny and metabolism in the uncultured marine bacteria, one cell at a time. Proc Natl Acad Sci U S A 104(21):9052–9057

Tanca A et al (2017) Potential and active functions in the gut microbiota of a healthy human cohort. Microbiome 5:79. https://doi.org/10.1186/s40168-017-0293-3

Tartaglia M, Ranauda MA, Falzarano A, Maisto M, Postiglione A, Prigioniero A, Scarano P, Zuzolo D, Sciarrillo R, Guarino C (2023) Metatranscriptomics of pastures under drought stress show a rhizospheric meta-organism reshape, 100687. Rhizosphere 26. https://doi.org/10.1016/j.rhisph.2023.100687

Tatusov RL, Galperin MY, Natale DA, Koonin EV (2000) The COG database: a tool for genome-scale analysis of protein functions and evolution. Nucleic Acids Res 28:33–36

Tedjo DI, Smolinska A, Savelkoul PH, Masclee AA, van Schooten FJ, Pierik MJ, Penders J, Jonkers DM (2016) The fecal microbiota as a biomarker for disease activity in Crohn's disease. Sci Rep 6:35216. https://doi.org/10.1038/srep35216

Thursby E, Juge N (2017) Introduction to the human gut microbiota. Biochem J 474:1823–1836. https://doi.org/10.1042/BCJ20160510

Torsvik V, Ovreas L (2002) Microbial diversity and function in soil: from genes to ecosystems. Curr Opin Microbiol 5(3):240–245

Tremaroli V, Backhed F (2012) Functional interactions between the gut microbiota and host metabolism. Nature 489:242–249

Tringe SG, Hugenholtz P (2008) A renaissance for the pioneering 16S rRNA gene. Curr Opin Microbiol 11:442–446

Tringe SG, Rubin EM (2005) Metagenomics: DNA sequencing of environmental samples. Nat Rev Genet 6(11):805–814

Tringe SG, von Mering C, Kobayashi A, Salamov AA, Chen K, Chang HW, Podar M, Short JM, Mathur EJ, Detter JC, Bork P, Hugenholtz P, Rubin EM (2005) Comparative metagenomics of microbial communities. Science 308(5721):554–557

Turnbaugh PJ, Gordon JI (2008) An invitation to the marriage of metagenomics and metabolomics. Cell 134:708–713

Turnbaugh PJ, Ley RE, Hamady M, Fraser-Liggett CM, Knight R, Gordon JI (2007) The human microbiome project. Nature 449(7164):804–810

Tyson GW, Chapman J, Hugenholtz P, Allen EE, Ram RJ, Richardson PM, Solovyev VV, Rubin EM, Rokhsar DS, Banfield JF (2004) Community structure and metabolism through reconstruction of microbial genomes from the environment. Nature 428:37–43

van Leeuwenhoek A (1684) An abstract of a Letter from Antonie van Leeuwenhoek, Sep. 12, 1683. About animals in the scruff of the teeth. Philos Trans R Soc Lond 14:568–574

Vemuri GN, Aristidou AA (2005) Metabolic engineering in the -omics era: elucidating and modulating regulatory networks. Microbiol Mol Biol Rev 69(2):197–216

Venter JC, Remington K, Heidelberg JF, Halpern AL, Rusch D, Eisen JA et al (2004) Environmental genome shotgun sequencing of the Sargasso Sea. Science 304:66–74. https://doi.org/10.1126/science.1093857

Ventura M, O'Flaherty S, Claesson MJ, Turroni F, Klaenhammer TR, van Sinderen D, O'Toole PW (2009) Genome-scale analyses of health-promoting bacteria: probiogenomics. Nat Rev Microbiol 7(1):61–71

Verberkmoes NC, Russell AL, Shah M, Godzik A et al (2009) Shotgun metaproteomics of the human distal gut microbiota. ISME J 3:179–189
Villas-Boas SG, Mas S, Akesson M, Smedsgaard J, Nielsen J (2005) Mass spectrometry in metabolome analysis. Mass Spectrom Rev 24(5):613–646
Vogel C, Marcotte EM (2012) Insights into the regulation of protein abundance from proteomic and transcriptomic analyses. Nat Rev Genet 13:227–232
Wang X, Buhimschi CS, Temoin S, Bhandari V et al (2013) Comparative microbial analysis of paired amniotic fluid and cord blood from pregnancies complicated by preterm birth and early-onset neonatal sepsis. PLoS One 8:e56131
Wang Y, Zhou Y, Xiao X, Zheng J, Zhou H (2020) Metaproteomics: a strategy to study the taxonomy and functionality of the gut microbiota. J Proteome 219:103737. https://doi.org/10.1016/j.jprot.2020.103737
Wei X, Jiang S, Chen Y, Zhao X, Li H, Lin W, Li B, Wang X, Yuan J, Sun Y (2016) Cirrhosis related functionality characteristic of the fecal microbiota as revealed by a metaproteomic approach. BMC Gastroenterol 16(1):121. https://doi.org/10.1186/s12876-016-0534-0
Whitman WB, Coleman DC, Wiebe WJ (1998) Prokaryotes: the unseen majority. Proc Natl Acad Sci U S A 95(12):6578–6583. https://doi.org/10.1073/pnas.95.12.6578
Wilmes P, Bond PL (2006) Metaproteomics: studying functional gene expression in microbial ecosystems. Trends Microbiol 14:92–97
Wilmes P, Bowen BP, Thomas BC, Mueller RS, Denef VJ, Verberkmoes NC, Hettich RL, Northen TR, Banfield JF (2010) Metabolome-proteome differentiation coupled to microbial divergence. MBio 1(5):e00246-10
Wilmes P, Heintz-Buschart A, Bond PL (2015) A decade of metaproteomics: where we stand and what the future holds. Proteomics 15:3409–3417
Wishart DS, Knox C, Guo AC, Eisner R, Young N, Gautam B, Hau DD, Psychogios N, Dong E, Bouatra S, Mandal R, Sinelnikov I, Xia J, Jia L, Cruz JA, Lim E, Sobsey CA, Shrivastava S, Huang P, Liu P, Fang L, Peng J, Fradette R, Cheng D, Tzur D, Clements M, Lewis A, De Souza A, Zuniga A, Dawe M, Xiong Y, Clive D, Greiner R, Nazyrova A, Shaykhutdinov R, Li L, Vogel HJ, Forsythe I (2009) HMDB: a knowledgebase for the human metabolome. Nucleic Acids Res 37(Database issue):D603–D610
Wiwanitkit V (2013) Utilization of multiple "omics" studies in microbial pathogeny for microbiology insights. Asian Pac J Trop Biomed 3(4):330–333. https://doi.org/10.1016/S2221-1691(13)60073-8
Xiao M, Yang J, Feng Y et al (2017) Metaproteomic strategies and applications for gut microbial research. Appl Microbiol Biotechnol 101:3077–3088
Xiong W, Abraham PE, Li Z, Pan C, Hettich RL (2015) Microbial metaproteomics for characterizing the range of metabolic functions and activities of human gut microbiota. Proteomics 15(20):3424–3438. https://doi.org/10.1002/pmic.201400571. Epub 2015 May 28
Xiong W, Brown CT, Morowitz MJ, Banfield JF, Hettich RL (2017) Genome-resolved metaproteomic characterization of preterm infant gut microbiota development reveals species-specific metabolic shifts and variabilities during early life. Microbiome 5(1):72. https://doi.org/10.1186/s40168-017-0290-6
Xue MY, Xie YY, Zhong Y et al (2022) Integrated meta-omics reveals new ruminal microbial features associated with feed efficiency in dairy cattle. Microbiome 10:32. https://doi.org/10.1186/s40168-022-01228-9
Young JC, Pan C, Adams RM, Brooks B, Banfield JF, Morowitz MJ, Hettich RL (2015) Metaproteomics reveals functional shifts in microbial and human proteins during a preterm infant gut colonization case. Proteomics 15(20):3463–3473. https://doi.org/10.1002/pmic.201400563
Zhang X et al (2017) Deep metaproteomics approach for the study of human microbiomes. Anal Chem 89:9407–9415. https://doi.org/10.1021/acs.analchem.7b02224
Zhang X et al (2018) Metaproteomics reveals associations between microbiome and intestinal extracellular vesicle proteins in pediatric inflammatory bowel disease. Nat Commun 9:2873. https://doi.org/10.1038/s41467-018-05357-4

Zhang X et al (2020) Widespread protein lysine acetylation in gut microbiome and its alterations in patients with Crohn's disease. Nat Commun 11:4120. https://doi.org/10.1038/s41467-020-17916-9

Zhang M, Wang X, Wang Z, Mao S, Zhang J, Li M, Pan H (2023) Metatranscriptomic analyses reveal important roles of the gut microbiome in primate dietary adaptation. Genes 14(1):228. https://doi.org/10.3390/genes14010228

Zhao Y et al (2018) Insights into the gut microbiota of freshwater shrimp and its associations with the surrounding microbiota and environmental factors. J Microbiol Biotechnol 28:946–956

Metagenomics Insight into the Mycobiome Analysis

7

Anshu Mathuria, Anushika Chaudhary, Harshita Sharma, and Indra Mani

7.1 Introduction

The mycobiome refers to the fungal segment of the microbiome. It was initially coined in 2010, specifically concerning the human oral mycobiome (Ghannoum et al. 2010). Fungi, constituting a significant kingdom within the Eukaryotic domain, hold economic significance by producing antibiotics, fermenting foods like bread and beer, and aiding in cellulose degradation. Despite an estimated multitude of fungal species, only a fraction has been identified due to factors like vast phenotypic diversity, genome adaptability, and challenges in culturing most of these species (Blackwell 2011; Roper et al. 2011; O'Brien et al. 2005).

The human gut microbiome, vital for health, mainly focuses research on prokaryotes and viruses, neglecting the complex relationships of fungi (Barrera-Vazquez and Gomez-Verjan 2020; Gregory et al. 2020; Camarillo-Guerrero et al. 2021; Pasolli et al. 2019). Although composing less than 1% of the microbiome of gut (Huffnagle and Noverr 2013), fungi profoundly impact health, influencing disease onset and the immune system. *Candida albicans*, for instance, affects immunocompromised individuals (Dai et al. 2021) and changes in the gut mycobiome are linked to various human diseases (Das et al. 2021; Frau et al. 2021). While specific fungal markers correlate with certain traits, the general composition of the mycobiome in the gut and its variability among individuals lack comprehensive understanding. In the microbiome of human gut, fungi coexist and interact with the prokaryotic microbiome, engaging in mutualistic, commensal, or competitive relationships (Richard and Sokol 2019; Gupta et al. 2022). Certain fungi–bacteria interactions are associated with human diseases, such as the favorable association between the bacterial species *Escherichia coli* and *Serratia marcescens* and *Candida*

A. Mathuria · A. Chaudhary · H. Sharma · I. Mani (✉)
Department of Microbiology, Gargi College, University of Delhi, New Delhi, India
e-mail: indra.mani@gargi.du.ac.in

I. Mani, V. Singh (eds.), *Multi-Omics Analysis of the Human Microbiome*,
https://doi.org/10.1007/978-981-97-1844-3_7

tropicalis in Crohn's disease (CD) (Hoarau et al. 2016). The availability and impact of fungi in the human lung, called the lung mycobiome, is increasingly recognized as important for respiratory health. Limited research suggests fungi exist in both healthy and diseased lungs, influencing chronic respiratory conditions (Hawksworth 1991).

Numerous human illnesses, from minor infections to potentially fatal invasive ones, can be brought on by fungi, especially in immunocompromised individuals. Despite the significant impact of invasive fungal infections, research in this field lags other microbial pathogens, hampering the development of new treatments and diagnostic tools. A fungus, *C. albicans*, that typically resides as a commensal in humans' normal microflora. However, it can turn pathogenic if the immune system of the host is compromised. *Cryptococcus neoformans, Histoplasma capsulatum*, and *Aspergillus fumigatus* are environmental pathogens. Inhaled spores or yeast form cells of these fungi can result in invasive infections, particularly affecting immunocompromised individuals (Reedy et al. 2007). *C. neoformans* is a dangerous fungal pathogen causing severe pneumonia and meningitis in immunocompromised individuals. In healthy individuals, *A. fumigatus* conidia are effectively eliminated by the immune system. However, these spores can evade the immune response in immunocompromised individuals, causing life-threatening invasive aspergillosis. *H. capsulatum* is the fungal pathogen responsible for causing histoplasmosis, a disease that shares some clinical and pathogenic characteristics with tuberculosis (Kim 2016). Recent studies have demonstrated the important role of microbiome in cancer, with emerging interest in the role of fungi like *Candida* and *Malassezia* in gut microbiota and their potential link to inflammation associated with certain cancers. Advanced metagenomics techniques offer avenues to study microbial diversity, aiding in identifying cancer-related microorganisms and potential targets for intervention across various cancer types.

Metagenomics evaluates diversity of species in contexts such as environmental ecosystems or the microbiome of humans using the technique of nucleic acid sequencing. Traditionally, 16S rRNA amplicon sequencing has been used for the identification of bacteria. Yet, the declining sequencing costs have spurred a shift from amplicon analyses to shotgun metagenomic sequencing. While shotgun metagenomics can detect a broad spectrum of species, it has been less commonly utilized for fungal identification (Donovan et al. 2018). In a mycobiome, two basic sequence-based approaches are used to identify species of fungi. The prevalent approach involves amplification by PCR of the internal transcribed spacer (ITS) regions, especially ITS2 found between the 5.8S and 28S rRNA genes, followed by sequencing. ITS2 sequences, which are very varied and are regarded as the standard fungal barcode sequence, provide a foundation for species identification (Schoch et al. 2012). Schematic representation of factors influencing human mycobiome and analysis and bias sources are shown in Fig. 7.1 (Thielemann et al. 2022). Numerous pipelines such as Plutof, PIPITS, Clotu, and CloVR-ITS have emerged for specific fungal species identification and frequency calculation using ITS data (Abarenkov et al. 2010; Kumar et al. 2011; Gweon et al. 2015; White et al. 2013). BioMaS,

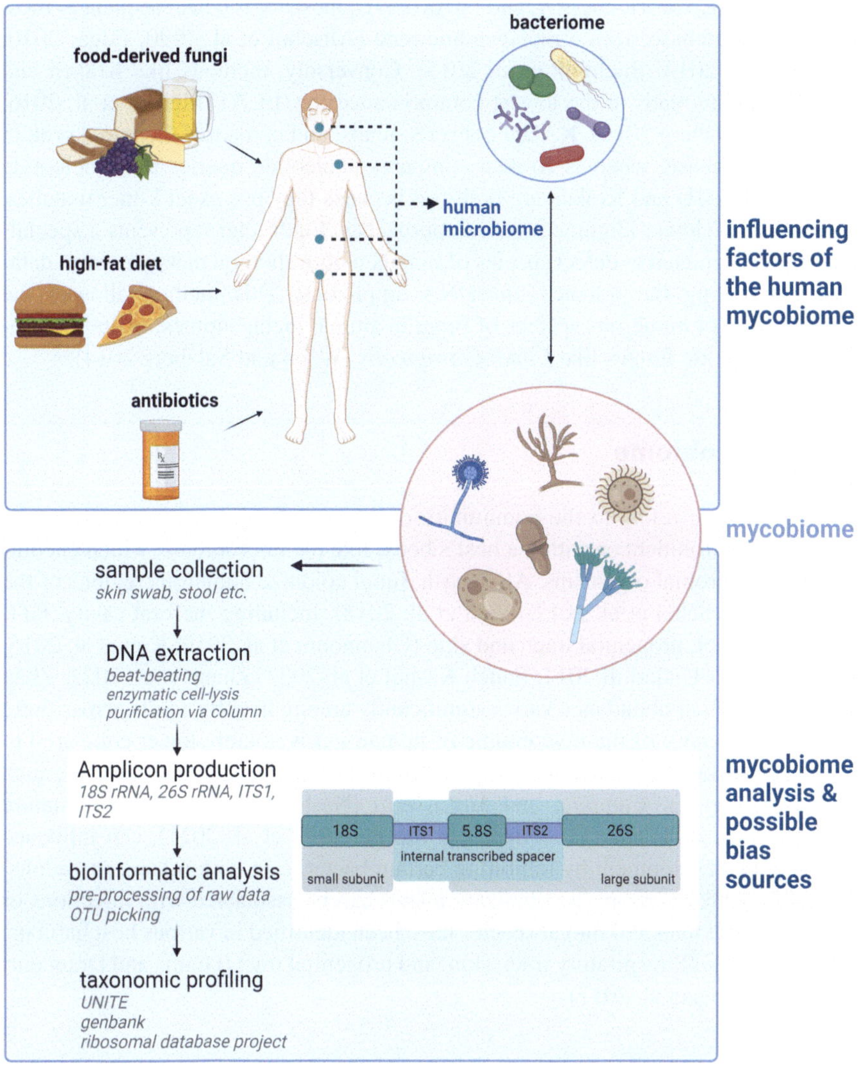

Fig. 7.1 Factors shaping mycobiota analysis and potential sources of bias. Various factors, including diet, antibiotics, environmental exposure, and interactions with the human bacteriome, shape the human mycobiome. Successful mycobiome analysis involves extracting fungal DNA from samples, using amplicon production strategies targeting the rRNA gene locus (e.g., 18S, ITS1, and ITS2). Bioinformatic analysis assigns taxonomy by comparing identified OTUs with reference sequence databases. (Adapted from Thielemann et al. 2022)

Mothur, and Qiime are versatile tools applicable to both fungal–bacterial and amplicon reads (Fosso et al. 2015; Schloss et al. 2009; Caporaso et al. 2010).

The alternative approach involves identifying species through shotgun metagenomes. Utilizing custom-built databases and search algorithms such as BLAST,

UBLAST, USEARCH, GhostX, and DIAMOND, these tools detect sequences most akin to the database from a metagenome read (Altschul et al. 1990; Edgar 2010; Suzuki et al. 2014; Buchfink et al. 2015). Conversely, methods like Kraken and KAIJU allocate reads to the lowest common ancestor (LCA) (Menzel et al. 2016; Wood and Salzberg 2014). KAIJU converts, reads, and compares them to a protein reference database, whereas Kraken compares nucleotide queries to a nucleotide reference. KAIJU and Kraken are both fast because they use exact k-mer matches, as opposed to slower alignment-based approaches. FindFungi represents a specialized pipeline created to detect species of fungus within shotgun metagenomics datasets, eliminating the reliance on rDNA amplicons. This method allowed the identification of numerous species of fungi in animal metagenomes, including prospective zoonotic fungus like *Candida tropicalis* (Wood and Salzberg 2014).

7.2 Mycobiome

The mycobiome refers to the community of fungi or fungal species present in a particular environment or within a host's body, like the microbiome, which encompasses all microbial organisms. After birth, fungi colonize numerous regions of the human body (Schei et al. 2017; Ward et al. 2018), including the oral cavity, GIT, respiratory tract, urogenital tract, and skin (Ghannoum et al. 2010; Kim et al. 2015; Ackerman and Underhill 2017; Saheb Kashaf et al. 2022; Zhang et al. 2022; Zhao et al. 2023). Their abundance varies significantly among healthy adults across these sites. The diversity of the mycobiome of human gut is notably lower compared to the bacterial communities in the gut. It is primarily yeast genera that dominate such as *Saccharomyces, Candida,* and *Malassezia* (Nash et al. 2017). In premature infants, fungi, especially *C. albicans* in the gut (Rao et al. 2021), can influence microbiome development by inhibiting certain bacteria. In both infants and adults, certain fungal expansions at colonization sites can be associated with infections or diseases. More than 390 fungal species have been identified in various host habitats, including the GIT, respiratory tract, skin, and urogenital tract (Gouba and Drancourt 2015; Halwachs et al. 2017).

7.2.1 Mycobiome in Gut

Fungi like *Candida, Saccharomyces,* and *Cladosporium* are prevalent in healthy adult fecal samples and play crucial roles in immune balance (Nash et al. 2017). These fungi interact intensively with the immune system, with antibodies like IgA and IgG targeting and providing protection against *Candida* (Doron et al. 2021; Ost et al. 2021). *Saccharomyces boulardii* is a common probiotic yeast, primarily employed in preventing and treating infectious antibiotic-associated diarrhea and managing preterm infants (Sivananthan and Petersen 2018; Guo et al. 2019). Studies suggest that a balanced gut fungal community is vital for immune health, and imbalances (dysbiosis) in this community are associated with diseases like inflammatory bowel disease (IBD), alcohol-associated liver disease, colorectal cancer (CRC), and allogeneic hematopoietic cell transplantation (allo-HCT).

Recent developments in research technologies have provided insight into the gut microbiome's significance, specifically focusing on host-fungal interactions in IBD. *Candida* and *Malassezia* fungal taxa have been linked to IBD's cellular and molecular pathways. Additionally, studies indicate that certain disease-associated fungi and bacteria may influence IBD not directly on the host but through their interactions within the fungal–bacterial ecosystem (Underhill and Braun 2022).

Despite this variability, certain fungi have consistently shown alterations in individuals with IBD across various studies. However, the exact mechanisms through which these commensal organisms might impact an individual's vulnerability to developing IBD or the disease's severity remain incompletely understood. In CD, there's often a rise in *Candida*, primarily *C. albicans*, in the fecal mycobiome across diverse regions (Richard and Sokol 2019; Hoarau et al. 2016; Nelson et al. 2021; Chehoud et al. 2015). Factors like antibiotic use influence *Candida* levels. *Candida*, especially *C. albicans*, can cause nosocomial infections in immunocompromised patients. Studies associate higher *Candida* levels in CD patients undergoing surgery and link reduced *Candida* after fecal transplants to favorable outcomes in ulcerative colitis. However, gaps exist in understanding causation and Candida's specific role in these contexts. Recent studies using ITS sequencing found strong links between *Malassezia* species (*M. globosa* and *M. restricta*) and CD (Limon et al. 2019). These fungi are primarily located in the intestinal mucosa rather than the gut lumen. *Malassezia*, usually a skin microbe, lacks essential enzymes for lipid metabolism and relies on environmental lipids for growth. The prevalence of *Malassezia* in CD patients is linked with the CARD9 IBD risk allele, impacting immune responses. Research in mouse models suggests *Malassezia* exacerbates colitis, requiring functional CARD9. More research is required to fully comprehend the potential role of *Malassezia* in immune responses and inflammatory pathways in CD.

The gut mycobiome's impact on metabolic diseases suggests new treatment strategies, including fungal probiotics, antifungal drugs, dietary interventions, and fecal microbiota transplantation (FMT) (Fig. 7.2) (Wang et al. 2023). *Saccharomyces boulardii* and *S. cerevisiae* are potential probiotics for metabolic diseases. Kefir, with over 50 yeast and lactic acid bacteria (LAB) species, alleviates obesity and hepatic steatosis in high-fat diet-fed mice (Kim et al. 2017). Antifungal drugs like amphotericin B, caspofungin, and fluconazole exhibit potential in mitigating hepatic steatosis and obesity (Demir et al. 2022). Studies in mice reveal reductions in specific fungi, leading to improved liver health. These findings suggest a therapeutic role for antifungal drugs in addressing metabolic disorders by modulating the gut mycobiome. FMT transfers donor fungi to recipients, affecting efficacy against *Clostridium difficile* (Zuo et al. 2018). Individuals who recovered after FMT treatment had a high abundance of *Saccharomyces* and *Aspergillus*, whereas those who did not recovered had a high abundance of *Candida* (Zuo et al. 2018). Furthermore, FMT therapy lowered *C. albicans* prevalence and severity of illness in ulcerative colitis patients (Leonardi et al. 2020).

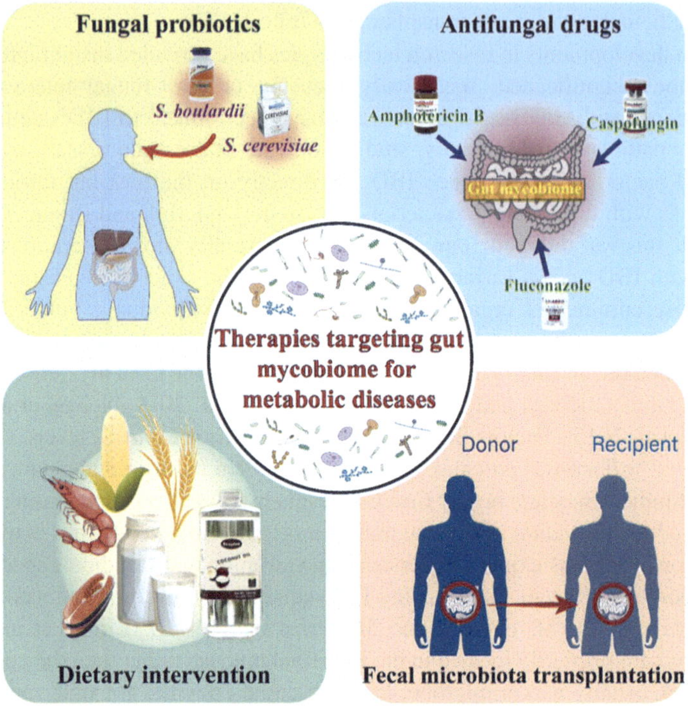

Fig. 7.2 Optimizing gut mycobiome for metabolic health. Strategies addressing the gut mycobiome for metabolic diseases encompass fungal probiotics, antifungal drugs, dietary interventions, and fecal microbiota transplantation. (Adapted with permission from Wang et al. 2023)

7.2.2 Mycobiome of Skin

Malassezia species dominate the mycobiome of skin, with *M. globosa, M. auris,* and *M. restricta,* commonly observed in healthy individuals (Saheb Kashaf et al. 2022). Atopic dermatitis patients show increased CCR6+ memory T cell frequency due to *Malassezia* (Sparber et al. 2019). Alongside *Malassezia, Candida* species like *C. albicans* and multidrug-resistant (MDR) *Candida auris* can asymptomatically colonize the skin. *C. auris* poses a healthcare-associated infection risk, colonizing multiple body sites, particularly the nares (Huang et al. 2021). Monitoring the dynamics of the mycobiome of skin might prove critical for quick action in *C. auris* infections in high-risk people (Proctor et al. 2021).

7.2.3 Tumor Mycobiome

Studies on the intratumor microbiome reveal diverse fungal communities within tumors across various cancers, including gastrointestinal and non-GI tumors.

Dysbiosis in the intratumor mycobiota is detected in cancers such as pancreatic, gastric, oral, and breast cancers (Alam et al. 2022; Banerjee et al. 2021; Perera et al. 2017; Zhong et al. 2021). In pancreatic ductal adenocarcinoma tumors, *Alternaria* and *Malassezia* species promote a type-2 immune response, which speeds up the growth of the malignancy (Alam et al. 2022). Species like *S. cerevisiae, C. albicans, M. restricta, Blastomyces gilchristii*, and *M. globosa* are implicated in contributing to carcinogenesis (Narunsky-Haziza et al. 2022). GI tumors are significantly enriched in *Candida* and *Saccharomyces* species, whereas lung tumors are more common in *Blastomyces*, and breast tumors are more abundant in *Malassezia*. These findings imply that the mycobiome associated with tumors may function as prognostic indicators and new targets for anticancer treatments.

7.2.4 Mycobiomes in Various Body Sites

Within a sound oral cavity, common fungal taxa include *Candida, Saccharomyces, Aspergillus, Cladosporium,* and *Aureobasidium* species (Ghannoum et al. 2010). Studies comparing the mycobiome of those who have dental cavities and those who do not, found enrichment of *C. albicans* in advanced caries, correlating with the presence of pathogens like *Streptococcus mutans* and *Scardovia wiggsiae* (Baraniya et al. 2020). Transmission of *Candida* species from the oral cavity in the case to the gut has been proposed as a key source of *Candida* detected in healthy individuals' feces, while conclusive understanding of this transmission in healthy or diseased settings is lacking.

In lung mycobiome research, sampling methods like bronchoalveolar lavage fluid or sputum can affect outcomes of sequencing due to potential oral cavity fungal contamination. Despite this, in conditions such as cystic fibrosis (CF), *Candida* species stable colonization in the lungs has been confirmed through sequencing and culture-based techniques (Kim et al. 2015; Demers et al. 2018), in chronic obstructive pulmonary disease, *Candida* species dominate bronchoalveolar lavage fluid, warranting further investigation into underlying mechanisms.

Research on the vaginal mycobiome is evolving. Studies in vulvovaginal candidiasis patients identified two dominant fungal community types: one led by *Candida glabrata* (CST I) and the other by *Candida albicans* (CST II) (Zhao et al. 2023). These types were linked, respectively, to *Prevotella* and *Ureaplasma urealyticum* infections. Because of pH variations among human (near-neutral) and mouse (pH 4.5) vaginal environments, using murine models to investigate vaginal mycobiome-related disorders is difficult (Fidel 2021).

7.3 Mycobiome in Health and Disease

One crucial element of the microbiome of human is the mycobiome. The primary objective of research on the mycobiome of humans is to determine how fungal communities impact health. A comprehensive review of recent studies reveals that distinct fungal taxa are found at diverse body parts of human and such diversified mycobiome profiles are linked to different diseases (Cui et al. 2013). As a significant component of the biomass in the microbiome of an individual, fungus represents a tiny percentage of all species but could be an area of interest for metabolic modification during disease progression and mechanisms of disease. Fungal-produced metabolic by-products influence niches in hosts, boost immunity, and alter their numbers ahead of alterations linked to metabolic disorders (Begum et al. 2022). So mycobiome performs an essential role in both health and diseases by interacting with the host and other biological communities present in microbiome. This has helped to accelerate the development of fungus-associated illnesses (Cui et al. 2013). Because of the higher biomass of fungi in human microbiome and corresponding high occupancy of surface area, fungi produce more metabolic products than bacteria do, and their production may have a stronger effect on the human microbiota and human health (Qin et al. 2010; Baker et al. 2017; Huffnagle and Noverr 2013).

There are several applications of metabolic products of fungi, such as in preserving food, biotechnology, sustainable farming, pharmaceuticals, and agricultural additives (Huffnagle and Noverr 2013; Frey-Klett et al. 2011). Regular ingestion of these products has unavoidably altered and has led the human mycobiome to adapt. Several diseases have been linked to mycobiome dysbiosis (Hoarau et al. 2016; Enaud et al. 2018), which is a discrepancy in the microbial community's makeup resulting from pathogenic or opportunistic microbial overgrowth, loss of symbiotic relationship, disrupted competition between microbes, and diversity in microbial life. A considerable percentage of people experience the effects of fungal dysbiosis, which has been traced to CD, IBD, and IBS (irritable bowel syndrome) (Gu et al. 2019). For instance, compared to healthy individuals, those suffering from CD possess a higher diversity of fungus species, and the development of IBD is associated with *C. albicans* excessive growth in the intestinal tract (Sokol et al. 2017; Ott et al. 2008; Witherden et al. 2017; Kumamoto 2011; Gerard et al. 2015; Trojanowska et al. 2010). Moreover, alterations in the ratios of *Ascomycota* and *Basidiomycota* may be the cause of inflammation-related events associated with IBD (Sokol et al. 2017; Ott et al. 2008; Hoarau et al. 2016). Chronic immune-mediated inflammation is the distant result of mycobiome dysbiosis, and it seems to be connected to a CNS infection linked to neurological disorders (Severance et al. 2012, 2016; Enaud et al. 2018). There may be interactions between the mycobiomes found in various locations within the body. A thorough knowledge of this is gained by understanding the gastrointestinal-respiratory relationship that exists between *C. albicans* and *A. fumigatus* where *C. albicans*-induced disruption of the gut mycobiome affects the lung mycobiome in allergic pulmonary illness caused by *A. fumigatus* (Noverr et al. 2005).

Additionally, the relationship between the bacteriome and the mycobiome may also influence health and illness. There are instances where the prevalence of fungi and bacteria has a positive association; aspergillosis and *Mycobacterium* superinfection are two examples of such situations (Darling 1976). But in certain scenarios, bacteria and fungus engage in competition; in CF, *Pseudomonas aeruginosa's* dominance inhibits the proliferation of *Candida* and maybe of others as well (Kerr 1994). The contradictory character of this relationship has been explained by several theories. The commensalism of both bacteria and fungus in the context of *Mycobacterium* along with aspergillosis may work in unison to increase their resistance to external stresses like antimicrobial drugs (Adam et al. 2002). When it comes to *Candida* and *Pseudomonas*, the bacteria's ability to limit fungal growth could assist them in fulfilling their nutritional needs and acquire additional areas of attachment on the host cell (Harrison et al. 2012). Other disorders are also influenced by mycobiome dysbiosis (Hoarau et al. 2016; Enaud et al. 2018). From an *in vivo* study, *Malassezia* spp. were found to be more prevalent in pancreatic ductal adenocarcinoma (PDA), demonstrating how the mycobiome is linked to the onset and development of cancer. Conversely, growth of tumors developing more slowly or not at all is linked to *Candida tropicalis* (Aykut et al. 2019).

Mycobiome analysis of immune-suppressed persons are also necessary to potentially determine the mechanistic relationship between immune suppression and fungal pathogenesis. Immunosuppressed individuals, such as those who are HIV patients, typically suffer from opportunistic infections caused by protozoa and fungi than individuals who are healthy (Cui et al. 2013). Numerous studies have made attempts to describe the mycobiome role in the CF, a lung illness. According to these research studies, CF patients have less diversity in fungi than people in general (Harrison et al. 2012; Nelson et al. 2010; Delhaes et al. 2012). It is a common misconception that more serious illness cases should have a higher fungal variety. The course of certain illnesses relates to the multitude of fungus present (Chen et al. 2011), such as in atopic dermatitis, Hepatitis B cirrhosis (Zhang et al. 2011), chronic hepatitis B, and IBD (Ott et al. 2008). But in some the fungal diversity is inversely correlated with the course of various disease settings, such as the nasal mycobiome in allergy-related patients (Sellart-Altisent et al. 2007) and the pulmonary mycobiome in CF (Harrison et al. 2012; Delhaes et al. 2012).

Mycobiome-related infections require early identification and treatment to improve prognosis and reduce mortality, especially in patients with weakened immune systems (Perlin et al. 2017; Barnes 2008; Limper et al. 2017). Traditional culture techniques yield antifungal sensitivity data and species-level determination. But culturally oriented approaches lack sensitivity and slow down the diagnosing process (Clancy and Nguyen 2013; Simoneau et al. 2005; Tarrand et al. 2003). For the purpose of infection control and management, next-generation sequencing (NGS) is being utilized more and more in epidemiological research and during epidemics of infectious disease. NGS is used to distinguish between species, determine pathogenicity, and detect antibiotic resistance (Snitkin et al. 2012; Lewis et al. 2010; Weterings et al. 2015; Mani 2020a, b; Mani et al. 2023). Furthermore, NGS can be utilized to solve enigmatic epidemiological instances in which it has proven to be more effective than conventional diagnostic techniques in identifying uncommon infections (Wilson et al. 2014, 2018; Thoendel et al. 2018).

7.4 Metagenomics in Gut Mycobiome

The human gut microbiome comprises a minor but essential component called the mycobiome of the gut. Intestinal fungus controls the proliferation of the co-residing gut bacterial microbiome, pathological and physiological mechanisms, and host homoeostasis (Zhang et al. 2022). In order to preserve intestinal homeostasis and maintain immunity, fungi reside on the mucosal surface (Iliev and Leonardi 2017). The majority of studies documenting the gut microbiome of the human in both healthy as well pathological states have focused on the bacterial population found in gut. However, the mycobiome, or microbiome related to fungi, is starting to be acknowledged as an essential component of our microbiome (Nash et al. 2017). The existence of fungi in microbial ecosystems is currently explained in a mere 3% of the microbiome literature. Recent developments in technologies for sequencing have made it possible to describe the fungal aspect of the gut microbiota and helped in identifying pathogenic fungi (Soverini et al. 2019). The lack of thorough, meticulously curated, precise taxonomic annotation in fungal databases has made it more difficult to characterize the gut mycobiome. To reduce differences between studies, standardizing mycobiome analysis techniques is essential (Zhang et al. 2021).

Direct sequencing and study of DNA of microbial ecosystems, also termed as metagenomics, has quickly gained acceptance as a commonly used technique to assess the functional capacity of microbial populations (Taş et al. 2021). To identify the fungal species, present in a mycobiome, two techniques based on sequences are typically employed. The most popular method involves amplifying rRNA operons' ITS region using PCR (Schoch et al. 2012). The second method uses shotgun metagenomics to identify species. The workflow for studying fungi includes extraction, library building, sequencing, and bioinformatic analysis.

To achieve greater throughput and quicker turnaround times, NGS platforms are heavily utilized to form metagenomic DNA libraries. The Illumina MiSeq platform is by far the most often used 16S amplicon sequencer out of all the platforms. Illumina is a popular NGS platform that offers excellent quality, high throughput, and rapid speed for an extensive number of samples (Zhang et al. 2021). Illumina sequencers have the following characteristics in comparison to Nanopore, which is part of the method known as third-generation sequencing (TGS) technique: (1) Using the dual-index approach, it has successfully performed hundreds of metabarcodings of complicated samples each run, as opposed to fewer than 24 sequences in a single Nanopore run (Morrison et al. 2020). (2) At 99.99% precision, it has a superior accuracy rate than roughly 90% of Nanopore. However, long segments can be precisely reassembled and classified as a species in nanospore, reducing the amount of incorrectly identified species found in illumina. Additionally, because nanopore can offer PCR-free metagenomics techniques, researchers are employing it to identify fungi (Loit et al. 2019).

To examine the mycobiome of the gut or in general gut microbiome of humans, frequently and widely utilized samples are stool samples. In order to study the metabolic pathways that fecal bacteria encode, the human genome project (HGP) incorporated relatively deep whole genome shotgun sequencing (WGS) on stool samples, with a target of 10 gigabases (Gb) per sample (The Human Microbiome Project Consortium 2012). Roughly only 0.01% of the over 27 billion metagenomic sequences produced from this were aligned to fungal genomes (Nash et al. 2017).

According to the fungi-based studies on stool samples from Human Microbiome Project (HMP), Ascomycota was the phylum with the highest prevalence, followed by Basidomycota which is also found in large numbers. There have been reports of this phyla's supremacy in various other body parts of human, like the skin, vagina, and oral cavity. The most common genera of gut mycobiome put together on the basis of sequencing of ITS2 and 18S rRNA., include *Candida*, *Malassezia*, *Saccharomyces*, *Aspergillus*, *Pichia*, *Cyberlindnera*, *Debaryomyces*, *Penicillium*, *Clavispora*, *Cladosporium*, and *Galactomyces*. Out of these, *Saccharomyces* emerged as one of the most prevalent fungus found in fecal samples of healthy humans, with *Candida* and *Malassezia* following closely behind (Nash et al. 2017). Fungi also invade the mucosa. The mucous membrane of the intestinal tract had a more stable fungal population than the luminal-associated gut mycobiota (Sonnenburg et al. 2004; Luan et al. 2015). But samples are harder to get than stool samples, so there is a dearth of research detailing the fungal makeup of mucosa layers (Sam et al. 2017). Even though the fungus populations in healthy human feces vary greatly, a number of fungal species were found in a significant number in HMP samples. For example, *S. cerevisiae* and *M. restricta* are found in 96.8% and 88.3% of the test samples, respectively (Nash et al. 2017). The most common species of fungi found in mucosal samples were *Septoria epambrosiae*, *Rhodotorula mucilaginosa*, *C. albicans*, *Galactomyces geotrichum*, *C. dubliensi*, *Bullera crocea*, *Raciborskiomyces longisetosum*, *Cladosporium cladospoirioides*, and *Penicillium ialicum* (Ott et al. 2008).

7.5 Metagenomics in Lung Mycobiome

The colony of fungi that naturally inhabit the lungs is referred to as the lung mycobiome. The mycobiome contributes to immune system and respiratory health, just like the gut microbiome does. The microbiota of the lungs, that was formerly believed to be constant or ideally ephemeral in those in good health, has since been determined to be a polymicrobial factor that contributes to the pathogenesis of several illnesses. The fungi that are present in the respiratory tract of humans are primarily from the phyla *Ascomycota* and *Basidiomycota* under the Dikarya sub-kingdom. In actuality, the family *Davidiellaceae* and the genera *Cladosporium*, *Eurotium*, *Penicillium*, and *Aspergillus* are among the most frequently discovered species in specimens of typical lungs (Charlson et al. 2012). Other species observed in individuals in good health include *Pneumocystis*, *Candida*, *Neosartorya*, *Hyphodontia*, *Malassezia*, and *Kluyveromyces* (Delhaes et al. 2012). The variety of fungi developmental patterns, which include yeast-like and filamentous development, is represented by the 11 taxa.

More focus has been placed on the lung mycobiome in recent years. A growing pool of research reveals that the mycobiome of the lung encounters a significant part in the clinical manifestations of chronic respiratory diseases (CRD), including CF, asthma, bronchiectasis, and chronic obstructive pulmonary disease (COPD). With a projected one billion cases globally, 7.5 million fatalities annually, and a significant financial, medical, and social impact on society, CRDs are ranked as the fourth most frequent cause of sickness and fatalities globally (Paudel et al. 2021). Although *Pseudomonas aeruginosa* is known to have the biggest effect on lung function, there has long been debate regarding the involvement of fungi in CF lung illness (Stoltz et al. 2015; LiPuma 2010).

Research indicates that the populations of microbes found in the lungs of people having (CRD) distinct from those observed among people in excellent health. This difference can be attributed to dysbiosis or the disturbance of microbial homeostasis. These alterations include variations in the overall microbial load along with their distribution, in addition to those pertaining to the microbiome's constitution (Whiteside et al. 2021).

The most hopeful investigative approach in the context of CRD is to use NGS technologies, which enable multiple analyses of bacteria, viruses, and fungi collectively in a holistic study approach not needing any previous understanding of existing microorganisms. Cultivation of fungi from respiratory specimens is a traditional technique for segregation, detecting, and assessing the antifungal sensitivity of fungi linked to the pathophysiology of CRDs. Nevertheless, there are several drawbacks to this traditional method. First, there are no established procedures for fungal culture in some laboratories (Pashley 2014; Pashley et al. 2012). Additionally, the standard 5-day incubation periods and Sabouraud-chloramphenicol agar seeding methods employed in the majority of labs may fail to identify the possibility of fungi in the lower respiratory tract (Pashley 2014; Pashley et al. 2012; Kramer et al. 2015).

As of right now, it is known that NGS can identify a far greater variety of fungal species over traditional culture, including those that are challenging or hard to grow using traditional methods (e.g., *Malassezia* species) (Cuthbertson et al. 2021; Delhaes et al. 2012; Soret et al. 2020; Willger et al. 2014; van Woerden et al. 2013; Gaitanis et al. 2012). However, given the high sensitivity of these approaches, a large number of the fungus found may be only temporary colonists from continual intake of *conidia* from the surroundings (Kramer et al. 2015). Clarifying the involvement of fungi in the pathophysiology of CRDs is more challenging because of NGS's highly sensitive nature and the vast number of species it has discovered.

In a pair of studies published in 2012, deep-sequencing methods were used for the first time to investigate the lung mycobiota of sick participants (Charlson et al. 2012; Delhaes et al. 2012). While Charlson et al. (2012) examined the bronchoalveolar lavage (BAL) and oropharyngeal wash (OW) specimen of individuals with lung conditions with transplants who used to have CF, COPD, interstitial lung disease, and heart disease. Delhaes et al. (2012) examined CF patients' sputum samples. Despite minor variations in the DNA extraction and data analysis techniques employed in the two investigations, the predominant genera identified in the lung mycobiome were *Candida* species, *Aspergillus* spp., *Penicillium*, *Cryptococcus*, and *Eurotium*.

Research on the lung microbiome could be restricted due to the various obstacles which arise during each level, in addition to the field's relative youth. The difficulties start when the lung is sampled and extend throughout the sample processing. The difficulties include gathering and handling samples, processing and amplifying sequences, and the past of giving several names to the same species. While some problems might never be resolved, others might be with greater information and research on the lung mycobiome.

Metagenomic techniques as well as mathematical modeling may be able to forecast how antimicrobial therapies would affect the lung microbiome, which could result in the more efficient use of antibiotics and/or antifungals with various treatment plans. All things considered these results offer fresh perspectives on how to deal with the connection between microbial ecosystems and lung illnesses. Facilitating the creation of novel therapy techniques and models may aid in the evaluation of lung infections. We need to do more research in this area to gain more insight into the relationship that exists between the mycobiome, the lung microbiome, as well as chronic conditions like CRD.

7.6 Metagenomics in Cancer Mycobiome

A serious health issue in both industrialized and underdeveloped nations is cancer. The International Agency for Research on Cancer (IARC) published the international cancer study in 2012, and it approximated that there were 14.1 million fresh cases of cancer, 8.2 million fatalities due to cancer, and 32.6 million cancer survivors. Cancer can result from the interaction of multiple causes, including genetic, environmental, and lifestyle factors. One such factor that contributes to cancer is microbe-induced cancer, which accounts for 16.1% of all cancer cases worldwide (De Martel et al. 2012).

The collection of fungi that reside in and on our bodies is referred to as our "mycobiome." The purpose of the mycobiome in cancer is advancing area of research which is gaining attention. While a lot is known about the bacterial components of the microbiome and their potential links to cancer, the fungal aspect is still being explored. Higher cancer incidence has been linked to a number of chronic inflammatory disorders, including IBD, ulcerative colitis, pancreatitis, and chronic atrophic gastritis (Greten and Grivennikov 2019). These conditions all suggest that inflammation contributes in the initiation and proliferation of cancer. The more robust immunological reaction that follows fungus infection has been the subject of numerous investigations (Underhill and Iliev 2014).

Variations in the mycobiome composition among cancer patients and non-cancer participants have been linked to head and neck cancer, colorectal carcinoma, and pancreatic ductal adenocarcinoma. *Malassezia* in particular has been associated with the onset and spread of colorectal carcinoma and pancreatic adenocarcinoma, and *Schizophyllum,* a member of the oral mycobiome, is considered to have cancer preventing properties. Multi-omics technology will also help determine whether changes in the human mycobiome cause or are a result of particular cancer types.

When the gut is healthy or ill, the mycobiome is crucial in triggering the Type 17 innate immunity as well as B-cell mediated immunity. Across cohorts, trans-kingdom traits were lately associated with colorectal malignancies (Coker et al. 2018; Liu et al. 2022) and bioactivated amines and fungal toxins been associated with the development of cancer (Chang et al. 1992; Yang 1980).

The metagenomics technique has completely changed the way that tissue specimens from cancer patients are recognized, evaluated, and aimed in cancer research. To investigate tumor-related mycobiomes linked with different cancers, metagenomic analysis of WGS data from various tumor samples across different malignancies that are accessible in tumors from The Cancer Genome Atlas (TCGA), was employed. Tumor-related fungus and human cancers were found to be connected in both cancer-specific and pan-cancer contexts through investigation of tumor mycobiomes. Further investigation into the tumor microbiome revealed that *Saccharomyces* and *Candida* species may function as "keystone taxa," promoting biological interactions and general variety in the microbiological composition of different kingdoms.

Tumor-specific mycobiomes were identified by pan-cancer investigation of several body regions, with up to 1 fungus per 10^4 cells. *Blastomyces* was linked to tumor tissues in lung cancer. High levels of *Candida* were associated with the development of proinflammatory immunological pathways in stomach malignancies, while *Candida* decreased cellular adhesions and was predictive of metastatic illness in colon cancers. Both ITS sequencing of tumor samples and culturally specific analysis in a separate cohort verified the existence of *Candida* in human GI cancers. These findings imply that fungal DNA linked to tumors may function as diagnostic or prognostic biomarkers and link the mycobiota to the pathophysiology of GI malignancies.

Metagenomics has expanded the field of microorganism studies in their natural environments. This strategy has also been strengthened by the most recent developments regarding the domain of NGS methods. Since microorganisms cause a sizable portion of tumors, metagenomics research may aid in the fight against cancer by identifying the germs that cause the disease. Determining possible targets for cancer research will improve corrective actions and broaden the field's scope. Therefore, it is recommended that future research investigate the cancer microbiome for further cancer types.

7.7 Conclusion and Future Perspectives

The human mycobiome, comprising diverse fungal communities across different body sites, is increasingly recognized for its crucial role in health and disease. Fungi such as *Candida, Malassezia*, and others have been linked to conditions ranging from inflammatory bowel disease and dermatological issues to respiratory ailments and cancer. Research into the mycobiome's impact on immunity and its potential as a diagnostic and therapeutic target for various diseases is rapidly advancing. Understanding the intricate interplay between fungi and the human body, including their interactions with bacteria, has provided valuable insights into disease mechanisms. Advanced sequencing techniques, particularly metagenomic analysis, have unveiled associations between fungal dysbiosis and specific health conditions, paving the way for potential diagnostic biomarkers and treatment strategies. The human gut mycobiome, a vital part of the microbiome, influences health and immunity. Studies using Illumina sequencing in stool samples reveal prevalent fungi like *Candida* and *Malassezia*. Mucosal fungal research is limited but specific species like *S. cerevisiae* and *M. restricta* are consistently found in healthy human feces. Similarly, the lung mycobiome's role in respiratory conditions like CF and chronic obstructive pulmonary disease underscores the importance of understanding fungal communities in these contexts. Moreover, the link between mycobiome alterations and cancer, as observed in various cancers like head and neck, colorectal, and pancreatic cancers, presents promising avenues for cancer research. Analyzing fungal DNA in tumors through metagenomic approaches holds potential for cancer diagnosis and targeted therapies. In essence, the burgeoning field of mycobiome research offers significant promise for advancing our understanding of human health and disease. Further exploration and comprehensive studies into the mycobiome's impact across different body sites could yield critical insights into disease mechanisms and potential therapeutic interventions, ushering in a new era of personalized medicine and targeted treatments.

Acknowledgments The authors are thankful to the Principal, of Gargi College for providing the infrastructural support.

References

Abarenkov K, Tedersoo L, Nilsson RK, Vellak K, Saar I, Veldre V et al (2010) PlutoF—a web based workbench for ecological and taxonomic research, with an online implementation for fungal ITS sequences. Evol Bioinformatics Online 6:189–196

Ackerman AL, Underhill DM (2017) The mycobiome of the human urinary tract: potential roles for fungi in urology. Ann Transl Med 5:31

Adam B, Baillie GS, Douglas LJ (2002) Mixed species biofilms of *Candida albicans* and *Staphylococcus epidermidis*. J Med Microbiol 5:344–349

Alam A, Levanduski E, Denz P, Villavicencio HS, Bhatta M, Alhorebi L, Zhang Y, Gomez EC, Morreale B, Senchanthisai S et al (2022) Fungal mycobiome drives IL-33 secretion and type 2 immunity in pancreatic cancer. Cancer Cell 40:153–167.e111

Altschul SF, Gish W, Miller W, Myers EW, Lipman DJ (1990) Basic local alignment search tool. J Mol Biol 215(3):403–410. https://doi.org/10.1016/S0022-2836(05)80360-2

Aykut B, Pushalkar S, Chen R, Li Q, Abengozar R, Kim JI, Shadaloey SA, Wu D, Preiss P, Verma N, Guo Y (2019) The fungal mycobiome promotes pancreatic oncogenesis via activation of MBL. Nature 574:1–4. https://doi.org/10.1038/s41586-019-1608-2

Baker JL, Bor B, Agnello M, Shi W, He X (2017) Ecology of the oral microbiome: beyond bacteria. Trends Microbiol 25(5):362–374. https://doi.org/10.1016/j.tim.2016.12.012

Banerjee S, Wei Z, Tian T, Bose D, Shih NNC, Feldman MD, Khoury T, De Michele A, Robertson ES (2021) Prognostic correlations with the microbiome of breast cancer subtypes. Cell Death Dis 12:831

Baraniya D, Chen T, Nahar A, Alakwaa F, Hill J, Tellez M, Ismail A, Puri S, Al-Hebshi NN (2020) Supragingival mycobiome and interkingdom interactions in dental caries. J Oral Microbiol 12:1729305

Barnes RA (2008) Early diagnosis of fungal infection in immunocompromised patients. J Antimicrob Chemother 61(Suppl 1):i3–i6. https://doi.org/10.1093/jac/dkm424

Barrera-Vazquez OS, Gomez-Verjan JC (2020) The unexplored world of human virome, mycobiome, and archaeome in aging. J Gerontol A Biol Sci Med Sci 75:1834–1837

Begum N, Harzandi A, Lee S, Uhlen M, Moyes DL, Shoaie S (2022) Host-mycobiome metabolic interactions in health and disease. Gut Microbes 14(1):2121576. https://doi.org/10.1080/19490976.2022.2121576. PMID: 36151873; PMCID: PMC9519009

Blackwell M (2011) The fungi: 1, 2, 3 … 5.1 million species? Am J Bot 98(3):426–438. https://doi.org/10.3732/ajb.1000298

Buchfink B, Xie C, Huson DH (2015) Fast and sensitive protein alignment using DIAMOND. Nat Methods 12(1):59–60. https://doi.org/10.1038/nmeth.3176

Camarillo-Guerrero LF, Almeida A, Rangel-Pineros G, Finn RD, Lawley TD (2021) Massive expansion of human gut bacteriophage diversity. Cell 184:1098–1109.e1099

Caporaso JG, Kuczynski J, Stombaugh J, Bittinger K, Bushman FD, Costello EK et al (2010) QIIME allows analysis of high-throughput community sequencing data. Nat Methods 7(5):335–336. https://doi.org/10.1038/nmeth.f.303

Chang F, Syrjanen S, Wang L, Syrjanen K (1992) Infectious agents in the etiology of esophageal cancer. Gastroenterology 103:1336–1348

Charlson ES, Diamond JM, Bittinger K, Fitzgerald AS, Yadav A, Haas AR, Bushman FD, Collman RG (2012) Lung-enriched organisms and aberrant bacterial and fungal respiratory microbiota after lung transplant. Am J Respir Crit Care Med 186:536–545. [cited 2016 June 29]

Chehoud C et al (2015) Fungal signature in the gut microbiota of pediatric patients with inflammatory bowel disease. Inflamm Bowel Dis 21(8):1948–1956. https://doi.org/10.1097/MIB.0000000000000454

Chen Y, Chen Z, Guo R, Chen N, Lu H, Huang S, Wang J, Li L (2011) Correlation between gastrointestinal fungi and varying degrees of chronic hepatitis B virus infection. Diagn Micr Infec Dis 5:492–498. https://doi.org/10.1016/j.diagmicrobio.2010.04.005

Clancy CJ, Nguyen MH (2013) Finding the "missing 50%" of invasive candidiasis: how nonculture diagnostics will improve understanding of disease spectrum and transform patient care. Clin Infect Dis 56(9):1284–1292. https://doi.org/10.1093/cid/cit006

Coker OO, Nakatsu G, Dai RZ, Wu WKK, Wong SH, Ng SC, Chan FKL, Sung JJY, Yu J (2018) Enteric fungal microbiota dysbiosis and ecological alterations in colorectal cancer. Gut 68:654

Cui L, Morris A, Ghedin E (2013) The human mycobiome in health and disease. Genome Med 5(7):63. https://doi.org/10.1186/gm467. PMID: 23899327; PMCID: PMC3978422

Cuthbertson L, Felton I, James P, Cox MJ, Bilton D, Schelenz S, Loebinger MR, Cookson WOC, Simmonds NJ, Moffatt MF (2021) The fungal airway microbiome in cystic fibrosis and non-cystic fibrosis bronchiectasis. J Cyst Fibrosis 20:295–302. https://doi.org/10.1016/j.jcf.2020.05.013

Dai Q, Zhang FL, Feng T (2021) Sesquiterpenoids specially produced by fungi: structures, biological activities, chemical and biosynthesis (2015–2020). J Fungi (Basel) 7(12):1026

Darling WM (1976) Co-cultivation of mycobacteria and fungus. Lancet 5:740

Das A, O'Herlihy E, Shanahan F, O'Toole PW, Jeffery IB (2021) The fecal mycobiome in patients with irritable bowel syndrome. Sci Rep 11:124

De Martel C, Ferlay J, Franceschi S, Vignat J, Bray F, Forman D, Plummer M (2012) Global burden of cancers attributable to infections in 2008: a review and synthetic analysis. Lancet Oncol 13(6):607–615

Delhaes L, Monchy S, Fréalle E, Hubans C, Salleron J, Leroy S, Prevotat A, Wallet F, Wallaert B, Dei-Cas E (2012) The airway microbiota in cystic fibrosis: a complex fungal and bacterial community-implications for therapeutic management. PLoS One 5:e36313. https://doi.org/10.1371/journal.pone.0036313

Demers EG, Biermann AR, Masonjones S, Crocker AW, Ashare A, Stajich JE, Hogan DA (2018) Evolution of drug resistance in an antifungal-naive chronic Candida lusitaniae infection. Proc Natl Acad Sci U S A 115:12040–12045

Demir M, Lang S, Hartmann P, Duan Y, Martin A, Miyamoto Y, Bondareva M, Zhang X, Wang Y, Kasper P, Bang C, Roderburg C, Tacke F, Steffen HM, Goeser T, Kruglov A, Eckmann L, Starkel P, Fouts DE, Schnabl B (2022) The fecal mycobiome in non-alcoholic fatty liver disease. J Hepatol 76(4):788–799. https://doi.org/10.1016/j.jhep.2021.11.029

Donovan PD, Gonzalez G, Higgins DG, Butler G, Ito K (2018) Identification of fungi in shotgun metagenomics datasets. PLoS One 13(2):e0192898. https://doi.org/10.1371/journal.pone.0192898

Doron I, Mesko M, Li XV, Kusakabe T, Leonardi I, Shaw DG, Fiers WD, Lin WY, Bialt-DeCelie M, Román E et al (2021) Mycobiota-induced IgA antibodies regulate fungal commensalism in the gut and are dysregulated in Crohn's disease. Nat Microbiol 6:1493–1504

Edgar RC (2010) Search and clustering orders of magnitude faster than BLAST. Bioinformatics 26(19):2460–2461. https://doi.org/10.1093/bioinformatics/btq461

Enaud R, Vandenborght L-E, Coron N, Bazin T, Prevel R, Schaeverbeke T, Berger P, Fayon M, Lamireau T, Delhaes L (2018) The mycobiome: a neglected component in the microbiota-gut-brain axis. Microorganisms 6(1):22. https://doi.org/10.3390/microorganisms6010022

Fidel PL (2021) The neutral vaginal pH in mice that is typical of most mammalian species should not deter research using experimental murine models of Candida vaginitis. Infect Immun 89:e00624-00620

Fosso B, Santamaria M, Marzano M, Alonso-Alemany D, Valiente G, Donvito G et al (2015) BioMaS: a modular pipeline for Bioinformatic analysis of Metagenomic AmpliconS. BMC Bioinformatics. 16:203. https://doi.org/10.1186/s12859-015-0595-z

Frau A, Ijaz UZ, Slater R, Jonkers D, Penders J, Campbell BJ, Kenny JG, Hall N, Lenzi L, Burkitt MD et al (2021) Inter-kingdom relationships in Crohn's disease explored using a multi-omics approach. Gut Microbes 13:1930871

Frey-Klett P, Burlinson P, Deveau A, Barret M, Tarkka M, Sarniguet A (2011) Bacterial-fungal interactions: hyphens between agricultural, clinical, environmental, and food microbiologists. Microbiol Mol Biol Rev 75(4):583–609. https://doi.org/10.1128/MMBR.00020-11

Gaitanis G, Magiatis P, Hantschke M, Bassukas ID, Velegraki A (2012) The *Malassezia* genus in skin and systemic diseases. Clin Microbiol Rev 25:106–141. https://doi.org/10.1128/CMR.00021-11

Gerard R, Sendid B, Colombel JF, Poulain D, Jouault T (2015) An immunological link between Candida albicans colonization and Crohn's disease. Crit Rev Microbiol 41(2):135–139. https://doi.org/10.3109/1040841X.2013.810587

Ghannoum MA, Jurevic RJ, Mukherjee PK, Cui F, Sikaroodi M, Naqvi A et al (2010) Characterization of the oral fungal microbiome (mycobiome) in healthy individuals. PLoS Pathog 6(1):e1000713. https://doi.org/10.1371/journal.ppat.1000713

Gouba N, Drancourt M (2015) Digestive tract mycobiota: a source of infection. Med Mal Infect 45(1–2):9–16

Gregory AC, Zablocki O, Zayed AA, Howell A, Bolduc B, Sullivan MB (2020) The gut virome database reveals age-dependent patterns of virome diversity in the human gut. Cell Host Microbe 28:724–740.e728

Greten FR, Grivennikov SI (2019) Inflammation and cancer: triggers, mechanisms, and consequences. Immunity 51:27–41. https://doi.org/10.1016/j.immuni.2019.06.025

Gu C, Kim GB, Kim WJ, Kim HU, Lee SY (2019) Current status and applications of genome-scale metabolic models. Genome Biol 20(1):121. https://doi.org/10.1186/s13059-019-1730-3

Guo Q et al (2019) Probiotics for the prevention of pediatric antibiotic-associated diarrhea. Cochrane Database Syst Rev 4(4):Cd004827

Gupta A, Singh V, Mani I (2022) Dysbiosis of human microbiome and infectious diseases. Prog Mol Biol Transl Sci 192(1):33–51. https://doi.org/10.1016/bs.pmbts.2022.06.016

Gweon HS, Oliver A, Taylor J, Booth T, Gibbs M, Read DS et al (2015) PIPITS: an automated pipeline for analyses of fungal internal transcribed spacer sequences from the Illumina sequencing platform. Methods Ecol Evol 6(8):973–980. https://doi.org/10.1111/2041-210X.12399

Halwachs B, Madhusudhan N, Krause R, Nilsson RH, Moissl-Eichinger C, Högenauer C et al (2017) Critical issues in mycobiota analysis. Front Microbiol 8. https://doi.org/10.3389/fmicb.2017.00180

Harrison MJ, Twomey KB, McCarthy Y, O'Connell OJ, Alston M, Febrer M, Murphy DM, Ryan RP, Plant BJ (2012) Fungal microbiota in the adult cystic fibrosis (CF) airway: characterization by second-generation sequencing and correlation with standard culture-based methods and clinical phenotype. Ir J Med Sci 5:S369–S437

Hawksworth DL (1991) The fungal dimension of biodiversity: magnitude, significance, and conservation. Mycol Res 95:641–655

Hoarau G, Mukherjee PK, Gower-Rousseau C, Hager C, Chandra J, Retuerto MA, Neut C, Vermeire S, Clemente J, Colombel JF (2016) Bacteriome and mycobiome interactions underscore microbial dysbiosis in familial Crohn's disease. mBio 7(5):e01250-16. https://doi.org/10.1128/mBio.01250-16

Huang X, Hurabielle C, Drummond RA, Bouladoux N, Desai JV, Sim CK, Belkaid Y, Lionakis MS, Segre JA (2021) Murine model of colonization with fungal pathogen Candida auris to explore skin tropism, host risk factors and therapeutic strategies. Cell Host Microbe 29:210–221.e216

Huffnagle GB, Noverr MC (2013) The emerging world of the fungal microbiome. Trends Microbiol 21(7):334–341. https://doi.org/10.1016/j.tim.2013.04.002

Iliev ID, Leonardi I (2017) Fungal dysbiosis: immunity and interactions at mucosal barriers. Nat Rev Immunol 17:635–646

Kerr J (1994) Inhibition of fungal growth by *Pseudomonas aeruginosa* and *Pseudomonas cepacia* isolated from patients with cystic fibrosis. J Infect 5:305–310

Kim JY (2016) Human fungal pathogens: why should we learn? J Microbiol 54(3):145–148. https://doi.org/10.1007/s12275-016-0647-8

Kim SH, Clark ST, Surendra A, Copeland JK, Wang PW, Ammar R, Collins C, Tullis DE, Nislow C, Hwang DM et al (2015) Global analysis of the fungal microbiome in cystic fibrosis patients reveals loss of function of the transcriptional repressor Nrg1 as a mechanism of pathogen adaptation. PLoS Pathog 11:e1005308

Kim DH, Kim H, Jeong D, Kang IB, Chon JW, Kim HS, Song KY, Seo KH (2017) Kefir alleviates obesity and hepatic steatosis in high-fat diet-fed mice by modulation of gut microbiota and mycobiota: targeted and untargeted community analysis with correlation of biomarkers. J Nutr Biochem 44:35–43. https://doi.org/10.1016/j.jnutbio.2017.02.014

Kramer R, Sauer-Heilborn A, Welte T, Guzman CA, Abraham W-R, Höfle MG (2015) Cohort study of airway mycobiome in adult cystic fibrosis patients: differences in community structure between fungi and bacteria reveal predominance of transient fungal elements. J Clin Microbiol 53:2900–2907. https://doi.org/10.1128/JCM.01094-15

Kumamoto CA (2011) Inflammation and gastrointestinal Candida colonization. Curr Opin Microbiol 14(4):386–391. https://doi.org/10.1016/j.mib.2011.07.015

Kumar S, Carlsen T, Mevik BH, Enger P, Blaalid R, Shalchian-Tabrizi K et al (2011) CLOTU: an online pipeline for processing and clustering of 454 amplicon reads into OTUs followed by taxonomic annotation. BMC Bioinformatics 12:182. https://doi.org/10.1186/1471-2105-12-182

Leonardi I, Paramsothy S, Doron I, Semon A, Kaakoush NO, Clemente JC, Faith JJ, Borody TJ, Mitchell HM, Colombel JF, Kamm MA, Iliev ID (2020) Fungal trans-kingdom dynamics linked to responsiveness to fecal microbiota transplantation (FMT) therapy in ulcerative colitis. Cell Host Microbe 27(5):823–829. https://doi.org/10.1016/j.chom.2020.03.006

Lewis T, Loman NJ, Bingle L, Jumaa P, Weinstock GM, Mortiboy D et al (2010) High-throughput whole-genome sequencing to dissect the epidemiology of *Acinetobacter baumannii* isolates from a hospital outbreak. J Hosp Infect 75(1):37–41. https://doi.org/10.1016/j.jhin.2010.01.012

Limon JJ et al (2019) Malassezia is associated with Crohn's disease and exacerbates colitis in mouse models. Cell Host Microbe 25(3):377–388. https://doi.org/10.1016/j.chom.2019.01.007

Limper AH, Adenis A, Le T, Harrison TS (2017) Fungal infections in HIV/AIDS. Lancet Infect Dis 17(11):e334–e343. https://doi.org/10.1016/S1473-3099(17)30303-1

LiPuma JJ (2010) The changing microbial epidemiology in cystic fibrosis. Clin Microbiol Rev 23:299–323. https://doi.org/10.1128/CMR.00068-09

Liu NN, Jiao N, Tan JC, Wang Z, Wu D, Wang AJ, Chen J, Tao L, Zhou C, Fang W et al (2022) Multi-kingdom microbiota analyses identify bacterial-fungal interactions and biomarkers of colorectal cancer across cohorts. Nat Microbiol 7:238–250

Loit K, Adamson K, Bahram M et al (2019) Relative performance of MinION (Oxford Nanopore Technologies) versus Sequel (Pacific Biosciences) third-generation sequencing instruments in identification of agricultural and forest fungal pathogens. Appl Environ Microbiol 85:e01368–e01319

Luan C, Xie L, Yang X et al (2015) Dysbiosis of fungal microbiota in the intestinal mucosa of patients with colorectal adenomas. Sci Rep 5:7980

Mani I (2020a) Current status and challenges of DNA sequencing. In: Singh V (ed) Advances in synthetic biology. Springer, Singapore

Mani I (2020b) Metagenomics approach for bioremediation: challenges and perspectives. In: Pandey VC, Singh V (eds) Bioremediation of pollutants. Elsevier, pp 275–285., ISBN 9780128190258. https://doi.org/10.1016/B978-0-12-819025-8.00012-0

Mani I, Singh V, Alzahrani KJ, Chu DT (eds) (2023) Microbial genomic islands in adaptation and pathogenicity. Springer, Singapore, p 356. ISBN: 978-981-19-9342-8. https://doi.org/10.1007/978-981-19-9342-8

Menzel P, Ng KL, Krogh A (2016) Fast and sensitive taxonomic classification for metagenomics with Kaiju. Nat Commun 7:11257. https://doi.org/10.1038/ncomms11257

Morrison GA, Fu J, Lee GC et al (2020) Nanopore sequencing of the fungal intergenic spacer sequence as a potential rapid diagnostic assay. J Clin Microbiol 58:e01972-20

Narunsky-Haziza L, Sepich-Poore GD, Livyatan I, Asraf O, Martino C, Nejman D, Gavert N, Stajich JE, Amit G, González A et al (2022) Pancancer analyses reveal cancer-type-specific fungal ecologies and bacteriome interactions. Cell 185:3789–3806.e3717

Nash AK, Auchtung TA, Wong MC, Smith DP, Gesell JR, Ross MC, Stewart CJ, Metcalf GA, Muzny DM, Gibbs RA, Ajami NJ, Petrosino JF (2017) The gut mycobiome of the Human Microbiome Project healthy cohort. Microbiome. 5(1):153. https://doi.org/10.1186/s40168-017-0373-4. PMID: 29178920; PMCID: PMC5702186

Nelson A, De Soyza A, Bourke S, Perry J, Cummings S (2010) Assessment of sample handling practices on microbial activity in sputum samples from patients with cystic fibrosis. Lett Appl Microbiol 5:272–277. https://doi.org/10.1111/j.1472-765X.2010.02891.x

Nelson A et al (2021) The impact of NOD2 genetic variants on the gut mycobiota in Crohn's disease patients in remission and in individuals without gastrointestinal inflammation. J Crohns Colitis 15(5):800–812. https://doi.org/10.1093/ecco-jcc/jjaa220

Noverr MC, Falkowski NR, McDonald RA, McKenzie AN, Huffnagle GB (2005) Development of allergic airway disease in mice following antibiotic therapy and fungal microbiota increase: role of host genetics, antigen, and interleukin-13. Infect Immun 5:30–38. https://doi.org/10.1128/IAI.73.1.30-38.2005

O'Brien HE, Parrent JL, Jackson JA, Moncalvo JM, Vilgalys R (2005) Fungal community analysis by large-scale sequencing of environmental samples. Appl Environ Microbiol 71(9):5544–5550. https://doi.org/10.1128/AEM.71.9.5544-5550.2005

Ost KS, O'Meara TR, Stephens WZ, Chiaro T, Zhou H, Penman J, Bell R, Catanzaro JR, Song D, Singh S et al (2021) Adaptive immunity induces mutualism between commensal eukaryotes. Nature 596:114–118

Ott SJ, Kühbacher T, Musfeldt M, Rosenstiel P, Hellmig S, Rehman A, Drews O, Weichert W, Timmis KN, Schreiber S (2008) Fungi and inflammatory bowel diseases: alterations of composition and diversity. Scand J Gastroenterol 43(7):831–841. https://doi.org/10.1080/00365520801935434

Pashley CH (2014) Fungal culture and sensitisation in asthma, cystic fibrosis and chronic obstructive pulmonary disorder: what does it tell us? Mycopathologia 178:457–463. https://doi.org/10.1007/s11046-014-9804-y

Pashley CH, Fairs A, Morley JP, Tailor S, Agbetile J, Bafadhel M, Brightling CE, Wardlaw AJ (2012) Routine processing procedures for isolating filamentous fungi from respiratory sputum samples may underestimate fungal prevalence. Med Mycol 50:433–438. https://doi.org/10.3109/13693786.2011.615762

Pasolli E, Asnicar F, Manara S, Zolfo M, Karcher N, Armanini F, Beghini F, Manghi P, Tett A, Ghensi P et al (2019) Extensive unexplored human microbiome diversity revealed by over 150,000 genomes from metagenomes spanning age, geography, and lifestyle. Cell 176:649–662.e620

Paudel KR, Jha SK, Allam VSRR, Prasher P, Gupta PK, Bhattacharjee R, Jha NK, Vishwas S, Singh SK, Shrestha J et al (2021) Recent advances in chronotherapy targeting respiratory diseases. Pharmaceutics 13:2008. https://doi.org/10.3390/pharmaceutics13122008

Perera M, Al-Hebshi NN, Perera I, Ipe D, Ulett GC, Speicher DJ, Chen T, Johnson NW (2017) A dysbiotic mycobiome dominated by Candida albicans is identified within oral squamous-cell carcinomas. J Oral Microbiol 9:1385369

Perlin DS, Rautemaa-Richardson R, Alastruey-Izquierdo A (2017) The global problem of antifungal resistance: prevalence, mechanisms, and management. Lancet Infect Dis 17(12):e383–e392. https://doi.org/10.1016/S1473-3099(17)30316-X

Proctor DM, Dangana T, Sexton DJ, Fukuda C, Yelin RD, Stanley M, Bell PB, Baskaran S, Deming C, Chen Q et al (2021) Integrated genomic, epidemiologic investigation of Candida auris skin colonization in a skilled nursing facility. Nat Med 27:1401–1409

Qin J, Li R, Raes J, Arumugam M, Burgdorf KS, Manichanh C, Nielsen T, Pons N, Levenez F, Yamada T et al (2010) A human gut microbial gene catalogue established by metagenomic sequencing. Nature 464(7285):59–65. https://doi.org/10.1038/nature08821

Rao C, Coyte KZ, Bainter W, Geha RS, Martin CR, Rakoff-Nahoum S (2021) Multi-kingdom ecological drivers of microbiota assembly in preterm infants. Nature 591:633–638

Reedy JL, Bastidas RJ, Heitman J (2007) The virulence of human pathogenic fungi: notes from the South of France. Cell Host Microbe 2:77–83

Richard ML, Sokol H (2019) The gut mycobiota: insights into analysis, environmental interactions and role in gastrointestinal diseases. Nat Rev Gastroenterol Hepatol 16(6):331–345

Roper M, Ellison C, Taylor JW, Glass NL (2011) Nuclear and genome dynamics in multinucleate ascomycete fungi. Curr Biol 21(18):R786–R793. https://doi.org/10.1016/j.cub.2011.06.042

Saheb Kashaf S, Proctor DM, Deming C, Saary P, Hölzer M, Taylor ME, Kong HH, Segre JA, Almeida A, Finn RD (2022) Integrating cultivation and metagenomics for a multi-kingdom view of skin microbiome diversity and functions. Nat Microbiol 7:169–179

Sam QH, Chang MW, Chai LYA (2017) The fungal mycobiome and its interaction with gut bacteria in the host. Int J Mol Sci 18:330

Schei K, Avershina E, Øien T, Rudi K, Follestad T, Salamati S, Ødegård RA (2017) Early gut mycobiota and mother-offspring transfer. Microbiome 5:107

Schloss PD, Westcott SL, Ryabin T, Hall JR, Hartmann M, Hollister EB et al (2009) Introducing mothur: open-source, platform-independent, community-supported software for describing and comparing microbial communities. Appl Environ Microbiol 75(23):7537–7541. https://doi.org/10.1128/AEM.01541-09

Schoch CL, Seifert KA, Huhndorf S, Robert V, Spouge JL, Levesque CA et al (2012) Nuclear ribosomal internal transcribed spacer (ITS) region as a universal DNA barcode marker for fungi. Proc Natl Acad Sci U S A 109(16):6241–6246. https://doi.org/10.1073/pnas.1117018109

Sellart-Altisent M, Torres-Rodríguez JM, Gómez de Ana S, Alvarado-Ramírez E (2007) Nasal fungal microbiota in allergic and healthy subjects. Rev Iberoam Micol 5:125–130. https://doi.org/10.1016/S1130-1406(07)70027-X

Severance EG, Alaedini A, Yang S, Halling M, Gressitt KL, Stallings CR, Origoni AE, Vaughan C, Khushalani S, Leweke FM (2012) Gastrointestinal inflammation and associated immune activation in schizophrenia. Schizophr Res 138(1):48–53. https://doi.org/10.1016/j.schres.2012.02.025

Severance EG, Gressitt KL, Stallings CR, Katsafanas E, Schweinfurth LA, Savage CL, Adamos MB, Sweeney KM, Origoni AE, Khushalani S, Leweke FM (2016) Candida albicans exposures, sex specificity and cognitive deficits in schizophrenia and bipolar disorder. NPJ Schizophr 2(1):1–7. https://doi.org/10.1038/npjschz.2016.18

Simoneau E, Kelly M, Labbe AC, Roy J, Laverdiere M (2005) What is the clinical significance of positive blood cultures with *Aspergillus* sp. in hematopoietic stem cell transplant recipients?

A 23 year experience. Bone Marrow Transpl 35(3):303–306. https://doi.org/10.1038/sj.bmt.1704793

Sivananthan K, Petersen AM (2018) Review of Saccharomyces boulardii as a treatment option in IBD. Immunopharmacol Immunotoxicol 40(6):465–475. https://doi.org/10.1080/08923973.2018.1469143

Snitkin ES, Zelazny AM, Thomas PJ, Stock F, Group NCS, Henderson DK et al (2012) Tracking a hospital outbreak of carbapenem-resistant *Klebsiella pneumoniae* with whole-genome sequencing. Sci Transl Med 4(148):148ra16. https://doi.org/10.1126/scitranslmed.3004129

Sokol H, Leducq V, Aschard H, Pham H-P, Jegou S, Landman C, Cohen D, Liguori G, Bourrier A, Nion-Larmurier I (2017) Fungal microbiota dysbiosis in IBD. Gut 66(6):1039–1048. https://doi.org/10.1136/gutjnl-2015-310746

Sonnenburg JL, Angenent LT, Gordon JI (2004) Getting a grip on things: how do communities of bacterial symbionts become established in our intestine? Nat Immunol 5:569–573

Soret P, Vandenborght LE, Francis F, Coron N, Enaud R, Avalos M, Schaeverbeke T, Berger P, Fayon M, Thiebaut R et al (2020) Respiratory mycobiome and suggestion of inter-kingdom network during acute pulmonary exacerbation in cystic fibrosis. Sci Rep 10:3589. https://doi.org/10.1038/s41598-020-60015-4

Soverini M, Turroni S, Biagi E et al (2019) HumanMycobiomeScan: a new bioinformatics tool for the characterization of the fungal fraction in metagenomic samples. BMC Genomics 20:496

Sparber F, De Gregorio C, Steckholzer S, Ferreira FM, Dolowschiak T, Ruchti F, Kirchner FR, Mertens S, Prinz I, Joller N et al (2019) The skin commensal yeast Malassezia triggers a Type 17 response that coordinates anti-fungal immunity and exacerbates skin inflammation. Cell Host Microbe 25:389–403.e386

Stoltz DA, Meyerholz DK, Welsh MJ (2015) Origins of cystic fibrosis lung disease. N Engl J Med 372:1574–1575. https://doi.org/10.1056/NEJMra1300109

Suzuki S, Kakuta M, Ishida T, Akiyama Y (2014) GHOSTX: an improved sequence homology search algorithm using a query suffix array and a database suffix array. PLoS One 9(8):e103833. https://doi.org/10.1371/journal.pone.0103833

Tarrand JJ, Lichterfeld M, Warraich I, Luna M, Han XY, May GS et al (2003) Diagnosis of invasive septate mold infections. A correlation of microbiological culture and histologic or cytologic examination. Am J Clin Pathol 119(6):854–858. https://doi.org/10.1309/EXBVYAUPENBM285Y

Taş N, de Jong AE, Li Y, Trubl G, Xue Y, Dove NC (2021) Metagenomic tools in microbial ecology research. Curr Opin Biotechnol 67:184–191. ISSN 0958-1669. https://doi.org/10.1016/j.copbio.2021.01.019

The Human Microbiome Project Consortium (2012) A framework for human microbiome research. Nature 486:215–221. https://doi.org/10.1038/nature11209

Thielemann N, Herz M, Kurzai O, Martin R (2022) Analyzing the human gut mycobiome—a short guide for beginners. Comput Struct Biotechnol J 20:608–614. https://doi.org/10.1016/j.csbj.2022.01.008

Thoendel MJ, Jeraldo PR, Greenwood-Quaintance KE, Yao JZ, Chia N, Hanssen AD et al (2018) Identification of prosthetic joint infection pathogens using a shotgun metagenomics approach. Clin Infect Dis 67(9):1333–1338. https://doi.org/10.1093/cid/ciy303

Trojanowska D, Zwolińska-Wcisło M, Tokarczyk M, Kosowski K, Mach T, Budak A (2010) The role of Candida in inflammatory bowel disease. Estimation of transmission of C. albicans fungi in gastrointestinal tract based on genetic affinity between strains. Med Sci Monit Int Med J Exp Clin Res 16:CR451–CR457

Underhill DM, Braun J (2022) Fungal microbiome in inflammatory bowel disease: a critical assessment. J Clin Invest 132(5):e155786. https://doi.org/10.1172/JCI155786

Underhill DM, Iliev ID (2014) The mycobiota: interactions between commensal fungi and the host immune system. Nat Rev Immunol 14:405–416. https://doi.org/10.1038/nri3684

van Woerden HC, Gregory C, Brown R, Marchesi JR, Hoogendoorn B, Matthews IP (2013) Differences in fungi present in induced sputum samples from asthma patients and non-atopic controls: a community based case control study. BMC Infect Dis 13:69. https://doi.org/10.1186/1471-2334-13-69

Wang L, Zhang K, Zeng Y, Luo Y, Peng J, Zhang J, Kuang T, Fan G (2023) Gut mycobiome and metabolic diseases: the known, the unknown, and the future. Pharmacol Res 193:106807. https://doi.org/10.1016/j.phrs.2023.106807. Epub 2023 May 25

Ward T, Dominguez-Bello M, Heisel T, Al-Ghalith G, Knights D, Gale C (2018) Development of the human mycobiome over the first month of life and across body sites. mSystems 3:e00140-00117

Weterings V, Zhou K, Rossen JW, van Stenis D, Thewessen E, Kluytmans J et al (2015) An outbreak of colistin-resistant *Klebsiella pneumoniae* carbapenemase-producing *Klebsiella pneumoniae* in the Netherlands (July to December 2013), with inter-institutional spread. Eur J Clin Microbiol Infect Dis 34(8):1647–1655. https://doi.org/10.1007/s10096-015-2401-2

White JR, Maddox C, White O, Angiuoli SV, Fricke WF (2013) CloVR-ITS: automated internal transcribed spacer amplicon sequence analysis pipeline for the characterization of fungal microbiota. Microbiome 1(1):6. https://doi.org/10.1186/2049-2618-1-6

Whiteside SA, McGinniss JE, Collman RG (2021) The lung microbiome: progress and promise. J Clin Invest 131:e150473. https://doi.org/10.1172/JCI150473

Willger SD, Grim SL, Dolben EL, Shipunova A, Hampton TH, Morrison HG, Filkins LM, O'Toole GA, Moulton LA, Ashare A et al (2014) Characterization and quantification of the fungal microbiome in serial samples from individuals with cystic fibrosis. Microbiome. 2:40. https://doi.org/10.1186/2049-2618-2-40

Wilson MR, Naccache SN, Samayoa E, Biagtan M, Bashir H, Yu G et al (2014) Actionable diagnosis of neuroleptospirosis by next-generation sequencing. N Engl J Med 370(25):2408–2417. https://doi.org/10.1056/NEJMoa1401268

Wilson MR, O'Donovan BD, Gelfand JM, Sample HA, Chow FC, Betjemann JP et al (2018) Chronic meningitis investigated via metagenomic next-generation sequencing. JAMA Neurol 75(8):947–955. https://doi.org/10.1001/jamaneurol.2018.0463

Witherden EA, Moyes DL, Bruce KD, Ehrlich SD, Shoaie S (2017) Using systems biology approaches to elucidate cause and effect in host–microbiome interactions. Curr Opin Syst Biol 3:141–146. https://doi.org/10.1016/j.coisb.2017.05.003

Wood DE, Salzberg SL (2014) Kraken: ultrafast metagenomic sequence classification using exact alignments. Genome Biol 15(3):R46. https://doi.org/10.1186/gb-2014-15-3-r46

Yang CS (1980) Research on esophageal cancer in China: a review. Cancer Res 40:2633–2644

Zhang E, Tanaka T, Tajima M, Tsuboi R, Nishikawa A, Sugita T (2011) Characterization of the skin fungal microbiota in patients with atopic dermatitis and in healthy subjects. Microbiol Immunol 5:625–632. https://doi.org/10.1111/j.1348-0421.2011.00364.x

Zhang L, Zhan H, Xu W, Yan S, Ng SC (2021) The role of gut mycobiome in health and diseases. Ther Adv Gastroenterol 14:17562848211047130. https://doi.org/10.1177/17562848211047130. PMID: 34589139; PMCID: PMC8474302

Zhang F, Aschenbrenner D, Yoo JY, Zuo T (2022) The gut mycobiome in health, disease, and clinical applications in association with the gut bacterial microbiome assembly. Lancet Microbe 3(12):e969–e983. https://doi.org/10.1016/S2666-5247(22)00203-8. Epub 2022 Sep 28. PMID: 36182668

Zhao C, Li Y, Chen B, Yue K, Su Z, Xu J, Xue W, Zhao G, Zhang L (2023) Mycobiome study reveals different pathogens of vulvovaginal candidiasis shape characteristic vaginal bacteriome. Microbiol Spectr 11:e0315222

Zhong M, Xiong Y, Zhao J, Gao Z, Ma J, Wu Z, Song Y, Hong X (2021) Candida albicans disorder is associated with gastric carcinogenesis. Theranostics 11:4945–4956

Zuo T, Wong SH, Cheung CP, Lam K, Lui R, Cheung K, Zhang F, Tang W, Ching JYL, Wu JCY, Chan PKS, Sung JJY, Yu J, Chan FKL, Ng SC (2018) Gut fungal dysbiosis correlates with reduced efficacy of fecal microbiota transplantation in Clostridium difficile infection. Nat Commun 9(1):3663. https://doi.org/10.1038/s41467-018-06103-6

Fluxomics and Metabolic Flux Analysis

8

Manoj Kumar Mishra and Ajay Kumar

8.1 Overview of Fluxomics

Fluxomics is a part of systems biology that focuses on the rates of molecule exchange or transfer (flux) within the context of metabolic reactions in biological systems. These systems can encompass cells, tissues, organisms (both prokaryotic and eukaryotic), or entire plants (Emwas et al. 2022). It provides a quantitative understanding of how cells and organisms allocate and utilize resources for growth, maintenance, and various cellular functions (Winter and Krömer 2013; Cortassa et al. 2015). Fluxomics establishes a connection between the enzymatic reactions forming the metabolic network, linking metabolites to subcellular trafficking processes such as transporters and vesicles at the cellular level. This network aims to ascertain reaction rates, proposing a flux map that translates into insights about enzymatic processes and regulation (Cascante and Marin 2008; Cortassa et al. 2015; Heux et al. 2017). These metabolic networks, assumed to be composed of nearly identical cells, have been investigated at the levels of individual cells, tissues, and organs (Heux et al. 2017). Fluxomics, which is associated with the rate of exchange between organs, may be utilized in the plant system to assess nutrient use efficiency (NUE), such as the reallocation of minerals and/or carbon through xylem and phloem (Salon et al. 2017), as well as modulation in, the content of a given component in the many organs in metabolic flux analysis. Metabolic pathways are interconnected series of chemical reactions that occur within a cell to convert nutrients into energy, biomass, and other molecules essential for cellular functions. Examples include glycolysis, the TCA cycle, and amino acid biosynthesis. Molecule fluxes are carried by physical force rather than enzymes. Water and minerals move up the root osmotically to the endodermis, where they are actively pumped into apoplast

M. K. Mishra (✉) · A. Kumar
Department of Biotechnology, Faculty of Engineering and Technology, Rama University, Kanpur, Uttar Pradesh, India

I. Mani, V. Singh (eds.), *Multi-Omics Analysis of the Human Microbiome*,
https://doi.org/10.1007/978-981-97-1844-3_8

(tracheid's) of xylem (Crown and Antoniewicz 2013; Salon et al. 2017). The ascent of sap in xylem cells is propelled by the transpiration of water from shoot organs. Substances like carbohydrates and amino acids travel from their origin points to the phloem in plant shoots. The flow of photosynthates and nutrients in the phloem is driven by osmotic pressure disparities between source organs and heterotrophic sinks. The metabolites transported are subsequently utilized for storage, maintenance, or growth (Crown and Antoniewicz 2013). Fluxomics can be used to discover how cells respond to various conditions and stimuli, how metabolic pathways are controlled, and how metabolic fluxes are linked to other cellular processes. It typically entails measuring and modeling metabolic fluxes using a combination of experimental and computational techniques (Fig. 8.1).

Experimental techniques encompass nuclear magnetic resonance (NMR) spectroscopy, mass spectrometry, and stable isotope labeling (Ong et al. 2003; Guo et al. 2013; Emwas et al. 2019; Giraudeau 2020). Computational approaches include flux balance analysis, dynamic flux balance analysis, and constraint-based modeling

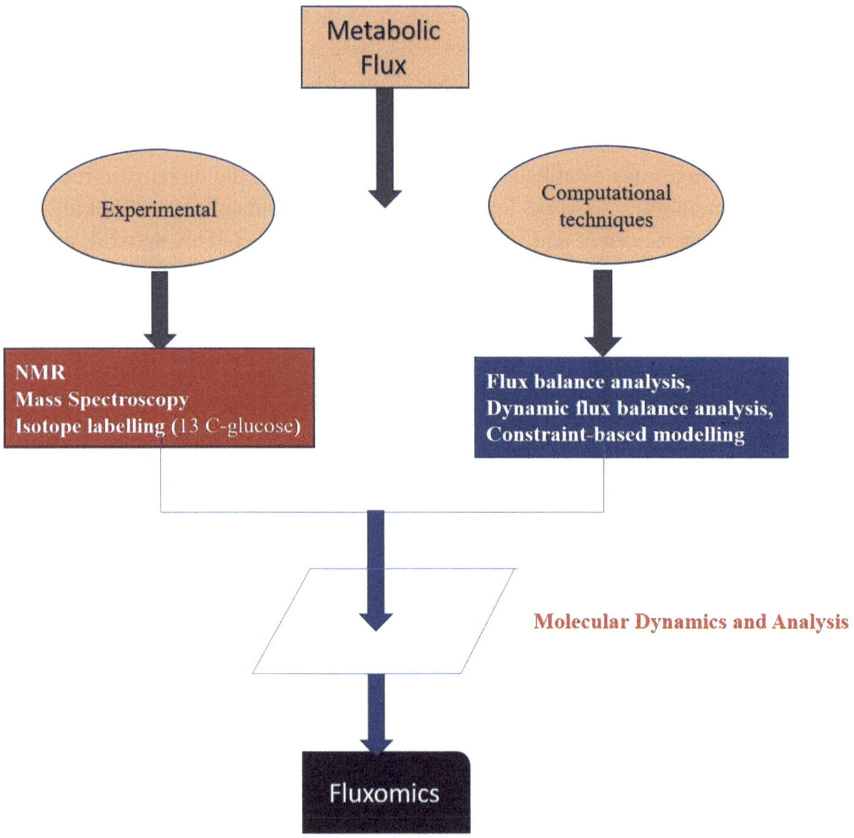

Fig. 8.1 Process of measuring and modeling metabolic fluxes

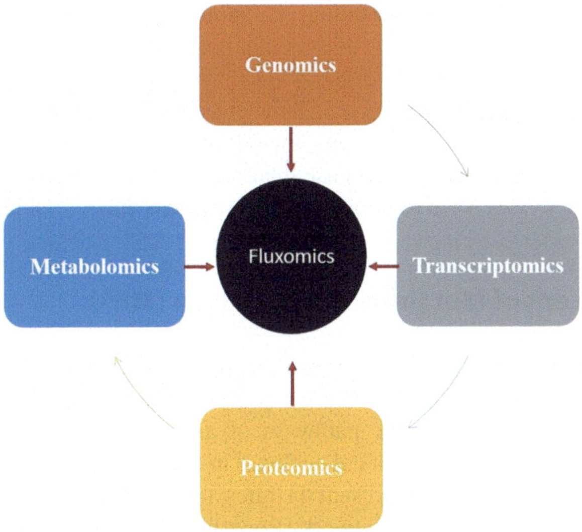

Fig. 8.2 The figure represents the relationships of fluxomic with each "-omics." The arrow present in the figure in a definite direction showed the influence of "-omics" in each other

(Emwas et al. 2022). The fluxome, representing the set of metabolic fluxes, provides a dynamic portrayal of the phenotype as it illustrates the functional connections of the metabolome with the environment and the genome (Salon et al. 2017). By integrating data from various cellular processes, the fluxome emerges as a distinctive phenotypic signature of cells. The substantial advantages of fluxomics over genomics and proteomics lie in its foundation on metabolite data, which are more abundant than genes or proteins (Cortassa et al. 2015; Giraudeau 2020; Salon et al. 2017). Although fluxomics is a valuable tool for metabolomics, there are certain challenges with this method. To develop a thorough comprehension of cellular function using fluxomics data, it is essential to integrate this information with other "omics" datasets, such as transcriptomics, proteomics, and genomics (Fig. 8.2) (Winter and Krömer 2013). Modeling metabolic networks can be challenging due to their complexity, requiring advanced mathematical and computational approaches. Additionally designing experiments that accurately capture metabolic fluxes can be complex and resource-intensive.

8.2 Applications of Fluxomics

Biotechnology: The application of fluxomics in metabolic engineering aims to enhance the production of biofuels, pharmaceuticals, and other valuable compounds by modifying cellular pathways.

Health and Medicine: Understanding metabolic fluxes is essential in studying diseases such as cancer, diabetes, and metabolic disorders, offering insights into potential treatment strategies.

Microbial Ecology: Fluxomics can help elucidate the metabolic interactions between microorganisms in ecosystems, including the human gut microbiome.

Biological Research: Researchers use fluxomics to gain insights into fundamental biological processes, such as photosynthesis in plants and microbial metabolism.

8.3 Concept of Metabolic Flux Analysis (MFA)

Within the realms of systems biology and bioengineering, metabolic flux analysis (MFA) proves to be a valuable method for quantifying and scrutinizing the flow of metabolites within the metabolic pathways of a cell or organism (Antoniewicz 2015). This technique is used to calculate fluxes in different organism or physiological and other pathways. MFA represent the rates at which metabolites are produced or consumed within a metabolic network (Bonarius et al. 1997). They are typically measured in units like mmol/gDW/h (millimoles dry weight in gram per hour) and describe how fast specific reactions occur. (Toya and Shimizu 2013; Antoniewicz 2015) It provides a comprehensive view of how nutrients are converted into various metabolites and energy in biological systems. Providing an isotopically labeled substrate is a common requirement for MFA. MFA helps researchers map out the intricate network of metabolic reactions within a cell. MFA relies on the assumption that a cell's metabolic state is in a steady state (Antoniewicz 2015; Emwas et al. 2022). This implies that throughout a specified time period, the concentrations of metabolites remain relatively stable, enabling the investigation of metabolic fluxes (Dauner 2010; Antoniewicz 2018). Experimental determination of metabolic fluxes often involves the prevalent use of stable isotope labeling techniques. These experiments involve feeding the cells with isotopically labeled substrates (e.g., 13C-glucose molecule) and tracking the integration of labeled atoms into downstream metabolites (Wiechert et al. 2001; Antoniewicz 2015, 2018). Analytical techniques such as nuclear magnetic resonance (NMR) spectrometry and mass spectrometry are commonly combined with isotope labeling experiments to assess the isotopic composition of metabolites and compute fluxes (Sriram et al. 2004; Emwas et al. 2022). Mathematical models, such as metabolic flux balance models or stoichiometric models, are employed to depict the complete metabolic network of an organism (Bonarius et al. 1997; Wiechert et al. 2001; Antoniewicz 2015). Metabolic flux models take into account the stoichiometry of reactions, constraints on nutrient uptake, and other relevant factors to calculate flux distributions. In metabolic flux analysis, an objective function is defined to represent the cell's goal. Common objectives include maximizing biomass production, ATP synthesis, or specific product yields. The flux distribution is then optimized to achieve this objective (Wiechert et al. 2001; Antoniewicz 2015, 2018). MFA involves solving a system of linear equations subject to constraints, which reflect the steady-state condition and other physiological limits. Constraint-based optimization algorithms, like flux balance analysis (FBA), are often used to calculate flux distributions that satisfy these constraints (Antoniewicz 2015). Metabolic flux analysis has various applications, including optimizing microbial strains for biofuel or pharmaceutical production, understanding the metabolic basis of diseases, and exploring the metabolic responses of organisms to environmental changes. Overall observation suggested the MFA can be challenging due to the complex metabolic networks and the need for precise measurements (Crown and Antoniewicz 2013; Emwas et al. 2022). Additionally, it assumes steady-state conditions, which may not always hold in dynamic systems.

8.4 Observation of Steady and Non-steady-State Flux Analysis

Fluxomics can be used in both steady-state and non-steady-state environments. Steady-state flux analysis presupposes that the metabolic system remains in a consistent and unchanging state, while non-steady-state analysis explores transient metabolic changes, which can be crucial in understanding responses to environmental perturbations (Sanford et al. 2002).

8.4.1 MFA at Steady State

In a living organism or cell operating at steady state, the analysis of biological system requires treating it as an integrated biochemical network model to compute metabolic flux. Metabolic flux analysis (MFA) relies on the equilibrium of intracellular metabolite fluxes within a theoretical metabolic network model. In this model, initially depicted as a stoichiometric matrix, columns symbolize metabolic fluxes, and rows symbolize balanced intracellular metabolites (Provost and Bastin 2004). In the stoichiometric matrix, the "biomass reaction" is defined as the consumption of precursor metabolites essential for cell growth (Long and Antoniewicz 2014). This method can be used to calculate metabolic fluxes in overdetermined system (contain a surplus of external rate measurements) and fully determined (contain all the required external rate measurements). The primary benefit of MFA at steady-state calculation is its accessibility to a wide range of researchers due to its minimal requirement of basic linear algebra and reliance on extracellular metabolite measurements, which are generally regarded as reliable. As an alternative, the flux in a system with unknown flux can be determined using constraint-based optimization and flux balance analysis (FBA). We use inequality constraints, such as upper and lower bounds on the flux, to execute constraints from measured extracellular rates and impose a biological objective on the model, such as maximum growth rate or maximum ATP production.

8.4.2 Isotopic, Steady-State Flux Analysis (13C-MFA)

Metabolic flux analysis based on isotopic 13C tracing (13C-MFA) represents a cutting-edge approach for determining metabolic flux in systems characterized by metabolic steady state (Antoniewicz 2018). In this method isotopically labeled compounds such as 13C-labeled glucose can be introduced into the system and allowed to incorporate into the metabolites and labeled and unlabeled metabolites analyzed through tandem mass spectrometry, NMR, and mass spectrometry (Sriram et al. 2004; Nöh et al. 2006; Antoniewicz 2018). In a 13C-MFA investigation employing tracer methodology, cells undergo cultivation with a precisely labeled 13C-tracer glucose for an extended duration, typically exceeding 3 h. This extended period leads to the integration of 13C atoms into various metabolic intermediates and products. Consequently, the assessments of 13C-labeling serve as a means to constrain flux estimations (He et al. 2014). This flux analysis addresses the nonlinear least squares regression challenge in measuring isotopic steady-state flux by determining a set of feasible intracellular metabolic fluxes that minimize the discrepancies between extracellular rate measurements (*r*) and the variance-weighted sum of squared residuals (SSR) derived from the disparity between predicted and measured 13C-labeling measurements. This method assumes the existence of constant isotopic labeling and metabolic fluxes over time, as well as the presence of both isotopic and metabolic steady states (Antoniewicz 2015). In an experimental setting, glycolytic metabolites, when exposed to 13C-labeled glucose in mammalian cells, can achieve isotopic steady state within 3 h, as demonstrated by Antoniewicz in 2018. In contrast, TCA cycle metabolites may necessitate more than 24 h to reach isotopic steady state under similar conditions. However, when [13C-labeled glutamine] is employed, TCA cycle metabolites typically attain isotopic steady state within 3 h, whereas glycolytic metabolites may exhibit a longer time requirement (Ahn and Antoniewicz 2013; Antoniewicz 2015, 2018). On the other hand, a pivotal benefit of 13C-MFA for quantifying flux is the abundance of redundant measurements, providing a robust basis for accurate flux estimation.

8.4.3 Isotopic, Non-steady State (13C-NMFA) Flux Analysis

It has been demonstrated that using 13C-MFA to study the metabolism of some living or biological systems is restricted by the postulation of an isotopic steady state. For instance, isotopic steady-state labeling measurements in autotrophic systems do not reveal flux information, whereas in mammalian cell cultures, it can take long times (hours or days) to achieve isotopic steady state. An innovative technique known as isotopic non-stationary 13C-MFA (or 13, C-NMFA) was created to address these problems (Yang et al. 2005; Antoniewicz 2018). The methods used in 13, C-NMFA and 13, C-MFA for parameter estimation using nonlinear least squares regression are similar. In addition to metabolic flux estimation, 13, C-NMFA fits metabolite pool sizes to explain the observed the labeling transients. Compared to 13C-MFA, 13C-NMFA requires a significantly longer computation time. The implementation of the EMU modeling framework has led to a substantial decrease in computation time for a specific metabolic network model encompassing all central carbon metabolism, reducing it from orders of magnitude to less than an hour (Choi and Antoniewicz 2019).

8.4.4 Dynamic Metabolic Flux Analysis (DMFA)

In recent years, multiple novel methods for dynamic metabolic flux analysis (DMFA) have been developed (Niklas et al. 2010; Ahn and Antoniewicz 2013; Antoniewicz 2015), all of which seek to use time-course data to identify temporal transients in metabolic fluxes. DMFA aims to comprehend metabolic alterations in systems experiencing non-steady metabolic states, employing time series data encompassing extracellular concentration and rate measurements.

Unlike intracellular metabolism, which usually takes a few seconds to minutes to reach a pseudo-steady state in terms of metabolism. It is generally assumed by DMFA methods that flux transients in cell culture occur relatively slowly, usually within hours. With this presumption, the three steps listed below can be used to complete DMFA analysis:

1. There are distinct time intervals within the experiment.
2. Each period's external rates (average data) are determined by taking the derivative of the external concentration measurements.
3. Using classical MFA, fluxes (average data) are calculated for every interval.

The evaluations conducted at different time intervals throughout the cultural process are aggregated to generate a temporal representation of flux transients in steady-state models (Niklas et al. 2011).

A recently introduced and improved approach to DMFA has been revealed, enabling the direct fitting of complete time series data. This method measures flow dynamics for the entire culture in a single step, eliminating the requirement for data pre-processing. An advantage of this approach is that it allows for the description of flux transients without manual selection of time intervals by the user. Instead, rigorous statistical criteria are employed to automatically discern the extent of dynamic information within the data. The key dominance of DMFA over stationary MFA lies in its capacity to capture information about metabolic transients, which classical MFA cannot observe, requiring minimal additional experimental or computational effort (Leighty and Antoniewicz 2011; Antoniewicz 2015).

8.5 Conclusion

The consensus is that metabolic flux analysis stands as a valuable analytical tool in the field of metabolic engineering. Presently, the tracer technique 13C-MFA is commonly employed to assess fluxes in various prokaryotic systems (Tang et al. 2009).

However, due to difficulties associated with metabolite compartmentation and metabolic reactions, conducting 13C-MFA in eukaryotic systems remains a challenging endeavor (Wahrheit et al. 2011). Analyzing flux in mammalian systems poses challenges due to the slow labeling dynamics involving the exchange of intra- and extracellular metabolites. Consequently, maintaining metabolic steady states for extended durations proves to be a formidable task in this system. However, recent advancements have made it feasible to construct comprehensive metabolic flux maps for "CHO cells" by employing both non-stationary 13C-NMFA and stationary 13C-MFA isotopic methods (Ahn and Antoniewicz 2013; Wiechert and Nöh 2013). The difficulties associated with compartmentalization in flux analysis have not received enough attention. No well-defined procedures for calculating compartment-specific fluxes have been established in eukaryotic systems and the effectiveness of 13, C tracers in addressing cellular compartmentation problems remains to be determined. Establishing parallel pathways across multiple compartments with current methodologies can be challenging when reliable extraction techniques for isolating compartment-specific metabolite pools and 13C-labeling are unavailable. Additionally, new analytical methods to supplement available data for flux estimation in compartmentalized network frameworks and new computational methods for designing the most efficient tracer experiments are required. Ultimately, enhanced techniques are essential for obtaining dynamic flux maps. Previous studies have introduced several DMFA techniques by analyzing dynamic extracellular metabolite data, effectively modeling dynamic behaviors and metabolic changes in cell cultures. To elucidate intracellular metabolic flux maps amidst dynamically changing culture conditions, novel 13C-DMFA techniques that comprehensively integrate dynamic metabolite concentrations and 13C-labeling data will be necessary.

8.6 Future Directions

Fluxomics is a field that is constantly changing as technology and computational techniques advance. Future studies might concentrate on single-cell fluxomics, dynamic fluxomics to record real-time metabolic responses, and the fusion of multi-omics data to produce comprehensive models of cellular function.

References

Ahn WS, Antoniewicz MR (2013) Parallel labeling experiments with [1, 2-13C] glucose and [U-13C] glutamine provide new insights into CHO cell metabolism. Metab Eng 15:34–47

Antoniewicz MR (2015) Methods and advances in metabolic flux analysis: a mini-review. J Ind Microbiol Biotechnol 42(3):317–325

Antoniewicz MR (2018) A guide to 13C metabolic flux analysis for the cancer biologist. Exp Mol Med 50(4):1–13

Bonarius HP, Schmid G, Tramper J (1997) Flux analysis of underdetermined metabolic networks: the quest for the missing constraints. Trends Biotechnol 15(8):308–314

Cascante M, Marin S (2008) Metabolomics and fluxomics approaches. Essays Biochem 45:67–82

Choi J, Antoniewicz MR (2019) Tandem mass spectrometry for 13C metabolic flux analysis: methods and algorithms based on EMU framework. Front Microbiol 10:31

Cortassa S, Caceres V, Bell LN, O'Rourke B, Paolocci N, Aon MA (2015) From metabolomics to fluxomics: a computational procedure to translate metabolite profiles into metabolic fluxes. Biophys J 108(1):163–172

Crown SB, Antoniewicz MR (2013) Publishing 13C metabolic flux analysis studies: a review and future perspectives. Metab Eng 20:42–48

Dauner M (2010) From fluxes and isotope labeling patterns towards in silico cells. Curr Opin Biotechnol 21(1):55–62

Emwas AH, Roy R, McKay RT, Tenori L, Saccenti E, Gowda GAN, Raftery D, Alahmari F, Jaremko L, Jaremko M, Wishart DS (2019) NMR spectroscopy for metabolomics research. Metabolites 9:123

Emwas AH, Szczepski K, Al-Younis I, Lachowicz JI, Jaremko M (2022) Fluxomics-new metabolomics approaches to monitor metabolic pathways. Front Pharmacol 13:805782

Giraudeau P (2020) NMR-based metabolomics and fluxomics: developments and future prospects. Analyst 145(7):2457–2472

Guo J, Zhang M, Elmore CS, Vishwanathan K (2013) An integrated strategy for in vivo metabolite profiling using high-resolution mass spectrometry based data processing techniques. Anal Chim Acta 780:55–64

He L, Xiao Y, Gebreselassie N, Zhang F, Antoniewicz MR, Tang YJ, Peng L (2014) Central metabolic responses to the overproduction of fatty acids in Escherichia coli based on 13C-metabolic flux analysis. Biotechnol Bioeng 111(3):575–585

Heux S, Bergès C, Millard P, Portais JC, Letisse F (2017) Recent advances in high-throughput 13C-fluxomics. Curr Opin Biotechnol 43:104–109

Leighty RW, Antoniewicz MR (2011) Dynamic metabolic flux analysis (DMFA): a framework for determining fluxes at metabolic non-steady state. Metab Eng 13(6):745–755

Long CP, Antoniewicz MR (2014) Metabolic flux analysis of Escherichia coli knockouts: lessons from the Keio collection and future outlook. Curr Opin Biotechnol 28:127–133

Niklas J, Schneider K, Heinzle E (2010) Metabolic flux analysis in eukaryotes. Curr Opin Biotechnol 21(1):63–69

Niklas J, Schräder E, Sandig V, Noll T, Heinzle E (2011) Quantitative characterization of metabolism and metabolic shifts during growth of the new human cell line AGE1. HN using time resolved metabolic flux analysis. Bioprocess Biosyst Eng 34:533–545

Nöh K, Wahl A, Wiechert W (2006) Computational tools for isotopically instationary 13C labeling experiments under metabolic steady state conditions. Metab Eng 8(6):554–577

Ong SE, Foster LJ, Mann M (2003) Mass spectrometric-based approaches in quantitative proteomics. Methods 29(2):124–130

Provost A, Bastin G (2004) Dynamic metabolic modelling under the balanced growth condition. J Process Control 14(7):717–728

Salon C, Avice JC, Colombié S, Dieuaide-Noubhani M, Gallardo K, Jeudy C, Ourry A, Prudent M, Voisin AS, Rolin D (2017) Fluxomics links cellular functional analyses to whole-plant phenotyping. J Exp Bot 68(9):2083–2098

Sanford K, Soucaille P, Whited G, Chotani G (2002) Genomics to fluxomics and physiomics—pathway engineering. Curr Opin Microbiol 5(3):318–322

Sriram G, Fulton DB, Iyer VV, Peterson JM, Zhou R, Westgate ME, Spalding MH, Shanks JV (2004) Quantification of compartmented metabolic fluxes in developing soybean embryos by employing biosynthetically directed fractional 13C labeling, two-dimensional [13C, 1H] nuclear magnetic resonance, and comprehensive isotopomer balancing. Plant Physiol 136(2):3043–3057

Tang YJ, Martin HG, Myers S, Rodriguez S, Baidoo EE, Keasling JD (2009) Advances in analysis of microbial metabolic fluxes via 13C isotopic labeling. Mass Spectrom Rev 28(2):362–375

Toya Y, Shimizu H (2013) Flux analysis and metabolomics for systematic metabolic engineering of microorganisms. Biotechnol Adv 31(6):818–826

Wahrheit J, Nicolae A, Heinzle E (2011) Eukaryotic metabolism: measuring compartment fluxes. Biotechnol J 6(9):1071–1085

Wiechert W, Nöh K (2013) Isotopically non-stationary metabolic flux analysis: complex yet highly informative. Curr Opin Biotechnol 24(6):979–986

Wiechert W, Möllney M, Petersen S, De Graaf AA (2001) A universal framework for 13C metabolic flux analysis. Metab Eng 3(3):265–283

Winter G, Krömer JO (2013) Fluxomics–connecting 'omics analysis and phenotypes. Environ Microbiol 15(7):1901–1916

Yang J, Wongsa S, Kadirkamanathan V, Billings SA, Wright PC (2005) Metabolic flux distribution analysis by 13C-tracer experiments using the Markov chain–Monte Carlo method. Biochem Soc Trans 33(Pt 6):1421–1422

Multi-omics in Gut Microbiome

9

Anshu Mathuria, Anushika Chaudhary, Harshita Sharma, and Indra Mani

9.1 Introduction

Joshua Lederberg, a molecular biologist, coined the term "microbiome" in 2001 to define the ecological system of microorganisms that resides inside the human body and are commensal, symbiotic, or possibly harmful (Lederberg 2001). These days, the phrase "microbiome" usually refers to a cluster of the genomes of microorganisms that inhabit a certain niche; however, these kinds of microorganisms, which include bacteria, fungi, viruses, archaea, and parasites (Gupta et al. 2022). The term "microbiota" refers to all of these together (Liu 2016). About 10–100 trillion microbial cells live in a symbiotic relationship with each individual, mainly gut bacteria, and are known as the human microbiota. The genes that these cells carry are known as the microbiome of humans (Turnbaugh et al. 2007). Microbiota produce specific vitamins and amino acids, boost immunity, and degrade potentially harmful dietary ingredients (den Besten et al. 2013). Every individual has a distinct microbiota network that is initially shaped by their genetic makeup. In the 1680s, Antonie van Leeuwenhoek was the first to examine and compare human fecal and oral microbiota, noting variations between these environments and among individuals in different health states (Dobell 1920; van Leeuwenhoek 1684). Traditional studies of the microbiomes of gut have mostly relied on cultivation methods. However, only 10–30% of the gut microbiota can be studied using conventional methods (Suau et al. 1999; Sokol and Seksik 2010; Tannock 2001). Recent advances in research have revolutionized the ability to assess microbial communities across settings using a single phylogenetic framework, thanks to culture-independent approaches (Pace 1997). Through the integration of the data analytic methodologies such as multi-omics with bioinformatic and environmental DNA sequencing techniques,

A. Mathuria · A. Chaudhary · H. Sharma · I. Mani (✉)
Department of Microbiology, Gargi College, University of Delhi, New Delhi, India
e-mail: indra.mani@gargi.du.ac.in

I. Mani, V. Singh (eds.), *Multi-Omics Analysis of the Human Microbiome*,
https://doi.org/10.1007/978-981-97-1844-3_9

researchers have initiated the process of deciphering the makeup, variety, and role of microbiomes (The Human Microbiome Project Consortium 2012; Sunagawa et al. 2015; Thompson et al. 2017). Multi-omics data of microbiome refers to the collection of several high-dimensional biological data from a microbiome sample, such as 16S rRNA, metagenomic, metatranscriptomic, and metabolomics data (Rodrigues et al. 2018; Wang et al. 2019; Mani 2020a).

There are four pivotal omics methodologies employed in investigating cellular functions: Metagenomics, which deals with DNA; proteins are the main focus of proteomics; minuscule particles or metabolic products are the focus of metabolomics and RNA is the target of transcriptomics. To delve into the intricate workings of the microscopic ecosystem, meta-omics techniques are utilized to scrutinize extensive gene or protein expressions as well as the compositions of metabolites (Xiao et al. 2017). Leveraging multi-omics methodologies can help pinpoint bacterial enzymes and little substances that impact the biological functioning of the host.

The term "metagenomics" refers to the direct genetic examination of genomes included in environmental samples (Handelsman et al. 1998). Compared to phylogenetic surveys, which frequently focus solely toward the variation within a particular gene, like the conserved 16S rRNA gene, studies like metagenomics offer a glimpse into the function-based gene profile of microbial population and as a result provides a far broader description. Utilizing a variety of genomic technologies along with bioinformatics tools, it gives direct access to the genetic makeup of entire microorganism communities (Thomas et al. 2012). Experimental workflow for metagenomics and metatranscriptomics analysis of the human microbiome is given in Fig. 9.1 (Bikel et al. 2015).

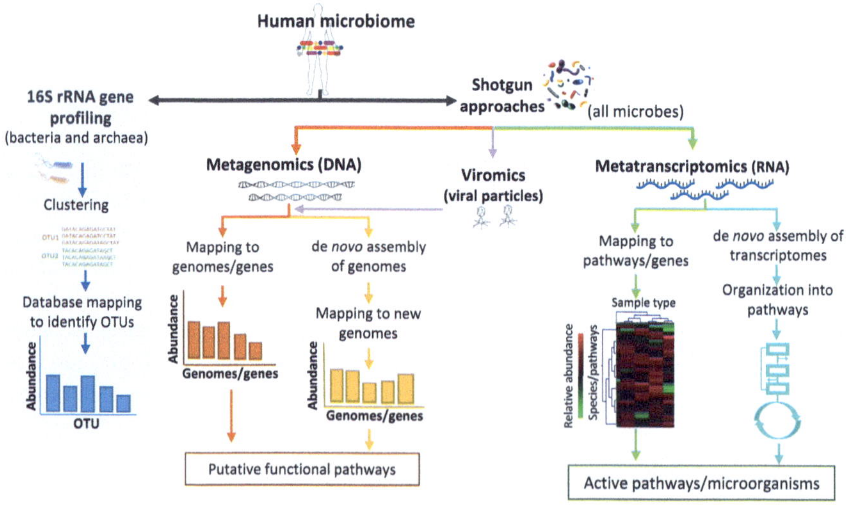

Fig. 9.1 Experimental workflow for metagenomics and metatranscriptomics analysis of the human microbiome. (Adapted from Bikel et al. 2015)

Metagenomics can provide genetic information on unique or potentially catalysts or enzymes produced by gut microbiota, as well as evolutionary histories of community function and structure and genomic connections between function and phylogeny of uncultured organisms. To define expressed activities in the gut, metatranscriptomic or metaproteomic techniques are also used in addition to it (Wilmes and Bond 2006; Gilbert et al. 2008). Metatranscriptomics involves examination of the combined transcriptomes within a specific environment or habitat (Filiatrault 2011). To generate data for metatranscriptomics, a series of steps is followed, including the collection of samples from the habitat, extraction of RNA, preparation of sequencing libraries, and subsequent sequencing of these libraries. After sequencing, the data is analyzed using bioinformatics techniques, as depicted in Fig. 9.2. Ultimately, the data that has been analyzed can be used to examine the expression patterns of genes within a population and determine what parts of that community are active (Filiatrault 2011; Shakya et al. 2019; Zhang et al. 2021).

By contrasting how genes express themselves across time, through clinical parameters, and under various circumstances, it is possible to uncover changes in microbial responses and activities. When the transcriptome profiling in human is also profiled simultaneously, it becomes feasible to unravel the dynamics between the organism and the community of microbes through a direct comparison of their expression networks. However, like many emerging technologies, metatranscriptomics has issues that need to be resolved in terms of both technology and samples. Metaproteomics, a category proteomics, has now emerged as a vital technique for

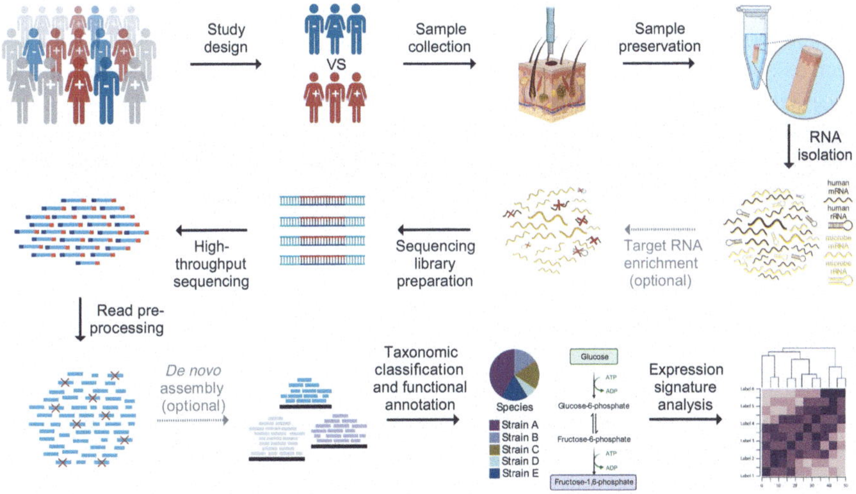

Fig. 9.2 Metatranscriptomics research involves selecting a study design, collecting samples, isolating RNA, and then inferring the target RNA through shotgun sequencing. Immediate sample processing is crucial due to the instability of microbial RNA. Enrichment steps, such as reducing rRNA or depleting host mRNA, can be employed. Computational analysis includes pre-processing, taxonomic and functional annotation, and optional de novo assembly and statistical analysis. Various bioinformatics tools are available for these tasks. (Adapted from Ojala et al. 2023)

the global functional assessment of a microbiome. It is increasingly being used in the research of digestive and metabolic illnesses, and it gives rich information to determine dysbiosis of the human gut microbiome in relation to well or disease states, in order to clarify the molecular mechanisms driving host-microbiome interactions. Despite major technological obstacles remaining, this emerging method has been shown to offer critical data in investigating functional modifications in the human microbiome, in addition to metagenomics and metatranscriptomics (Pan and Chen 2020). Metabolomics is the complete evaluation of metabolites present in a biological sample, and it is an emerging field of research with potential to influence precision medicine. Metabolomics has the potential to present a more precise depiction of the valid physiological state of the cell by examining the entire set of metabolites in a microbe and tracking the worldwide result of relationships among its growth processes and the surroundings (Clish 2015). The metabolome can be studied and analyzed thanks to recent technological advancements and post-genomic discoveries. As a result of this distinctive contribution, metabolomics has become one of the "omics" platforms used by numerous scientific disciplines.

Bacterial cells that reside in the gastrointestinal tract (GIT) of humans exceed host cells ten times over, and the genes expressed by these bacteria outweigh host genes beyond a hundred times. These bacteria or microbes associated with digestive tract of the human are referred to as the gut microbiome. A thorough investigation was conducted on the human gut microbiome and its relevance in both health and disease, showing its contribution to metabolic processes, nutrition, physiological processes, and in immunity. Imbalances in the typical gut microbiota might be linked to gastrointestinal disorders such as inflammatory bowel disease (IBD) and irritable bowel syndrome (IBS), along with broader systemic symptoms of ailment such as obesity, type 2 diabetes mellitus (T2M), as well as atopy. Gut bacteria produce vital components impacting health. Metabolite profiles from the gut offer insights into how diet affects illnesses. Studying biofluids' metabolites helps identify, quantify, and understand microbial-produced molecules. Mass spectrometry (MS) and nuclear magnetic resonance (NMR) are key in this research, fostering new biomarker creation for personalized therapies.

9.2 Gut Microbiome

A typical description of the microbiome of the gut is the entirety of protozoa, viruses, bacteria, fungi, and other microorganisms as well as their combined genetic composition that are found within the gut. All microorganisms that live in the GIT—both commensal and pathogenic—make up the microbiota of the gut. It has been investigated over the years for possible interactions between gut microbes and hosts, including impacts on immunological, metabolic, and neuroendocrine responses. Russian scientist Eli Metchnikoff was first to link the longer lifespan of Bulgarians living in rural areas to the use of products made after fermentation of milk in the early 1900s. He postulated that consuming fermented dairy products could introduce beneficial bacteria into the gut, hence preventing the growth of dangerous microbes which leads to extended lifespan. He also proposed that beneficial bacteria may be substituted for bad bacteria in order to alter the gut microbiome (Cresci and Bawden 2015).

The majority of biomass and diversity in the environments linked to humans is made up of bacteria, however, eukaryotes and archaea are also present in lower quantities and should not be disregarded (Marchesi 2010; Breitbart et al. 2008). So bacteria are found throughout the body, encompassing the GIT, skin, vagina, mouth cavity, and placenta, a significant portion of bacteria are found in the GIT, with the colon housing the most of the bacteria, which are mainly anaerobic. Approximately one trillion microorganisms are found in the gut of humans, making it habitat of a large and diverse microscopic ecosystem (Thursby and Juge 2017). Based on estimates, the colon contains between 10^{11} and 10^{12} bacterial cells per milliliter, making it one of the planet's most dense known environments for microbes (Ley et al. 2006). The Human Gut Microbiome Project, which preceded the Human Microbiome Project, has increased the understanding of microbial ecology which exists in the human GIT (Backhed et al. 2005). There are over three million prokaryotic genes that are expressed by the gut microbiota, which shows scope of bacterial presence and possible host impacts. The four primary phyla that make up the gut microbiota are Bacteriodetes, Firmicutes, Proteobacteria, and Actinobacteria (Belizário and Napolitano 2015). Other than these, *Prevotella* and *Ruminococcus* are also present in the gut. In the colon, bacteria like *Lactobacillus*, *Clostridium*, *Bifidobacterium*, and *Peptostreptococcus* are present, which are anaerobic in nature (Arumugam et al. 2011).

Viruses are also present in large number in gut and the overall count of RNA and DNA viruses present in gut represent the virome of the gut, which is at least equal to that of bacterial cells (Reyes et al. 2012), yet they might surpass the quantity of bacterial cells by a factor of 20 on the surfaces and in the mucous layers of the gut (Barr et al. 2013). An estimated 10^8–10^9 virus-like particles (VLPs) are present in every gram of intestinal material. In addition to the genomic variation within the host, viruses can alter the phenotypic traits of their host, which in the case of humans, through interactions across different kingdoms. These virus-induced phenotypic alterations may confer benefits to the host or raise the danger of illness (Hunter 2013; Villarreal 2012). The majority of the virome has not yet been described; less than 1% of it is believed to have been sequenced (Virgin 2014).

9.2.1 Variability in Gut Microbiome

An average healthy person's GI is home to trillions of microbial lives. However, the microbiomes of two healthy individuals could differ greatly (Ding and Schloss 2014). Both internal and extrinsic microbiome elements are to blame for this. The former are contingent upon the condition of the microbiome, for example, interactions between taxa. The different environmental tiers that affect or interact with the microbiome of the gut are recognized as extrinsic factors (Schmidt et al. 2018).

From the analysis of samples with different lifestyles, relationships between life-history traits and the microbiome makeup have been found (Ding and Schloss 2014). Numerous elements, including the way an infant is born and fed, the aging process, the makeup of the diet, the place of residence, the use of medications, and stress, all have an impact on the microbiome (Cresci and Bawden 2015). It was once thought the gut was uncontaminated while the fetus was still inside the mother and that delivery marked the initial establishment and emergence of the microbiome in the gut (Salazar et al. 2014). Recent research, however, has refuted this notion by showing that the amniotic fluid, the placenta, and the umbilical cord all contain microbes (DiGiulio 2012; Jiménez et al. 2005; Aagard et al. 2014). It is contemplation that the fetus colonizes the growing GIT by ingesting amniotic fluid and its microorganisms while still in the uterus (Moles et al. 2013). After birth, breast milk is crucial for the development of the gut microbiota. Breast milk contains up to 600 distinct bacterial species, including the health-promoting *Bifidobacterium adolescentis, Bifidobacterium longum, B dentium,* and *B breve*, and not at all sterile (Martín et al. 2009). Through their ability to combat pathogens, *Bifidobacteria* have been associated with enhancing the protection of mucosa of the gut (Fukuda et al. 2011). It has been demonstrated that *bifidobacteria* enhance immunoglobulin A synthesis, which is associated with gut immune system modulation (Ouwehand et al. 2002).

The gut microbiota's diversity alters during the early stages of human development and is impacted by nutrition, as revealed by metagenomic research (Koenig et al. 2011). Also, while it is likely that host-associated microorganisms are obtained from their surroundings, the makeup of the microbiota of mammals, particularly of the intestine, differs significantly from those of microbial communities that exist freely. This implies that over hundreds of millions of years, coevolution among microbial groups and their vertebrates has been chosen for a specific microbial population that thrives in the gut's warm, steady, and eutrophic conditions (Ley et al. 2008a, b).

9.2.2 Importance of Gut Microbiome

It carries out a few fundamental tasks in the human body's immunological, metabolic, structural, and neurological systems (Adak and Khan 2019). It significantly influences the absorption of nutrients and minerals, the production of short-chain fatty acids (SCFAs), vitamins, enzymes, and amino acids. Fermentation byproducts including acetate, propionate, and butyrate are essential for gut health, because they give epithelial cells energy, improve the integrity of the barrier of epithelium, and support immune modulation and pathogen defense. It plays a crucial immunological function by preventing the colonization of the pathogenic bacteria by making use of available resources like bacteriocins. Additionally, by preserving the structural strength of the gut epithelium, gut microbiota inhibits the infiltration by bacteria (Khosravi and Mazmanian 2013). It also uses a range of competing strategies, including food consumption, pH regulation, the release of antimicrobial peptides, and impacts on pathways for cell signaling to prevent pathogenic bacteria.

An individual's microbiota of gut has a major influence on their physical and emotional well-being (Adak and Khan 2019). It is now acknowledged to have a part in a number of disorders. These comprise diseases related to autoimmunity like multiple sclerosis and autism spectrum disorder, as well as cancer. A thorough knowledge of the function of the intestinal microbiotas in health has produced some very promising advances in therapies, including medicines, probiotics, prebiotics, as well as fecal transplantation (Adak and Khan 2019).

Advances in genome sequencing technology, bioinformatics, and culturomics have now made it possible for researchers to investigate the microbiota, and specifically its roles, at a deeper level than they were previously able to.

9.3 Metagenomics in Gut Microbiome

The microbiome of the human gut, which is made up of about 1000 rare and roughly 200 common bacterial species, functions normally like a multicellular organ (Ley et al. 2008a, b). In mutually advantageous ways, an organism's gut microbiota enhances its biology (Backhed et al. 2005). The incapacity to cultivate the majority of gut microbial species, however, presents a significant obstacle to the research of gut microbiota (Siezen and Kleerebezem 2011). There have been a number of prior attempts in this area. However, considering the recent introduction of several sequencing technology, it has become easier to analyze a vast range of microbes in diverse environments (Venter et al. 2004; Tyson et al. 2004; Tringe et al. 2005) and parts of the human body (Ding and Schloss 2014), including gut of human (Human Microbiome Project Consortium 2012; Tyakht et al. 2013). The study of uncultivated gut microbial communities has made use of two efficient DNA sequencing techniques like sequencing of 16S rDNA and shotgun metagenomics. The former has shown a number of novel linkages between the gut microbiota configuration and disease (Blaser et al. 2013; Cho and Blaser 2012; Ren et al. 2013) and focuses on sequencing the universally preserved 16S rDNA gene found in all microorganisms (Cole et al. 2007; Woese and Fox 1977). 16S rDNA-based research aims to identify "who's there" within a specific community of microbes, whereas shotgun metagenomic may tackle the corresponding query of "what can they do?" (Lepage et al. 2013; Handelsman 2004).

The analysis of genetic material that has been directly isolated from environmental samples such as gut, soil, and water is known as metagenomics (Ley et al. 2008a, b). Handelsman and Rodon (Handelsman et al. 1998; Rondon et al. 1999) were the first to report on metagenomics in 1998, and it has since grown to be another DNA sequencing method for examining the intricate gut microbial ecology. By randomly sequencing every piece of DNA that is taken out of the sample, it seeks to catalog every gene from a community (Gill et al. 2006; Qin et al. 2010; Turnbaugh et al. 2009). It has helped in understanding the richness within the human gut microbiota, finding new genes, and identifying the origin of dysbiosis (Wang et al. 2015).

Metagenomic studies of two types can be used to understand gut microbiota: descriptive metagenomics and functional metagenomics. The measure of degree of microbial prevalence depending on several biological and external variables is known as descriptive metagenomics, and it is used to show the microbiome's variety and community structure. Whereas research on the interactions between microbes and hosts with the goal of developing a reliable evolving ecosystem model to represent a relationship between a microbe's or a community's identity is known as functional metagenomics (Xia et al. 2011; Garmendia et al. 2012; Faust and Raes 2012; Chistoserdovai 2010). More and more research of human gut microbiome has been centered on functional metagenomics in recent years (Qin et al. 2010; Turnbaugh et al. 2009; Yatsunenko et al. 2012; Koenig et al. 2011). Given that numerous types of bacterial genes can be expressed in *Escherichia coli* (*E. coli*), this host is most frequently utilized for functional metagenomics (Handelsman 2004). Additionally, for the gene expression, Gram-positive bacterial species like *Lactococcus lactis, Streptomyces,* and *Bacillus subtilis* are used (Gabor et al. 2004).

9.3.1 Sequencing Strategies Used in Metagenomics

In general, two types of sequencing methodologies are included in the field of metagenomics: shotgun sequencing, which records all the DNA in a sample, and amplicon sequencing, which often uses phylogenetic marker like 16S rRNA gene (Morgan and Huttenhower 2012).

For microbiome analysis, the gene encoding 16S ribosomal RNA has been a successful and economical phylogenetic marker that even permits the estimation of functional content based on species abundances. However, mistakes in preparing samples and sequencing might introduce bias into PCR-based phylogenetic marker methods. Technical discrepancies in microbiome composition appear to be mostly caused by the selection of 16S rRNA gene hypervariable regions to focus on for sequencing (Hiergeist et al. 2015; Tremblay et al. 2015; Gohl et al. 2016). Moreover, with regard to the database and classifiers employed, amplicon sequencing of 16S rRNA gene is restricted to genus-level classification by taxon (Walsh et al. 2018) and yields very little functional information. There is now more interest in creating more dependable techniques for preparation of amplicon-based library as well as sequencing due to these acknowledged constraints of microbiological analysis based on amplicon, which have questioned the accuracy and consistency of studies using 16S rRNA as phylogenetic markers (Faith et al. 2013; Gohl et al. 2016).

On the other hand, the benefit of classifying microbe at the species and strain levels is provided by shotgun metagenomics (or whole genome shotgun sequencing). Furthermore, by explicitly identifying the functional content of samples, it permits researchers to investigate the functional interactions between hosts and microbes (Walsh et al. 2018; Jovel et al. 2016) and to explore as-yet-undiscovered microbial life that might have been remained unclassifiable (Rinke et al. 2013). Shotgun metagenomics has not, however, been widely used for microbiome investigation due to its relatively high costs and more stringent bioinformatic criteria (Morgan and Huttenhower 2012; Walsh et al. 2018).

In response to the need for more affordable and quicker metagenomic sequencing methods many sequencing methods have been evolved so far. For example, first generation Sanger platforms, such as Applied Biosystems, was followed by the second generation of 454 Life Sciences Roche platforms and Illumina platforms. Lastly, single-molecule real-time (SMRT), a third-generation sequencing method was introduced by Pacific Bioscience (Markowitz et al. 2012). The second and third generation sequencing are part of next generation sequencing (NGS), which has developed into an effective method for studying the microbiome, enabling the characterization of microbial communities at an previously unattainable level of detail devoid of necessity for prior culturing (Escobar-Zepeda et al. 2015; Mani 2020b).

9.3.2 How Is Metagenomics Performed?

First, all DNA of different microorganisms present in the gut is retrieved in whole from fecal samples. Acquired DNA samples are then sheared arbitrarily using a "shotgun" method prior to sequencing. Following that, the extensive sequences are examined to produce whole genome-based genomic profiling (Tringe et al. 2005) or species analyses determined by phylogenetic markers (16S rDNA) (Sunagawa et al. 2013). By using metagenomics, the shotgun sequence reads undergo screening to produce high-quality sequences of the entire genomic profile. The obtained sequences are then put together to create contigs of genomic sequences based on sequence overlaps. Continuous sequence coding requires computational techniques. Next, advanced algorithms for data analysis and database searches are applied to annotate genes (Thomas et al. 2012).

Numerous bioinformatic techniques have been created to manage sequence data to better assess taxonomic diversity and functional capacity. The main functions of genome databases appropriate for taxonomic categorization include annotation, mining, and preservation of single-sequenced bacterial genomes. The International Nucleotide Sequence Database Collaboration (INSDC www.insdc.org), which is composed of the European Nucleotide Archive (ENA), the National Center for Biotechnology Information (NCBI), and the DNA Data Bank of Japan (DDBJ), is the primary worldwide resource for genome sequencing data from all realms of life (Yen and Johnson 2021). Because substantial genome sharing and interrelationships allow for variable levels of curation among various resources, metagenomic reference databases are fundamentally built on this form of linkage. For instance, the NCBI RefSeq database (Haft et al. 2018; O'Leary et al. 2015) has been put together using GenBank to offer a unique collection of top-quality genomes with clear classification and comprehensive annotations (Sayers et al. 2020), an INSDC-supported database. There is an extensive correlation of sequencing data throughout annotated databases which is a helpful tool for connecting disparate complementary data sources.

9.3.3 Challenges Faced During Metagenomics

The characterization of the gut metagenome using metagenomics is fundamentally hampered by two issues. The first is the size of ever-expanding sequencing datasets. The second is the material that in metagenomes is most frequently mentioned as its "dark matter." Because of their immense size and complexity, the human gut metagenomes have not yet been fully defined despite enormous attempts to sequence and examine them.

Additionally, as existing sequencing techniques are limited, the sequences derived from metagenomic research are incomplete and fragmented, ranging from 20 to 700 base pairs. A sample contains a wide variety of species, most of which lack a complete genome. As a result, it is not feasible to identify the species from which a given genome sequence originated. Furthermore, compared to sequencing a single genome, the number of sequence-based data acquired from metagenomic sequencing is significantly bigger (Wooley et al. 2010). Also because of variations in the underlying bioinformatic techniques used, taking alternative paths to obtain the same information may expose one to unnoticed biases. A further problem is with functional annotation of metagenomic datasets, as many higher-order biological processes documented in databases require manual curation (Yen and Johnson 2021).

9.3.4 Application of Metagenomics

With the help of metagenomics many aspects of the microbiome and microbiota of the gut have come to light. (1) The gut microbiota in healthy individuals is mostly composed of bacteria, primarily Firmicutes and *Bacteroides* (Fernandes et al. 2014). Studies suggest distinct enterotypes marked by *Prevotella, Ruminococcus*, and *Bacteroides* spp., identified through comprehensive metagenomic examination of healthy individuals' stool samples (Arumugam et al. 2011). (2) The gut virome, uncovered by metagenomic sequencing, impacts microorganism behavior but its role in human diseases remains unclear (Kelder et al. 2014). (3) The gut microbiome of humans, studied by the MetaHIT team, revealed 3.3 million genes using metagenomics, surpassing the human gene count (Blottière et al. 2013; Qin et al. 2010). Various elements such as medication, diet, age, geography, and environment impact intestinal microbiome diversity (Wang et al. 2015). (4) Metagenomic research has unearthed an array of carbohydrate-active enzymes (CAZymes) (Gill et al. 2006; Turnbaugh et al. 2009, 2010a, b; Li et al. 2009) and discovered a comprehensive genetic inventory of enzymes like bile-salt hydrolase found in the human gut microbiota (Jones et al. 2008). (5) Metagenomics helps reveal the extensive presence of genes for antibiotic resistance (ARGs) within the human gut microbiome (Salyers et al. 2004; Sommer et al. 2009), providing crucial insights into resistance dynamics of antibiotic (D'Costa et al. 2006). (6) Metagenomics helps identify dysbiosis, a reduction in microbial diversity linked to health issues like IBS, colorectal cancer, liver cirrhosis, and obesity (Wang et al. 2015). (7) Metagenomics unveils changes in microbial functions. In a study on type 2 diabetes patients, differences in gut microbes showed higher functions related to oxidative stress resistance and sulfate reduction in patients compared to healthy individuals (Qin et al. 2010).

Thus genetic makeup of the entire microbiota can be obtained from metagenomic sequencing data, yet it tells little about how gene expression functions. While functional metagenomics could be helpful, the current goal is to find non-abundant, functionally-important genes. Combining metagenomics information with metatranscriptomics, metaproteomics, and metabolomics data is necessary to get glimpses into the host–microbiota relationship's cellular as well as molecular dynamics (Mandal et al. 2015).

9.4 Metatranscriptomics in Gut Microbiome

The microbiota of the human gut is the most diverse microbial ecosystem in humans, consisting of 4600 different bacterial species and 13.9 million distinct gastrointestinal protein clusters with a protein identity criterion of 90% (Almeida et al. 2021). This microbiota is established either before birth (Blaser et al. 2021) or shortly after birth (Rodriguez et al. 2015), undergoes its most significant development in the first 3–5 years of life (Tamburini et al. 2016), and typically maintains a largely steady composition throughout maturity and into old age (Rodriguez et al. 2015). Nonetheless, the makeup of the microbiota in the gut can be altered by a variety of circumstances, including antibiotic usage, dietary changes, and xenobiotic exposure. Following such temporary perturbations, it usually returns to a composition close to its previous form (David et al. 2014a, b; Chen et al. 2021). While it remains stable within individual hosts over time, it exhibits considerable variation between different individuals and populations (Falony et al. 2016; Vangay et al. 2018). Furthermore, the gut microbiota can undergo significant alterations in response to various diseases, either as a causal factor or as a consequence of the disease itself. Studies utilizing metatranscriptomics to explore the human gut microbiota have yielded a comprehensive knowledge of gut communities in various populations, including healthy infants (Vatanen et al. 2022), adults (Franzosa et al. 2014; Gosalbes et al. 2011; Hatch et al. 2019; Abu-Ali et al. 2018; Mehta et al. 2018; Maurice et al. 2013; David et al. 2014a, b; Wang et al. 2021; Li et al. 2022), and twins (Turnbaugh et al. 2010a, b; McNulty et al. 2011). These investigations have not only illuminated the temporal dynamics of transcription (Abu-Ali et al. 2018; Mehta et al. 2018) but have also provided insights into how the microbiota of gut responds to dietary changes (David et al. 2014a, b; Wang et al. 2021; Li et al. 2022; McNulty et al. 2011) and exposure to xenobiotics (Maurice et al. 2013) in a healthy gut. Crucially, these studies have revealed instances where certain functions and taxa are over transcribed when aligned to their genomic abundance. However, in research that integrate metagenomics and metatranscriptomics, there is often a high connection between transcript and genomic abundances (Franzosa et al. 2014; Abu-Ali et al. 2018; Mehta et al. 2018).

Metatranscriptomics has found extensive application in the investigation of disease-associated gut microbiota, with several recent studies addressing conditions such as IBD (Heintz-Buschart et al. 2016; Lloyd-Price et al. 2019; Schirmer et al. 2018; Ma et al. 2021; Li et al. 2021; Gallardo-Becerra et al. 2020; Granata et al. 2020). Two recent studies specifically centered on IBD (Lloyd-Price et al. 2019; Schirmer et al. 2018) have identified an elevation in the activity of transcription of certain obligate anaerobic microorganisms during IBD. Additionally, there have been noteworthy endeavors examining the microbiota of gut in the context of diabetes (Heintz-Buschart et al. 2016) and chronic systemic inflammation (Ma et al. 2021).

9.4.1 Microbiome mRNA Isolation and Processing

A metatranscriptome experiment involves isolating total RNA from the microbiome, but mRNA of a prokaryote lacks a poly-A tail like eukaryotic mRNA, making traditional synthesis of cDNA unsuitable (Sultan et al. 2014; Wendisch et al. 2001; Sharma et al. 2010). Various approaches, like rRNA removal using magnetic beads attached to specific probes, have been developed (Sultan et al. 2014). This involves annealing probes to rRNA, removing them with a magnet, and using RNase inhibitors to prevent degradation. The enriched mRNA population, representative of active genes, is used to generate a library through cDNA synthesis, adapter ligation, amplification, and sequencing. To identify expressed genes, sequence readings are mapped to reference genomes (Giannoukos et al. 2012).

9.4.2 Metatransciptomics Data Computational Analysis

Millions of mRNA molecules are present in metatranscriptome datasets (RNA-seq reads), requiring automated, high-throughput analyses for meaningful biological insights. Tools like HUMAnN (Abubucker et al. 2012) and MG-RAST (Glass et al. 2010), combined with bioinformatic tools, e.g., BOWTIE (Langmead and Salzberg 2012), Trimmomatic (Bolger et al. 2014), and CuffDuff (Ghosh and Chan 2016), help infer expression levels of genes and changes from raw mRNA reads. Standard steps involve filtering non-mRNA and host reads, quality filtering, open-reading frame (ORF) identification, read mapping, normalization, and gene expression level calculations. Optional assembly of reads into contigs followed by mapping to reference genomes reveals additional gene expression details like gene relations and start-stop sites.

Assembly demands deeper sequencing and is crucial in the absence of reference genomes, common in nonmodel organisms. Annotation of transcripts without reference genomes relies on similarity to annotated proteins using tools like Blast2GO (Conesa et al. 2005). Suites like Trinity aid in full transcriptome reconstruction (Grabherr et al. 2011). Integrating RNA-seq and metagenomics data helps distinguish expressed genes from potentially existing genes. Post-normalization, relative gene expression values undergo statistical analyses similar to those in 16S rRNA gene and sequencing of metagenomes.

9.4.3 Metatranscriptomics in Disease and Health

Improving metatranscriptomics for microbiome analysis is crucial for understanding microbial roles in diseases and health, enabling clinical integration and deeper insights into mammalian physiology (Bashiardes et al. 2016).

9.4.3.1 Understanding Microbiome Impacts on Drug Metabolism and Health

Characterizing active bacteria in the microbiome highlights disease-driving agents. Methods like 16S rRNA detection reveal active phyla (e.g., Firmicutes, Bacteroidetes) and metabolic pathways in healthy individuals (Gosalbes et al. 2011). The combined metagenomics–metatranscriptomics approach uncovers the microbiota of gut impacts on metabolism of drugs, crucial for understanding drug efficacy and toxicity variations among individuals.

The microbiota of the gut plays a pivotal role in metabolization of proteins and carbohydrates through enzyme production. These activities can impact xenobiotic stability, where gut microbiota affect approximately 40 xenobiotics on the market without clearly known mechanisms (Haiser and Turnbaugh 2012). A tragic incident involved sorivudine, an antiviral drug converted by bacteria present in gut into a molecule that prevented the anticancer medication 5-fluorouracil from being metabolized (Sousa et al. 2008), causing its toxic accumulation and leading to the deaths of 18 patients. Despite acknowledging gut bacteria's role in drug metabolism, the specific bacteria and molecular pathways involved often remain unidentified.

Flow cytometry was used in an innovative study to isolate and characterize active bacterial communities in the gut. Characterization using 16S rRNA gene sequencing and metatranscriptomics revealed diverse sets of active gut bacteria, mostly Firmicutes. More than 50% of the microbiota in the gut were highly metabolically active, while the other one-third were inactive but identifiable using metagenomics. Exposure to xenobiotics influenced both the structure and expression of the gene profile of the active microbiome in the gut (Maurice et al. 2013). Digoxin, a cardiac drug, can be inactivated by the metabolism of the microbiome of gut, particularly by specific strains like *Eggerthella lenta*. Transcriptional profiling identified an upregulated cytochrome-encoding operon in these bacteria, predictive of digoxin inactivation. In gnotobiotic mice, higher dietary protein reduced digoxin microbial metabolism, elevating drug serum concentrations. This underscores the gut microbiome's significance in drug efficacy (Haiser and Turnbaugh 2012).

9.4.3.2 Microbiome Impact on Mucosal Immune System

Toll-like receptor 5 (TLR5) knockout (KO) mice studies showcase the interplay between metatranscriptomics, metagenomics, and 16S rRNA characterization (Vijay-Kumar et al. 2010; Carvalho et al. 2012) in understanding microbiota-immune interactions. TLR5-deficient mice exhibit altered microbiota and compromised mucosal barrier function. While 16S rRNA sequencing reveals dysbiosis, metatranscriptomics highlights upregulated flagella motility genes in their gut microbiota compared to wild-type mice. This absence of TLR5 affects antibody production, influencing bacterial motility and compromising the mucosal barrier (Cullender et al. 2013).

9.4.3.3 Antisense RNA in Microbiome

Traditional metatranscriptome analysis focuses on mRNA transcripts to discern activated metabolic pathways, but bacterial transcriptomes reveal complexity. Strand-specific RNA-seq revealed cis-encoded antisense RNAs, which are produced from the opposite DNA strand of genes. These antisense transcripts can interact with sense transcripts, regulating mRNA expression, as seen in various bacteria like *Synechocystis* sp. and *Listeria monocytogenes* (Georg et al. 2009; Toledo-Arana et al. 2007). While prior studies focused on single-cultured strains, implementing global microbiota-scale analyses of antisense RNAs complements traditional metatranscriptomics (Thomason and Storz 2010). In the human gut microbiota, analysis revealed dynamic presence and variation in antisense RNA transcription between individuals, impacting various genes like transposases, defense mechanisms, and housekeeping genes (Bao et al. 2015).

9.4.3.4 Metatranscriptomics in Diagnosis of Respiratory Conditions

Metatranscriptomics offers insights into respiratory conditions by studying diverse communities of microbes in the respiratory tract. In cystic fibrosis (CF) (Lim et al. 2013; Quinn et al. 2014), it reveals patient-specific polymicrobial communities responsive to treatments and transcriptionally unstable, impacting disease progression. For chronic obstructive pulmonary disease (COPD), it links peculiar bacterial activities in samples with severity of disease and host immune responses (Lee et al. 2016; Ren et al. 2018). In asthma, microbial activity in nasal epithelia and specific signatures in infants with bronchiolitis predict childhood asthma risk. Allergic rhinitis studies show altered host expression and microbial diversity, improved through allergen immunotherapy. Furthermore, research is being conducted on respiratory infections, such as COVID-19 and respiratory syncytial virus (Rajagopala et al. 2021), influenza (Zhang et al. 2020), and pneumonia (Shi et al. 2022; Spottiswoode et al. 2022), has benefited from metatranscriptomics, providing comprehensive insights into disease dynamics.

9.4.3.5 Metatranscriptomics in Cancer

Microbes intricately affect tumorigenesis and therapy outcomes, contributing to new cancer cases approximately 13% globally (de Martel et al. 2020). Infection-related cancers vary, including examples like hepatocellular carcinoma and noncardia gastric cancer. The microbiota of gut significantly influences responses to cancer treatment, impacting chemotherapy (Viaud et al. 2013) and immunotherapy efficacy (Routy et al. 2018; Sivan et al. 2015; Andrews et al. 2021; Smith et al. 2022). Intratumoral bacteria's metabolic activities influence tumor resistance to therapy, while microbial virulence factors and toxins can promote tumor growth. Metatranscriptomics studies have unveiled insights into cancer-related microbiotas, associating specific taxa and functions with survival outcomes in cancer patients. Analyses of tumor RNA-sequence data reveal microbial signatures correlating with cancer types, demonstrating potential predictive value for cancer diagnosis from whole blood RNA, yet requiring further exploration.

9.4.4 Challenges in Metatranscriptome Analysis

Numerous challenges are associated with metatranscriptome analysis and deserve attention. (1) *RNA Sample Quality*: Isolating high-quality RNA from certain biological samples (e.g., feces) remains challenging despite experimental strategies to improve isolation methods (Franzosa et al. 2014). (2) *Host RNA Contamination*: Biopsy samples may contain high levels of host RNA, complicating removal strategies and increasing processing costs, potentially affecting downstream analysis. (3) *mRNA Stability*: mRNA, with its short half-life, may not effectively capture rapid or responses that are short-lived to stimuli (Bikel et al. 2015). (4) *mRNA and Protein Correlation*: mRNA presence does not always directly correlate with protein presence or activity, requiring integration with other omics data (metagenomics, metabolomics, metaproteomics) for a comprehensive understanding. (5) *Variability in Analysis*: Even when using equivalent databases, distinct metagenomic analysis techniques can provide varying findings, demanding standardization of RNA isolation, processing, sequencing, and analysis for wider usage. (6) *Integration of Techniques*: Integrating multiple omics datasets could provide a holistic view of the composition of microbiome and function at different layers, enhancing insights into microbial communities.

Traditionally, extensive expression studies relied on methods like microarrays and serial analysis of gene expression (SAGE). These studies were validated using quantitative polymerase chain reaction (qPCR) as an independent technique, regarded as the gold standard (Jeanty et al. 2010; Felix et al. 2009). However, as large-scale expression methods have transitioned to NGS, particularly RNA-seq, it is important not to disregard the validation of findings using qPCR. This approach remains crucial, adding substantial value to observed expression patterns in studies utilizing NGS technologies (Fang and Cui 2011).

9.5 Metaproteomics in Gut Microbiome

The microbiome of the human gut is a vital and intricate environment that coexists in a symbiotic relationship with us, playing a vital role in digestion. Its components, functions, balance, and interactions with the host greatly impact human health. The microbiome is extremely intricate and varied, consisting largely of bacteria from the Bacteroides, Proteobacteria, Firmicutes, and Actinobacteria phyla, making up over 98% of the microbial population (Frank et al. 2007). Proteins drive physiological processes in the genome. Metaproteomics, alongside metagenomics and metatranscriptomics, reveals how microbial communities interact with host biology by globally characterizing proteins and their origins. This approach uncovers translational and post-translational changes, shedding light on functional aspects relevant to disease states and health (Mills et al. 2019; Verberkmoes et al. 2009).

Shotgun metaproteomics using mass spectrometry (MS) is the preferred method for detecting proteins in the microbiota of the gut. Current research focuses on isolating microorganisms from feces, colon biopsies or colonic lavage. Despite variations in study designs and sample diversity, a typical metaproteomic workflow includes bacterial protein extraction, protein/peptide preparation, liquid chromatography-mass spectrometry (LC-MS)/MS analysis, database search, and taxonomic/functional analysis (Pan and Chen 2020). Due to the diverse microbiota, methods like bead-beating, ultrasonication, freeze-thawing, heating, and detergents are combined to enhance the extraction of proteins. Notably, Gram-positive bacteria, with their resistant cell walls, require special attention. Post-translational modifications (PTMs) are also explored in metaproteomics, revealing their uneven distribution across microbial taxonomic and functional categories, influencing bacterial adaptation to environmental conditions (Tanca et al. 2014; Zhang et al. 2016, 2018a, b).

9.5.1 Applications

Metagenomics, metaproteomics, metatranscriptomics, and metabolomics are interconnected, with metaproteomics playing a pivotal role in unraveling the makeup and functions of microbial communities more efficiently. Recent advancements in MS and computational informatics have propelled metaproteomics into a noteworthy approach for characterizing the microbiome of the human gut. Combining the metagenomic analysis with metaproteomics has revealed unique sets of active microorganisms within the gut microbiome across individuals (Erickson et al. 2012). The applications of metaproteome studies include as follows:

9.5.1.1 Bioinformatics in the Study of Microbiomes in the Human Gut

Metaproteomic data processing involves searching databases to identify proteins. The challenge is creating accurate databases for all gut microbial genomes. Synthetic databases help but miss some genomes of the uncultivable microbiota of gut (Kolmeder et al. 2012; Erickson et al. 2012). Solutions include multi-step searches (e.g., MetaPro-IQ) (Zhang et al. 2016) and building comprehensive databases using Human Microbiome Project (HMP) data. Algorithms like SEQUEST (Mills et al. 2019), X!Tandem with MetaProteomeAnalyzer (Muth et al. 2015), Mascot (Li et al. 2016), and Myrimatch (Xiong et al. 2015) are used for metaproteomic database searches.

9.5.1.2 Dietary and Obesity Effects on the Gut Metaproteome

A combination of metabolomics, 16S rRNA gene sequencing, and proteome analysis was used to examine how dietary resistant starch (RS) affected the microbiome of the human gut (Erickson et al. 2012). A customized sequence database was used to identify >50,000 proteins from fecal samples of participants on high-RS and low-RS diets (Maier et al. 2017). The most abundant bacterial families were *Bacteroidaceae*, *Lachnospiraceae*, *Ruminococcaceae*, and *Prevotellaceae*. Human proteins associated with lipid metabolism were elevated and the Firmicutes/Bacteroidetes ratio increased with a high-RS diet.

9.5.1.3 Metaproteomic Research on the Functional Relationship Between the Host Microbiota and Type 1 Diabetes

An autoimmune condition, type 1 diabetes is impacted by both environmental and genetic variables. A metaproteomic study found that people with ongoing islet autoimmunity have decreased levels of exocrine pancreas proteins and increased levels of immune-modulating proteins compared to healthy controls. Integrative analysis revealed that *Alistipes* and *F. prausnitzii* exhibited positive correlations with adhesion molecule CEACAM5, mucous epithelial barrier proteins MUC2 and FCGBP, and a set of exocrine pancreas-derived proteins, while *Prevotella* was inversely correlated with *Bacteroides*, goblet cell-produced CLCA1, and two antibody components (Gavin et al. 2018).

9.5.1.4 Metaproteomic Studies in Gastric Cancer and IBD

The imbalance in the microbiota of gut has been linked to conditions like IBD, which includes Crohn's disease (CD) and ulcerative colitis (UC), as well as IBS (Lee et al. 2017; Matsuoka and Kanai 2015; Mondot and Lepage 2016). Metaproteomics has been used by researchers to investigate alterations in the metaproteome and host–microbiota interactions in relation to IBD. This analysis has been conducted using fecal samples (Mills et al. 2019; Erickson et al. 2012; Lehmann et al. 2019; Presley et al. 2012) or samples from the mucosal-luminal interface (MLI) (Zhang et al. 2018a, b; Li et al. 2016; Presley et al. 2012). A research on twins with or without Crohn's disease revealed reduced bacterial diversity, especially in the Firmicutes phylum, in CD patients' guts. *Faecalibacterium* was negatively associated with an inflammation marker. Metagenomic and metaproteomic data showed significant differences despite consistent taxonomic associations at the genus level (Mills et al. 2019).

In addition to increased expression of human proteins linked to antimicrobial activity, a metaproteomics study conducted on pediatric IBD patients found elevated microbial oxidative stress responses in IBD cases. Interestingly, the study found higher levels of *Faecalibacterium prausnitzii*, a major gut bacterium, in IBD patients compared to controls. This was attributed to specific strains of *Faecalibacterium* in the samples (Zhang et al. 2018a, b). Imbalanced gut microbiota is linked to GIT cancers, affecting cancer risk and treatment response. A metaproteomic study on patients with various GI conditions revealed shifts in microbial and host proteins, including changes in specific bacterial phyla and host proteins related to tumor-related inflammation in gastric carcinoma patients (Garrett 2015).

9.5.2 Advancements and Challenges in Human Gut Metaproteomics

Due to the intricate nature of samples, metaproteomics faces several challenges in studying the human gut. These include issues related to preparation of sample, and complexities in interpretation of data. However, significant strides have been made through notable advancements and innovations in the field. Advancements include optimized procedures for extraction, lysis, and purification of cells, enhancing the robustness of metaproteomic analysis. Moreover, improvements in LC methods have facilitated enhanced separation techniques. The utilization of rapid, precise, and sensitive MS has significantly expanded the analysis of metaproteomics. Additionally, the availability of tailored sequence databases generated from high-quality metagenomics data, coupled with the development of sophisticated bioinformatics tools and efficient workflow pipelines, has notably increased the number of identified proteins. Importantly, integrating metaproteomic data with other meta-omics datasets offers a comprehensive and complementary perspective on the functional status of the intestinal microbiome. This synergy provides a more exhaustive understanding of the microbial community's functionality and its function within the gut environment.

9.6 Metabolomics in Gut Microbiome

The microbiota of the digestive tract, a rich and ever-changing population of microorganisms found in the human GIT, has a crucial impact on the host during both health and disease. More than 100 trillion different microorganisms make up the rich and diverse microbial community found in the human GIT (Thursby and Juge 2017). Firmicutes and Bacteroidetes are the two phyla that comprise the bulk of the gut microbiota, along with Actinobacteria, Proteobacteria, Fusobacteria, and Verrucomicrobia (Arumugam et al. 2011).

Metabolomics is a rapidly developing field, which can be described as the thorough evaluation of all metabolites or reduced molecular weight metabolites in a biomedical sample. Metabolomics holds potential as a crucial objective lens in the molecular microscope for precision medicine since it facilitates the characterization of a far larger array of metabolites than are currently covered in typical diagnostic strategies, and consequently a broad understanding of biological activities and pathways associated with metabolism. Because the metabolome is thought to be the most precise depiction of phenotype, metabolomics can shed light on how cells react to certain interactions or stimuli. By characterizing and quantifying tiny compounds in a variety of biofluids (saliva, urine, serum, plasma, feces, and stools), metabolomics employs high throughput strategies to identify a distinct metabolic signature (Nicholson and Lindon 2008). Schematic representation of the use and workflow of multi-omics in interactions of food-gut microbiota are depicted in Fig. 9.3 (Farag et al. 2023).

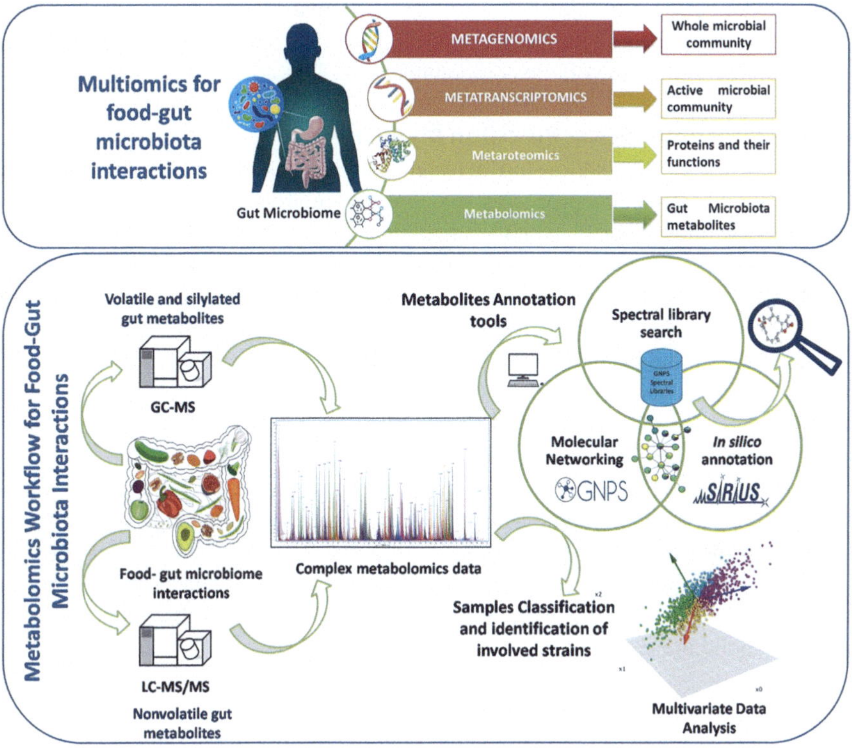

Fig. 9.3 Schematic representation of the use of multi-omics and metabolomics workflow for interactions of food-gut microbiota. (Adapted with permission from Farag et al. 2023)

9.6.1 Application

Relevant metabolomic innovations are far more willing than regular clinical strategies at accurately analyzing hundreds to thousands of metabolites. Thus, metabolomics provides precise phenotypic characterization and can facilitate precision medicine on several fronts: identifying novel targets for therapeutic intervention, characterizing the metabolic dysregulations underlying disease, and identifying biomarkers for potential use in disease diagnosis and treatment response monitoring.

9.6.1.1 Detecting Microbial Metabolites

For metabolomics analysis, MS is frequently utilized, primarily given its great sensitivity and ability to identify and measure a variety of molecules in intricate biological samples. In addition, MS has a wider dynamic range, shows excellent selectivity and sensitivity (Zhao and Hartung 2015), and lets one identify metabolite fingerprints to create metabolome libraries, that simplify recognition of metabolites (Martins-de-Souza 2014). MS-based metabolomics enables targeted and untargeted metabolome assessment and has developed into a vital tool in metabolome analysis

(Milne et al. 2013). In metabolomics, gas chromatography-mass spectrometry (GC)-MS) is regarded as the pinnacle of analysis (Harrigan and Goodacre 2003). However, it is limited to volatile compounds.

9.6.1.2 Metabolomics in Precision Medicine

Because of the close relationship between metabolite's abundance in biological specimens and pathogenic processes, metabolites are often referred to as the closest indicators of illness (Gerszten and Wang 2008).

We must expedite the identification of particular disease markers, drug pharmacodynamics, and metabolite profiles linked to the external environment and disease risk in order to fully realize the possibilities of precision medicine. The identification and verification of metabolic markers of disease can be accelerated by current metabolomics technologies. Metabolics methods like LC-MS are appropriate in human cohort discovery research because they can precisely measure up to hundreds of metabolites on a regular basis. Research on the ties between dietary habits and health has also been influenced by metabolomics studies. For example, elevated levels of branched chain amino acids and obesity have been linked to insulin resistance (Newgard et al. 2009). Considering the discovery method can serve as an initial basis and different technology may not be needed, it can be fairly easy to construct relevant tests for metabolites that are relevant for subsequent research after metabolic indicators of disease have been identified.

9.6.1.3 Microbial Metabolism Reconstruction

Metabolomics provides additional value to transcriptomics and proteomics. Metabolomes are produced through modifications and expressions of systems biology components and are located further downstream from hierarchical functional analyses. As such, it represents microbial phenotypes more accurately. Transcripts and proteins alone cannot be used to calculate or infer metabolic fluxes and interactions, which makes metabolomics data necessary for the elucidation of cellular networks (Hollywood et al. 2006).

Large collections of "omics"-related data, in conjunction with a variety of computational techniques or instruments, can be used to reconstruct metabolic networks in order to study cellular functions (Durot et al. 2009; Schellenberger et al. 2010). Predictive systems are constructed by integrating global data from the "omics" with prior understanding of microbial physiology through the use of statistical and mathematical techniques. Iteratively, experimental data are used to resolve any discrepancies and improve the models (Durot et al. 2009; Vemuri and Aristidou 2005).

9.6.2 Metabolomics and Its Challenges

The capability to recognize and quantify the complete set of intracellular and extracellular metabolites with masses less than 1000 Da is the primary technical challenge in metabolomics (Villas-Boas et al. 2005; Ohashi et al. 2008). Hydrophilic carbohydrates, volatile alcohols, ketones, organic acids that are amino or non-amino, hydrophobic lipids, and secondary metabolites like antibiotics, pigments, non-ribosomal peptides, and cofactors are present in the metabolome (Villas-Boas et al. 2005; Durot et al. 2009; Gill et al. 2006). The dynamic nature of these metabolites further complicates the analyses. Their compositions and concentrations are constantly fluctuating and adapt quickly to external stimuli (Villas-Boas et al. 2005; Hollywood et al. 2006; Vemuri and Aristidou 2005). Because there is not always an intimate correlation within genes and metabolites—for instance, microorganisms have fewer metabolites than genes—correlation of results might be difficult to interpret (Vemuri and Aristidou 2005; Forster et al. 2003).

Each of these elements adds to the intricacy and challenge of research on metabolomics. In closing, the development of novel gut microbiota biomarkers will present an opportunity to assess the causal link between metabolites and pathogenesis by linking intricate metabolic pathways to the genesis of various diseases. These novel biomarkers may also contribute to the emergence of mechanistic hypotheses that could be used to develop personalized therapy plans and nutritional supplements for early disease prediction in asymptomatic conditions. This would allow for a more precise prognosis of the disease's course.

9.7 Gut Microbiome in Health and Disease

The microbiota of the gut is ruled by stringent anaerobes, which surpass facultative anaerobes and aerobes by a hundredfold (Harris et al. 1976). While over 50 phyla were identified in the human intestinal tract (Schloss and Handelsman 2004), the microbiota is subjugated by only two phyla: the Bacteroidetes and the Firmicutes. While the proportional proportions of these phyla could differ, the Bacteroidetes and Firmicutes have remained constant in almost all people (Lozupone et al. 2012). When intestines were analyzed in the transverse section, a considerable number of differences in the makeup of the microbiota appeared due to microbial heterogeneity throughout the course of the GIT (Swidsinski et al. 2005a, b). There were several microenvironments throughout the gut, and luminal microbiome varied greatly in content from the bacteria nearby and connected to the epithelium. For instance, *Bacteroides*, *Bifidobacterium*, *Streptococcus*, *Enterococcus*, *Clostridium*, *Lactobacillus*, and *Ruminococcus*, were all found in the feces, indicating that the composition was indicative of the luminal community, whereas only *Clostridium*, *Lactobacillus*, and *Enterococcus* were found in the mucus layer and epithelial crypts of the small intestine (Swidsinski et al. 2005a, b).

The gut microbiota can perform an array of chemical and metabolic tasks that the humans are either incapable of performing entirely or can just execute to a restricted range. Gut bacteria can synthesize several vitamins, synthesize every amino acid, both vital and non-vital, and perform bile acid biotransformation (Vyas and Ranganathan 2012). Furthermore, the microbiome provides critical biochemical routes for the breakdown of indigestible carbohydrates, which include starches, cellulose, pectins, and gums; a few oligosaccharides which bypass breakdown; unabsorbed sugars along with alcohols (Cummings et al. 1987); and mucins originated in the host (Koropatkin et al. 2012). Many intestinal bacteria release antimicrobial agents and seek out assets and adhesion sites in the intestinal lining, preventing the entry of pathogens. This is referred to as the barrier or competitive-exclusion effect. Furthermore, bacteria can limit the developments of rivals by creating antibacterial compounds known as bacteriocins, which are extensively dispersed among gastrointestinal bacteria (Guarner and Malagelada 2003).

Human gut microbiota can be linked to a growing number of illnesses, syndromes, and functional abnormalities. IBS is described as a set of functional bowel diseases characterized by abdominal irritation or pain during excretion or an alteration in bowel motions, as well as symptoms of disordered defecation. The actual cause of it is uncertain. Its development is assumed to be impacted by familial variables, GIT motor dysfunction, visceral hypersensitivity, infection, inflammation, and immunology, including psychopathological causes (Ghoshal et al. 2012). Alteration in the gut microbial environment, in addition to these characteristics, is suggested to be a contributor to the syndrome's low-grade intestinal inflammation (Guinane and Cotter 2013).

CD and UC are included under IBD. Located along the GIT's length, CD is represented by an inflammatory pattern resembling cobblestones. Adjacent inflammation that only affects the intestinal wall's outer layers is the usual sign of UC. It usually starts at the rectum and is mainly contained in the colon (Baumgart and Sandborn 2007). Studies demonstrating that the administration of antibiotics can prevent or reduce inflammation in people as well as in mouse models of the disease have established the importance of the gut microbiota in the pathogenesis of IBD (Swidsinski et al. 2005a, b). Like IBS, IBD dysbiosis is characterized by extensive changes in the variety and number of the Firmicute population, whose overall abundance has been found to be significantly lower in people with IBD. Type 2 diabetes and obesity come under the examples of systemic metabolic diseases. The discovery that the gut microbiota of obese mice carrying a mutation in the leptin gene differed markedly from that of mice lacking the mutation provided early evidence that the gut microbiota plays a role in obesity (Ley et al. 2005).

Human gut microbiome undergoes changes over the course of a person's life and seems to be crucial to both health and illness. When a host is in good health, the gut microbiota performs a multitude of beneficial roles, such as regulating the immune system, recovering energy from the metabolism of indigestible food components, and shielding the host from pathogen invasion. Although it is unclear how specifically the gut microbiota contributes to these conditions, a dysbiotic state of the gut microbiota is increasingly understood to be an outside influence that affects the host's metabolic processes and has an impact in illness, both systemic—obesity, diabetes, and atopy—and gut-related IBS and IBD. Different microbes have been linked to the diverse etiology of gastrointestinal and metabolic diseases, however, not much has been established about the causal connection.

9.8 Conclusion and Future Perspectives

The exploration of the microbiome of humans has unveiled a large landscape of diversity of microbes and its profound influence on human health. Coined as the microbiome encompassing both genetic material and the microbial organisms themselves, this ecosystem within us has emerged as a pivotal player in shaping diseases and health. Advanced techniques like metagenomics, metatranscriptomics, metaproteomics, and metabolomics provide insights into various diseases such as IBS, IBD, obesity, and diabetes. Analyzing metabolites sheds light on diet-related health impacts. These multi-omics approaches offer promising prospects for precision medicine and shaping the future of healthcare. Metagenomics, a groundbreaking technique, has provided an unprecedented glimpse into the genetic richness of the gut microbiome, unraveling its diverse species, genetic elements, and their potential implications for health. The identification of crucial genes encoding enzymes, antibiotic resistance, and other functional elements has highlighted the intricate relationships among the microbiome and various health conditions, including dysbiosis-related disorders. Complementary approaches like metatranscriptomics and metaproteomics have furthered our understanding by delving into the dynamic responses and functional roles of gut microbes. These techniques, despite their challenges, offer valuable insights into microbial functionalities and their associations with diet, diseases, and therapeutic interventions. Their potential for personalized medicine in deciphering complex microbial interactions highlights their significance in comprehending the microbiome's impact on human health. Moreover, microbial metabolomics, while facing challenges, remains a promising avenue in studying microbial metabolism and its implications for human and environmental ecosystems. Collaborative efforts and technological advancements in this field are paving the way for a greater comprehension of the function of microbes, contributing significantly to the broader landscape of systems biology.

The human gut microbiota, primarily composed of Bacteroidetes and Firmicutes, serves multifaceted roles in nutrient metabolism, immune defense, and overall well-being. Its disruptions have been linked to various health conditions such as IBS, IBD, obesity, and diabetes. Unraveling the intricate interplay between gut microbes and human health holds immense promise for potential therapeutic interventions and precision medicine strategies. In essence, the burgeoning research in multi-omics approaches, alongside collaborative endeavors and technological advancements, promises a future where understanding the microbiome's impact on health is paramount. The comprehensive insights garnered from these diverse methodologies stand to revolutionize healthcare, offering novel perspectives for disease management and personalized interventions rooted in an in-depth comprehension of our symbiotic relationship with the microbiome.

Acknowledgments The authors are thankful to the principal, of Gargi College for providing the infrastructural support.

Conflict of Interest None.

References

Aagard K, Ma J, Antony K et al (2014) The placenta harbors a unique microbiome. Sci Transl Med 6:237–265

Abu-Ali GS et al (2018) Metatranscriptome of human faecal microbial communities in a cohort of adult men. Nat Microbiol 3:356–366

Abubucker S, Segata N, Goll J et al (2012) Metabolic reconstruction for metagenomic data and its application to the human microbiome. PLoS Comput Biol 8(6):e1002358

Adak A, Khan MR (2019) An insight into gut microbiota and its functionalities. Cell Mol Life Sci 76(3):473–493. https://doi.org/10.1007/s00018-018-2943-4. Epub 2018 Oct 13. PMID: 30317530

Almeida A et al (2021) A unified catalog of 204,938 reference genomes from the human gut microbiome. Nat Biotechnol 39:105–114

Andrews MC et al (2021) Gut microbiota signatures are associated with toxicity to combined CTLA-4 and PD-1 blockade. Nat Med 27:1432–1441

Arumugam M, Raes J, Pelletier E, Le Paslier D, Yamada T, Mende DR, Fernandes GR, Tap J, Bruls T, Batto JM et al (2011) Enterotypes of the human gut microbiome. Nature 473:174–180

Backhed F, Ley RE, Sonnenburg JL, Peterson DA, Gordon JI (2005) Host-bacterial mutualism in the human intestine. Nature 307:1915–1920

Bao G, Wang M, Doak TG, Ye Y (2015) Strand-specific community RNA-seq reveals prevalent and dynamic antisense transcription in human gut microbiota. Front Microbiol 6:896

Barr JJ, Auro R, Furlan M et al (2013) Bacteriophage adhering to mucus provide a non-host-derived immunity. Proc Natl Acad Sci U S A 110:10771–10776

Bashiardes S, Zilberman-Schapira G, Elinav E (2016) Use of metatranscriptomics in microbiome research. Bioinform Biol Insights 10:19–25. https://doi.org/10.4137/BBI.S34610

Baumgart DC, Sandborn WJ (2007) Inflammatory bowel disease: clinical aspects and established and evolving therapies. Lancet 369(9573):1641–1657

Belizário JE, Napolitano M (2015) Human microbiomes and their roles in dysbiosis, common diseases, and novel therapeutic approaches. Front Microbiol 6:1050. https://doi.org/10.3389/fmicb.2015.01050. PMID: 26500616; PMCID: PMC4594012

Bikel S, Valdez-Lara A, Cornejo-Granados F, Rico K, Canizales-Quinteros S et al (2015) Combining metagenomics, metatranscriptomics and viromics to explore novel microbial interactions: towards a systems-level understanding of human microbiome. Comput Struct Biotechnol J 13:390–401. https://doi.org/10.1016/j.csbj.2015.06.001

Blaser M, Bork P, Fraser C, Knight R, Wang J (2013) The microbiome explored: recent insights and future challenges. Nat Rev Microbiol 11:213–217

Blaser MJ et al (2021) Lessons learned from the prenatal microbiome controversy. Microbiome 9:8

Blottière HM, de Vos WM, Ehrlich SD, Doré J (2013) Human intestinal metagenomics: state of the art and future. Curr Opin Microbiol 16:232–239

Bolger AM, Lohse M, Usadel B (2014) Trimmomatic: a flexible trimmer for Illumina sequence data. Bioinformatics 30(15):2114–2120

Breitbart M, Haynes M, Kelley S, Angly F, Edwards RA, Felts B, Mahaffy JM, Mueller J, Nulton J, Rayhawk S et al (2008) Viral diversity and dynamics in an infant gut. Res Microbiol 159:367–373

Bull MJ, Plummer NT (2014) Part 1: the human gut microbiome in health and disease. Integr Med (Encinitas) 13(6):17–22. PMID: 26770121; PMCID: PMC4566439

Carvalho FA, Koren O, Goodrich JK et al (2012) Transient inability to manage proteobacteria promotes chronic gut inflammation in TLR5-deficient mice. Cell Host Microbe 12(2):139–152

Chen L et al (2021) The long-term genetic stability and individual specificity of the human gut microbiome. Cell 184:2302–2315.e2312

Chistoserdovai L (2010) Functional metagenomics: recent advances and future challenges. Biotechnol Genet Eng Rev 26:335–352

Cho I, Blaser MJ (2012) The human microbiome: at the interface of health and disease. Nat Rev Genet 13:260–270
Clish CB (2015) Metabolomics: an emerging but powerful tool for precision medicine. Cold Spring Harb Mol Case Stud 1(1):a000588. https://doi.org/10.1101/mcs.a000588
Cole JR, Chai B, Farris RJ, Wang Q, Kulam-Syed-Mohideen AS, McGarrell DM, Bandela AM, Cardenas E, Garrity GM, Tiedje JM (2007) The ribosomal database project (RDP-II): introducing myRDP space and quality controlled public data. Nucleic Acids Res 35:D169–D172
Conesa A, Götz S, García-Gómez JM, Terol J, Talón M, Robles M (2005) Blast2GO: a universal tool for annotation, visualization and analysis in functional genomics research. Bioinformatics 21(18):3674–3676
Cresci GA, Bawden E (2015) Gut microbiome: what we do and don't know. Nutr Clin Pract. 30(6):734–746. https://doi.org/10.1177/0884533615609899. Epub 2015 Oct 8. PMID: 26449893; PMCID: PMC4838018
Cullender TC, Chassaing B, Janzon A et al (2013) Innate and adaptive immunity interact to quench microbiome flagellar motility in the gut. Cell Host Microbe 14(5):571–581
Cummings JH, Pomare EW, Branch WJ, Naylor CP, Macfarlane GT (1987) Short chain fatty acids in human large intestine, portal, hepatic and venous blood. Gut 28(10):1221–1227
D'Costa VM, McGrann KM, Hughes DW, Wright GD (2006) Sampling the antibiotic resistome. Science 311:374–377
David LA et al (2014a) Host lifestyle affects human microbiota on daily timescales. Genome Biol 15:R89
David LA et al (2014b) Diet rapidly and reproducibly alters the human gut microbiome. Nature 505:559–563
de Martel C et al (2020) Global burden of cancer attributable to infections in 2018: a worldwide incidence analysis. Lancet Glob Health 8:e180–e190
den Besten G et al (2013) The role of short-chain fatty acids in the interplay between diet, gut microbiota, and host energy metabolism. J Lipid Res 54(9):2325–2340
DiGiulio DB (2012) Diversity of microbes in amniotic fluid. Semin Fetal Neonatal Med 17:2–11
Ding T, Schloss PD (2014) Dynamics and associations of microbial community types across the human body. Nature 509(7500):357–360
Dobell C (1920) The discovery of intestinal protozoa in man. Proc R Soc Med 13:1–15
Durot M, Bourguignon PY, Schachter V (2009) Genome-scale models of bacterial metabolism: reconstruction and applications. FEMS Microbiol Rev 33(1):164–190
Erickson AR, Cantarel BL, Lamendella R, Darzi Y, Mongodin EF, Pan C, Shah M, Halfvarson J, Tysk C, Henrissat B, Raes J, Verberkmoes NC, Fraser CM, Hettich RL, Jansson JK (2012) Integrated metagenomics/metaproteomics reveals human host-microbiota signatures of Crohn's disease. PLoS One 7:e49138
Escobar-Zepeda A, Vera-Ponce de León A, Sanchez-Flores A (2015) The road to metagenomics: from microbiology to DNA sequencing technologies and bioinformatics. Front Genet 6:348
Faith JJ, Guruge JL, Charbonneau M, Subramanian S, Seedorf H, Goodman AL, Clemente JC, Knight R, Heath AC, Leibel RL et al (2013) The long-term stability of the human gut microbiota. Science 341(6141):1237439
Falony G et al (2016) Population-level analysis of gut microbiome variation. Science 352:560–564
Fang Z, Cui X (2011) Design and validation issues in RNA-seq experiments. Brief Bioinform 12(3):280–287
Farag MA, Baky MH, von Bergen M, Hegazi NM (2023) The use of omics in monitoring food gut microbiota interaction outcomes: a review of novel trends and technologies. Curr Opin Food Sci 52:101064. https://doi.org/10.1016/j.cofs.2023.101064
Faust K, Raes J (2012) Microbial interactions: from networks to models. Nat Rev Microbiol 10:538–550
Felix RS, Colleoni GWB, Caballero OL et al (2009) SAGE analysis highlights the importance of p53csv, ddx5, mapkapk2 and ranbp2 to multiple myeloma tumorigenesis. Cancer Lett 278(1):41–48

Fernandes J, Su W, Rahat-Rozenbloom S, Wolever TM, Comelli EM (2014) Adiposity, gut microbiota and faecal short chain fatty acids are linked in adult humans. Nutr Diabetes 4:E121
Filiatrault MJ (2011) Progress in prokaryotic transcriptomics. Curr Opin Microbiol 14:579–586
Forster J, Famili I, Fu P, Palsson BO, Nielsen J (2003) Genome-scale reconstruction of the *Saccharomyces cerevisiae* metabolic network. Genome Res 13(2):244–253
Frank DN, St Amand AL, Feldman RA, Boedeker EC, Harpaz N, Pace NR (2007) Molecular-phylogenetic characterization of microbial community imbalances in human inflammatory bowel diseases. Proc Natl Acad Sci U S A 104:13780–13785
Franzosa EA, Morgan XC, Segata N et al (2014) Relating the metatranscriptome and metagenome of the human gut. Proc Natl Acad Sci USA 111(22):E2329–E2338
Fukuda S, Toh H, Hase K et al (2011) Bifidobacteria can protect from enteropathogenic infection through production of acetate. Nature 469:543–547
Gabor EM, Alkema WB, Janssen DB (2004) Quantifying the accessibility of the metagenome by random expression cloning techniques. Environ Microbiol 6:879–886
Gallardo-Becerra L et al (2020) Metatranscriptomic analysis to define the Secrebiome, and 16S rRNA profiling of the gut microbiome in obesity and metabolic syndrome of Mexican children. Microb Cell Factories 19:61
Garmendia L, Hernandez A, Sanchez MB, Martinez JL (2012) Metagenomics and antibiotics. Clin Microbiol Infect 18:27–31
Garrett WS (2015) Cancer and the microbiota. Science 348:80–86
Gavin PG, Mullaney JA, Loo D, Cao KL, Gottlieb PA, Hill MM, Zipris D, Hamilton-Williams EE (2018) Intestinal metaproteomics reveals host-microbiota interactions in subjects at risk for type 1 diabetes. Diabetes Care 41:2178–2186
Georg J, Voss B, Scholz I, Mitschke J, Wilde A, Hess WR (2009) Evidence for a major role of antisense RNAs in cyanobacterial gene regulation. Mol Syst Biol 5:305
Gerszten RE, Wang TJ (2008) The search for new cardiovascular biomarkers. Nature 451:949–952
Ghosh S, Chan C-KK (2016) Analysis of RNA-seq data using TopHat and Cufflinks. Methods Mol Biol 1374:339–361
Ghoshal UC, Shukla R, Ghoshal U, Gwee KA, Ng SC, Quigley FM (2012) The gut microbiota and irritable bowel syndrome: friend or foe? Int J Inflam 2012:151085
Giannoukos G, Ciulla DM, Huang K et al (2012) Efficient and robust RNA-seq process for cultured bacteria and complex community transcriptomes. Genome Biol 13(3):R23
Gilbert JA, Field D, Huang Y, Edwards R, Li W, Gilna P, Joint I (2008) Detection of large numbers of novel sequences in the metatranscriptomes of complex marine microbial communities. PLoS One 3(8):e3042. https://doi.org/10.1371/journal.pone.0003042
Gill SR, Pop M, Deboy RT, Eckburg PB, Turnbaugh PJ, Samuel BS, Gordon JI, Relman DA, Fraser-Liggett CM, Nelson KE (2006) Metagenomic analysis of the human distal gut microbiome. Science 312:1355–1359
Glass EM, Wilkening J, Wilke A, Antonopoulos D, Meyer F (2010) Using the metagenomics RAST server (MG-RAST) for analyzing shotgun metagenomes. Cold Spring Harb Protoc 2010(1):db.rot5368
Gohl DM, Vangay P, Garbe J, MacLean A, Hauge A, Becker A, Gould TJ, Clayton JB, Johnson TJ, Hunter R et al (2016) Systematic improvement of amplicon marker gene methods for increased accuracy in microbiome studies. Nat Biotechnol 34(9):942–949
Gosalbes MJ, Durbán A, Pignatelli M et al (2011) Metatranscriptomic approach to analyze the functional human gut microbiota. PLoS One 6(3):e17447
Grabherr MG, Haas BJ, Yassour M et al (2011) Full-length transcriptome assembly from RNA-Seq data without a reference genome. Nat Biotechnol 29(7):644–652
Granata I et al (2020) Duodenal metatranscriptomics to define human and microbial functional alterations associated with severe obesity: a pilot study. Microorganisms 8:1811
Guarner F, Malagelada JR (2003) Gut flora in health and disease. Lancet 361(9356):512–519
Guinane CM, Cotter PD (2013) Role of the gut microbiota in health and chronic gastrointestinal disease: understanding a hidden metabolic organ. Ther Adv Gastroenterol 6(4):295–308

Gupta A, Singh V, Mani I (2022) Dysbiosis of human microbiome and infectious diseases. Prog Mol Biol Transl Sci 192(1):33–51. https://doi.org/10.1016/bs.pmbts.2022.06.016

Haft DH et al (2018) RefSeq: an update on prokaryotic genome annotation and curation. Nucleic Acids Res 46:D851–D860. https://doi.org/10.1093/nar/gkx1068

Haiser HJ, Turnbaugh PJ (2012) Is it time for a metagenomic basis of therapeutics? Science 336(6086):1253–1255

Handelsman J (2004) Metagenomics: application of genomics to uncultured microorganisms. Microbiol Mol Biol Rev 68:669–685

Handelsman J, Rondon MR, Brady SF, Clardy J, Goodman RM (1998) Molecular biological access to the chemistry of unknown soil microbes: a new frontier for natural products. Chem Biol 5:R245–R249

Harrigan GG, Goodacre R (2003) Metabolic profiling: its role in biomarker discovery and gene function analysis. Springer, Boston, MA

Harris MA, Reddy CA, Carter GR (1976) Anaerobic bacteria from the large intestine of mice. Appl Environ Microbiol 31(6):907–912

Hatch A et al (2019) A robust metatranscriptomic technology for population-scale studies of diet, gut microbiome, and human health. Int J Genomics 2019:1718741

Heintz-Buschart A et al (2016) Integrated multi-omics of the human gut microbiome in a case study of familial type 1 diabetes. Nat Microbiol 2:16180

Hiergeist A, Glasner J, Reischl U, Gessner A (2015) Analyses of intestinal microbiota: culture versus sequencing. ILAR J 56(2):228–240

Hollywood K, Brison DR, Goodacre R (2006) Metabolomics: current technologies and future trends. Proteomics 6(17):4716–4723

Human Microbiome Project Consortium (2012) A framework for human microbiome research. Nature 486:215–221

Hunter P (2013) The secret garden's gardeners: research increasingly appreciates the crucial role of gut viruses for human health and disease. EMBO Rep 14:683–685

Jeanty C, Longrois D, Mertes P-M, Wagner DR, Devaux Y (2010) An optimized protocol for microarray validation by quantitative PCR using amplified amino allyl labeled RNA. BMC Genomics 11:542

Jiménez E, Fernándes L, Marín M et al (2005) Isolation of commensal bacteria from umbilical cord blood of healthy neonates born by cesarean section. Curr Microbiol 51:270–274

Jones BV, Begley M, Hill C, Gahan CG, Marchesi JR (2008) Functional and comparative metagenomic analysis of bile salt hydrolase activity in the human gut microbiome. Proc Natl Acad Sci U S A 105:13580–13585

Jovel J, Patterson J, Wang W, Hotte N, O'Keefe S, Mitchel T, Perry T, Kao D, Mason AL, Madsen KL et al (2016) Characterization of the gut microbiome using 16S or shotgun metagenomics. Front Microbiol 7(459):459

Kelder T, Stroeve JH, Bijlsma S, Radonjic M, Roeselers G (2014) Correlation network analysis reveals relationships between diet-induced changes in human gut microbiota and metabolic health. Nutr Diabetes 4:e122

Khosravi A, Mazmanian SK (2013) Disruption of the gut microbiome as a risk factor for microbial infections. Curr Opin Microbiol 16:221–227. https://doi.org/10.1016/j.mib.2013.03.009

Koenig JE, Spor A, Scalfone N, Fricker AD, Stombaugh J, Knight R, Angenent LT, Ley RE (2011) Succession of microbial consortia in the developing infant gut microbiome. Proc Natl Acad Sci USA 108(Suppl 1):4578–4585

Kolmeder CA, de Been M, Nikkila J, Ritamo I, Matto J, Valmu L, Salojarvi J, Palva A, Salonen A, de Vos WM (2012) Comparative metaproteomics and diversity analysis of human intestinal microbiota testifies for its temporal stability and expression of core functions. PLoS One 7:e29913

Koropatkin NM, Cameron EA, Martens EC (2012) How glycan metabolism shapes the human gut microbiota. Nat Rev Microbiol 10(5):323–335

Langmead B, Salzberg SL (2012) Fast gapped-read alignment with Bowtie 2. Nat Methods 9(4):357–359

Lederberg J (2001) Ome Sweet'Omics—a genealogical treasury of words. Scientist
Lee SW et al (2016) Metagenome and metatranscriptome profiling of moderate and severe COPD sputum in Taiwanese Han males. PLoS One 11:e0159066
Lee PY, Chin SF, Neoh HM, Jamal R (2017) Metaproteomic analysis of human gut microbiota: where are we heading? J Biomed Sci 24:36
Lehmann T, Schallert K, Vilchez-Vargas R, Benndorf D, Puttker S, Sydor S, Schulz C, Bechmann L, Canbay A, Heidrich B, Reichl U, Link A, Heyer R (2019) Metaproteomics of fecal samples of Crohn's disease and ulcerative colitis. J Proteome 201:93–103
Lepage P, Leclerc MC, Joossens M, Mondot S, Blottière HM, Raes J, Ehrlich D, Doré J (2013) A metagenomic insight into our gut's microbiome. Gut 62:146–158
Ley RE, Bäckhed F, Turnbaugh P, Lozupone CA, Knight RD, Gordon JI (2005) Obesity alters gut microbial ecology. Proc Natl Acad Sci USA 102(31):11070–11075
Ley RE, Turnbaugh PJ, Klein S, Gordon JI (2006) Microbial ecology: human gut microbes associated with obesity. Nature 444:1022–1023. https://doi.org/10.1038/4441022a
Ley RE, Lozupone CA, Hamady M, Knight R, Gordon JI (2008a) Worlds within worlds: evolution of the vertebrate gut microbiota. Nat Rev Microbiol 6:776–788
Ley RE, Hamady M, Lozupone C, Turnbaugh PJ, Ramey RR, Bircher JS et al (2008b) Evolution of mammals and their gut microbes. Science 320:1647–1651
Li LL, McCorkle SR, Monchy S, Taghavi S, van der Lelie D (2009) Bioprospecting metagenomes: glycosyl hydrolases for converting biomass. Biotechnol Biofuels 2:10
Li X, LeBlanc J, Elashoff D, McHardy I, Tong M, Roth B, Ippoliti A, Barron G, McGovern D, McDonald K, Newberry R, Graeber T, Horvath S, Goodglick L, Braun J (2016) Microgeographic proteomic networks of the human colonic mucosa and their association with inflammatory bowel disease. Cell Mol Gastroenterol Hepatol 2:567–583
Li P et al (2021) Gut microbiota dysbiosis is associated with elevated bile acids in Parkinson's disease. Meta 11:29
Li J et al (2022) Interplay between diet and gut microbiome, and circulating concentrations of trimethylamine N-oxide: findings from a longitudinal cohort of US men. Gut 71:724–733
Lim YW et al (2013) Metagenomics and metatranscriptomics: windows on CF-associated viral and microbial communities. J Cyst Fibros 12:154–164
Liu X (2016) Microbiome. Yale J Biol Med 89(3):275–276. PMCID: PMC5045136
Lloyd-Price J et al (2019) Multi-omics of the gut microbial ecosystem in inflammatory bowel diseases. Nature 569:655–662
Lozupone CA, Stombaugh JI, Gordon JI, Jansson JK, Knight R (2012) Diversity, stability and resilience of the human gut microbiota. Nature 489(7415):220–230
Ma W et al (2021) Dietary fiber intake, the gut microbiome, and chronic systemic inflammation in a cohort of adult men. Genome Med 13:102
Maier TV, Lucio M, Lee LH, Verberkmoes NC, Brislawn CJ, Bernhardt J, Lamendella R, McDermott JE, Bergeron N, Heinzmann SS, Morton JT, Gonzalez A, Ackermann G, Knight R, Riedel K, Krauss RM, Schmitt-Kopplin P, Jansson JK (2017) Impact of dietary resistant starch on the human gut microbiome, metaproteome, and metabolome. MBio 8
Mandal RS, Saha S, Das S (2015) Metagenomic surveys of gut microbiota. Genomics Proteomics Bioinformatics 13(3):148–158. https://doi.org/10.1016/j.gpb.2015.02.005
Mani I (2020a) Metagenomics approach for bioremediation: challenges and perspectives. In: Pandey VC, Singh V (eds) Bioremediation of pollutants. Elsevier, pp 275–285., ISBN 9780128190258. https://doi.org/10.1016/B978-0-12-819025-8.00012-0
Mani I (2020b) Current status and challenges of DNA sequencing. In: Singh V (ed) Advances in synthetic biology. Springer, Singapore. https://doi.org/10.1007/978-981-15-0081-7_5
Marchesi JR (2010) Prokaryotic and eukaryotic diversity of the human gut. Adv Appl Microbiol 72(72):43–62
Markowitz VM, Chen IM, Chu K, Szeto E, Palaniappan K, Grechkin Y et al (2012) IMG/M: the integrated metagenome data management and comparative analysis system. Nucleic Acids Res 40:D123–D129

Martín R, Jiménez E, Heilig H, Fernández L, Marín ML, Zoetendal EG, Rodríguez JM (2009) Isolation of bifidobacteria from breast milk and assessment of the bifidobacterial population by PCR-denaturing gradient gel electrophoresis and quantitative real-time PCR. Appl Environ Microbiol 75:965–969

Martins-de-Souza D (2014) Proteomics, metabolomics, and protein interactomics in the characterization of the molecular features of major depressive disorder. Dialogues Clin Neurosci 16:63–73

Matsuoka K, Kanai T (2015) The gut microbiota and inflammatory bowel disease. Semin Immunopathol 37:47–55

Maurice CF, Haiser HJ, Turnbaugh PJ (2013) Xenobiotics shape the physiology and gene expression of the active human gut microbiome. Cell 152(1–2):39–50

McNulty NP et al (2011) The impact of a consortium of fermented milk strains on the gut microbiome of gnotobiotic mice and monozygotic twins. Sci Transl Med 3:106ra106

Mehta RS et al (2018) Stability of the human faecal microbiome in a cohort of adult men. Nat Microbiol 3:347–355

Mills RH, Vazquez-Baeza Y, Zhu Q, Jiang L, Gaffney J, Humphrey G, Smarr L, Knight R, Gonzalez DJ (2019) Evaluating metagenomic prediction of the metaproteome in a 45-year study of a patient with Crohn's disease. mSystems 4(1):e00337-18

Milne SB, Mathews TP, Myers DS, Ivanova PT, Brown HA (2013) Sum of the parts: mass spectrometry-based metabolomics. Biochemistry (Mosc) 52:3829–3840. https://doi.org/10.1021/bi400060e

Moles L, Gomez M, Heilig H, Bustos G, Fuentes S, de Vos W, Fernández L, Rodríguez JM, Jiménez E (2013) Bacterial diversity in meconium of preterm neonates and evolution of their fecal microbiota during the first months of life. PLoS One 8:1–13

Mondot S, Lepage P (2016) The human gut microbiome and its dysfunctions through the meta-omics prism. Ann N Y Acad Sci 1372:9–19

Morgan XC, Huttenhower C (2012) Chapter 12: human microbiome analysis. PLoS Comput Biol 8(12):e1002808

Muth T, Behne A, Heyer R, Kohrs F, Benndorf D, Hoffmann M, Lehteva M, Reichl U, Martens L, Rapp E (2015) The MetaProteomeAnalyzer: a powerful open-source software suite for metaproteomics data analysis and interpretation. J Proteome Res 14:1557–1565

Newgard CB, An J, Bain JR, Muehlbauer MJ, Stevens RD, Lien LF, Haqq AM, Shah SH, Arlotto M, Slentz CA et al (2009) A branched-chain amino acid-related metabolic signature that differentiates obese and lean humans and contributes to insulin resistance. Cell Metab 9:311–326

Nicholson JK, Lindon JC (2008) Systems biology: metabonomics. Nature 455:1054–1056. https://doi.org/10.1038/4551054a

O'Leary NA et al (2015) Reference sequence (RefSeq) database at NCBI: current status, taxonomic expansion, and functional annotation. Nucleic Acids Res 44:D733–D745. https://doi.org/10.1093/nar/gkv1189

Ohashi Y, Hirayama A, Ishikawa T, Nakamura S, Shimizu K, Ueno Y, Tomita M, Soga T (2008) Depiction of metabolome changes in histidine-starved *Escherichia coli* by CE-TOF-MS. Mol BioSyst 4(2):135–114

Ojala T, Kankuri E, Kankainen M (2023) Understanding human health through metatranscriptomics. Trends Mol Med 29(5):376–389. https://doi.org/10.1016/j.molmed.2023.02.002

Ouwehand A, Isolauri E, Salminen S (2002) The role of the intestinal microflora for the development of the immune system in early childhood. Eur J Nutr 41(Suppl 1):I32–I37

Pace NR (1997) A molecular view of microbial diversity and the biosphere. Science 276:734–740

Pan S, Chen R (2020) Metaproteomic analysis of human gut microbiome in digestive and metabolic diseases. Adv Clin Chem 97:1–12. https://doi.org/10.1016/bs.acc.2019.12.002. Epub 2020 Feb 17

Presley LL, Ye J, Li X, LeBlanc J, Zhang Z, Ruegger PM, Allard J, McGovern D, Ippoliti A, Roth B, Cui X, Jeske DR, Elashoff D, Goodglick L, Braun J, Borneman J (2012) Host-microbe relationships in inflammatory bowel disease detected by bacterial and metaproteomic analysis of the mucosal-luminal interface. Inflamm Bowel Dis 18:409–417

Qin J, Li R, Raes J, Arumugam M, Burgdorf KS, Manichanh C, Nielsen T, Pons N, Levenez F, Yamada T et al (2010) A human gut microbial gene catalogue established by metagenomic sequencing. Nature 464:59–65

Quinn RA et al (2014) Biogeochemical forces shape the composition and physiology of polymicrobial communities in the cystic fibrosis lung. MBio 5:e00956–e00913

Rajagopala SV et al (2021) Metatranscriptomics to characterize respiratory virome, microbiome, and host response directly from clinical samples. Cell Rep Methods 1:10091

Ren Z, Cui G, Lu H, Chen X, Jiang J, Liu H, He Y, Ding S, Hu Z, Wang W et al (2013) Liver ischemic preconditioning (IPC) improves intestinal microbiota following liver transplantation in rats through 16s rDNA-based analysis of microbial structure shift. PLoS One 8:e75950

Ren L et al (2018) Transcriptionally active lung microbiome and its association with bacterial biomass and host inflammatory status. mSystems 3:e00199-18

Reyes A, Semenkovich NP, Whiteson K et al (2012) Going viral: next-generation sequencing applied to phage populations in the human gut. Nat Rev Microbiol 10:607–617

Rinke C, Schwientek P, Sczyrba A, Ivanova NN, Anderson IJ, Cheng J-F, Darling A, Malfatti S, Swan BK, Gies EA et al (2013) Insights into the phylogeny and coding potential of microbial dark matter. Nature 499(7459):431–437

Rodrigues RR, Shulzhenko N, Morgun A (2018) Transkingdom networks: a systems biology approach to identify causal members of host–microbiota interactions. Methods Mol Biol 1849:227–242. https://doi.org/10.1007/978-1-4939-8728-3_15

Rodriguez JM et al (2015) The composition of the gut microbiota throughout life, with an emphasis on early life. Microb Ecol Health Dis 26:26050

Rondon MR, Raffel SJ, Goodman RM, Handelsman J (1999) Toward functional genomics in bacteria: analysis of gene expression in Escherichia coli from a bacterial artificial chromosome library of Bacillus cereus. Proc Natl Acad Sci USA 96:6451–6455

Routy B et al (2018) Gut microbiome influences efficacy of PD-1-based immunotherapy against epithelial tumors. Science 359:91–97

Salazar N, Arboleya S, Valdés L et al (2014) The human intestinal microbiome at extreme ages of life. Front Genet 5:1–9

Salyers AA, Gupta A, Wang Y (2004) Human intestinal bacteria as reservoirs for antibiotic resistance genes. Trends Microbiol 12:412–416

Sayers EW, Cavanaugh M, Clark K, Ostell J, Pruitt KD, Karsch-Mizrachi I (2020) GenBank. Nucleic Acids Res 48:D84–D86. https://doi.org/10.1093/nar/gkz956

Schellenberger J, Park JO, Conrad TM, Palsson BO (2010) BiGG: a biochemical genetic and genomic knowledgebase of large scale metabolic reconstructions. BMC Bioinformatics 11:213

Schirmer M et al (2018) Dynamics of metatranscription in the inflammatory bowel disease gut microbiome. Nat Microbiol 3:337–346

Schloss PD, Handelsman J (2004) Status of the microbial census. Microbiol Mol Biol Rev 68(4):686–691

Schmidt TSB, Raes J, Bork P (2018) The human gut microbiome: from association to modulation. Cell 172(6):1198–1215. https://doi.org/10.1016/j.cell.2018.02.044. PMID: 29522742

Shakya M et al (2019) Advances and challenges in metatranscriptomic analysis. Front Genet 10:904

Sharma CM, Hoffmann S, Darfeuille F et al (2010) The primary transcriptome of the major human pathogen *Helicobacter pylori*. Nature 464(7286):250–255

Shi M et al (2022) Total infectome characterization of respiratory infections in pre-COVID-19 Wuhan, China. PLoS Pathog 18:e1010259

Siezen RJ, Kleerebezem M (2011) The human gut microbiome: are we our enterotypes? Microb Biotechnol 4:55053

Sivan A et al (2015) Commensal Bifidobacterium promotes antitumor immunity and facilitates anti-PD-L1 efficacy. Science 350:1084–1089

Smith M et al (2022) Gut microbiome correlates of response and toxicity following anti-CD19 CAR T cell therapy. Nat Med 28:713–723

Sokol H, Seksik P (2010) The intestinal microbiota in inflammatory bowel diseases: time to connect with the host. Curr Opin Gastroenterol 26:327–331

Sommer MO, Dantas G, Church GM (2009) Functional characterization of the antibiotic resistance reservoir in the human microflora. Science 325:1128–1131
Sousa T, Paterson R, Moore V, Carlsson A, Abrahamsson B, Basit AW (2008) The gastrointestinal microbiota as a site for the biotransformation of drugs. Int J Pharm 363(1–2):1–25
Spottiswoode N et al (2022) Pneumonia surveillance with culture-independent metatranscriptomics in HIV-positive adults in Uganda: a cross-sectional study. Lancet Microbe 3:e357–e365
Suau A, Bonnet R, Sutren M, Godon JJ, Gibson GR, Collins MD, Doré J (1999) Direct analysis of genes encoding 16S rRNA from complex communities reveals many novel molecular species within the human gut. Appl Environ Microbiol 65:4799–4807
Sultan M, Amstislavskiy V, Risch T et al (2014) Influence of RNA extraction methods and library selection schemes on RNA-seq data. BMC Genomics 15:675
Sunagawa S, Mende DR, Zeller G, Izquierdo-Carrasco F, Berger SA, Kultima JR, Coelho LP, Arumugam M, Tap J, Nielsen HB et al (2013) Metagenomic species profiling using universal phylogenetic marker genes. Nat Methods 10:1196–1199
Sunagawa S, Coelho LP, Chaffron S, Kultima JR, Labadie K, Salazar G et al (2015) Structure and function of the global ocean microbiome. Science 348:1261359. https://doi.org/10.1126/science.1261359
Swidsinski A, Loening-Baucke V, Lochs H, Hale LP (2005a) Spatial organization of bacterial flora in normal and inflamed intestine: a fluorescence in situ hybridization study in mice. World J Gastroenterol 11(8):1131–1140
Swidsinski A, Weber J, Loening-Baucke V, Hale LP, Lochs H (2005b) Spatial organization and composition of the mucosal flora in patients with inflammatory bowel disease. J Clin Microbiol 43(7):3380–3389
Tamburini S et al (2016) The microbiome in early life: implications for health outcomes. Nat Med 22:713–722
Tanca A, Palomba A, Pisanu S, Deligios M, Fraumene C, Manghina V, Pagnozzi D, Addis MF, Uzzau S (2014) A straightforward and efficient analytical pipeline for metaproteome characterization. Microbiome 2:49
Tannock GW (2001) Molecular assessment of intestinal microflora. Am J Clin Nutr 73:410S–414S
The Human Microbiome Project Consortium (2012) Structure, function and diversity of the healthy human microbiome. Nature 486:207–214. https://doi.org/10.1038/nature11234
Thomas T, Gilbert J, Meyer F (2012) Metagenomics—a guide from sampling to data analysis. Microb Inform Exp 2(1):3. https://doi.org/10.1186/2042-5783-2-3. PMID: 22587947; PMCID: PMC3351745
Thomason MK, Storz G (2010) Bacterial antisense RNAs: how many are there, and what are they doing? Annu Rev Genet 44(1):167–188
Thompson LR, Sanders JG, McDonald D, Amir A, Ladau J, Locey KJ et al (2017) A communal catalogue reveals Earth's multiscale microbial diversity. Nature 551:457–463. https://doi.org/10.1038/nature24621
Thursby E, Juge N (2017) Introduction to the human gut microbiota. Biochem J 474:1823–1836. https://doi.org/10.1042/BCJ20160510
Toledo-Arana A, Repoila F, Cossart P (2007) Small noncoding RNAs controlling pathogenesis. Curr Opin Microbiol 10(2):182–188
Tremblay J, Singh K, Fern A, Kirton ES, He S, Woyke T, Lee J, Chen F, Dangl JL, Tringe SG (2015) Primer and platform effects on 16S rRNA tag sequencing. Front Microbiol 6:771
Tringe SG, von Mering C, Kobayashi A, Salamov AA, Chen K, Chang HW, Podar M, Short JM, Mathur EJ, Detter JC et al (2005) Comparative metagenomics of microbial communities. Science 308:554–557
Turnbaugh PJ, Ley RE, Hamady M, Fraser-Liggett CM, Knight R, Gordon JI (2007) The human microbiome project. Nature 449:804–810
Turnbaugh PJ, Hamady M, Yatsunenko T, Cantarel BL, Duncan A, Ley RE, Sogin ML, Jones WJ, Roe BA, Affourtit JP et al (2009) A core gut microbiome in obese and lean twins. Nature 457:480–484

Turnbaugh PJ et al (2010a) Organismal, genetic, and transcriptional variation in the deeply sequenced gut microbiomes of identical twins. Proc Natl Acad Sci U S A 107:7503–7508

Turnbaugh PJ, Henrissat B, Gordon JI (2010b) Viewing the human microbiome through three-dimensional glasses: integrating structural and functional studies to better define the properties of myriad carbohydrate-active enzymes. Acta Crystallogr Sect F Struct Biol Cryst Commun 66:1261–1264

Tyakht AV, Kostryukova ES, Popenko AS, Belenikin MS, Pavlenko AV, Larin AK, Karpova IY, Selezneva OV, Semashko TA, Ospanova EA et al (2013) Human gut microbiota community structures in urban and rural populations in Russia. Nat Commun 4:2469

Tyson GW, Chapman J, Hugenholtz P, Allen EE, Ram RJ, Richardson PM, Solovyev VV, Rubin EM, Rokhsar DS, Banfield JF (2004) Community structure and metabolism through reconstruction of microbial genomes from the environment. Nature 428:37–43

van Leeuwenhoek A (1684) An abstract of a Letter from Antonie van Leeuwenhoek, Sep. 12, 1683. About animals in the scrurf of the teeth. Philos Trans R Soc Lond 14:568–574

Vangay P et al (2018) US immigration westernizes the human gut microbiome. Cell 175:962–972.e910

Vatanen T et al (2022) Transcription shifts in gut bacteria shared between mothers and their infants. Sci Rep 12:1276

Vemuri GN, Aristidou AA (2005) Metabolic engineering in the -omics era: elucidating and modulating regulatory networks. Microbiol Mol Biol Rev 69(2):197–216

Venter JC, Remington K, Heidelberg JF, Halpern AL, Rusch D, Eisen JA, Wu D, Paulsen I, Nelson KE, Nelson W et al (2004) Environmental genome shotgun sequencing of the Sargasso Sea. Science 304:66–74

Verberkmoes NC, Russell AL, Shah M, Godzik A, Rosenquist M, Halfvarson J, Lefsrud MG, Apajalahti J, Tysk C, Hettich RL, Jansson JK (2009) Shotgun metaproteomics of the human distal gut microbiota. ISME J 3:179–189

Viaud S et al (2013) The intestinal microbiota modulates the anticancer immune effects of cyclophosphamide. Science 342:971–976

Vijay-Kumar M, Aitken JD, Carvalho FA et al (2010) Metabolic syndrome and altered gut microbiota in mice lacking toll-like receptor 5. Science 328(5975):228–231

Villarreal L (2012) Viruses and host evolution: virus-mediated self identity. In: Lopez-Larrea C (ed) Self and nonself. Springer, New York, pp 185–217

Villas-Boas SG, Mas S, Akesson M, Smedsgaard J, Nielsen J (2005) Mass spectrometry in metabolome analysis. Mass Spectrom Rev 24(5):613–646

Virgin HW (2014) The virome in mammalian physiology and disease. Cell 157:142–150

Vyas U, Ranganathan N (2012) Probiotics, prebiotics, and synbiotics: gut and beyond. Gastroenterol Res Pract 2012:872716

Walsh AM, Crispie F, O'Sullivan O, Finnegan L, Claesson MJ, Cotter PD (2018) Species classifier choice is a key consideration when analysing low-complexity food microbiome data. Microbiome 6(1):50

Wang WL, Xu SY, Ren ZG, Tao L, Jiang JW, Zheng SS (2015) Application of metagenomics in the human gut microbiome. World J Gastroenterol 21(3):803–814. https://doi.org/10.3748/wjg.v21.i3.803. PMID: 25624713; PMCID: PMC4299332

Wang Q, Wang K, Wu W, Giannoulatou E, Ho JWK, Li L (2019) Host and microbiome multi-omics integration: applications and methodologies. Biophys Rev 11:55–65. https://doi.org/10.1007/s12551-018-0491-7

Wang DD et al (2021) The gut microbiome modulates the protective association between a Mediterranean diet and cardiometabolic disease risk. Nat Med 27:333–343

Wendisch VF, Zimmer DP, Khodursky A, Peter B, Cozzarelli N, Kustu S (2001) Isolation of *Escherichia coli* mRNA and comparison of expression using mRNA and total RNA on DNA microarrays. Anal Biochem 290(2):205–213

Wilmes P, Bond PL (2006) Metaproteomics: studying functional gene expression in microbial ecosystems. Trends Microbiol 14(2):92–97. https://doi.org/10.1016/j.tim.2005.12.006

Woese CR, Fox GE (1977) Phylogenetic structure of the prokaryotic domain: the primary kingdoms. Proc Natl Acad Sci U S A 74:5088–5090

Wooley JC, Godzik A, Friedberg I (2010) A primer on metagenomics. PLoS Comput Biol 6:e1000667

Xia LC, Cram JA, Chen T, Fuhrman JA, Sun F (2011) Accurate genome relative abundance estimation based on shotgun metagenomic reads. PLoS One 6:e27992

Xiao M, Yang J, Feng Y, Zhu Y, Chai X, Wang Y (2017) Metaproteomic strategies and applications for gut microbial research. Appl Microbiol Biotechnol 101(8):3077–3088. https://doi.org/10.1007/s00253-017-8215-7

Xiong W, Giannone RJ, Morowitz MJ, Banfield JF, Hettich RL (2015) Development of an enhanced metaproteomic approach for deepening the microbiome characterization of the human infant gut. J Proteome Res 14:133–141

Yatsunenko T, Rey FE, Manary MJ, Trehan I, Dominguez-Bello MG, Contreras M, Magris M, Hidalgo G, Baldassano RN, Anokhin AP et al (2012) Human gut microbiome viewed across age and geography. Nature 486:222–227

Yen S, Johnson JS (2021) Metagenomics: a path to understanding the gut microbiome. Mamm Genome 32(4):282–296. https://doi.org/10.1007/s00335-021-09889-x. Epub 2021 Jul 14

Zhang A, Sun H, Wang X (2012) Saliva metabolomics opens door to biomarker discovery, disease diagnosis, and treatment. Appl Biochem Biotechnol 168:1718–1727

Zhang W, Sun J, Cao H, Tian R, Cai L, Ding W, Qian PY (2016) Post-translational modifications are enriched within protein functional groups important to bacterial adaptation within a deep-sea hydrothermal vent environment. Microbiome 4:49

Zhang X, Li L, Mayne J, Ning Z, Stintzi A, Figeys D (2018a) Assessing the impact of protein extraction methods for human gut metaproteomics. J Proteome 180:120–127

Zhang X, Deeke SA, Ning Z, Starr AE, Butcher J, Li J, Mayne J, Cheng K, Liao B, Li L, Singleton R, Mack D, Stintzi A, Figeys D (2018b) Metaproteomics reveals associations between microbiome and intestinal extracellular vesicle proteins in pediatric inflammatory bowel disease. Nat Commun 9:2873

Zhang L et al (2020) Characterization of antibiotic resistance and host-microbiome interactions in the human upper respiratory tract during influenza infection. Microbiome 8:39

Zhang Y et al (2021) Metatranscriptomics for the human microbiome and microbial community functional profiling. Annu Rev Biomed Data Sci 4:279–311

Zhao L, Hartung T (2015) Metabonomics and toxicology. Methods Mol Biol 1277:209–231. https://doi.org/10.1007/978-1-4939-2377-9_15

10 Multi-omics in Study of Skin Microbiome

Krishna Joshi, Mansi Panchasara, Hetsi Goswami,
Ashok Kumar Bishoyi, and Gaurav Sanghvi

10.1 Introduction

Human health is being affected by communities of microorganisms residing on various human body parts. Human microbiome being diversified affects and supports human health as it has a mutual relations with the human host, and in turn affecting the status of illness in the human being. Microorganism including bacteria, viruses, and fungi invade on human skin, mucosal surfaces, gut, and other bodily parts. Microorganism present and affect human health exhibit potential diversity based on the factors such as age, life style, diet, medication, and other physical and external variables throughout the life span (Roux et al. 2022). Skin is considered as the human body's largest organ covering the outermost surface of an individual. Essentially, there are three layers that make up the skin. The epidermis layer is further divided into five layers on palm and soles, the other body portion only have four layers; it lacks the stratum lucidum. The layers are stratum corneum, stratum lucidum, stratum spinosum, and stratum granulosum. The dermis, which lies beneath the epidermidis, is made up of sweat glands, blood vessels, lymphatic vessels, connective tissue, and hair follicles. The third and most profound layer is the subcutaneous tissues. (Oakley et al. 2020)

Skin as an outermost layer functions as an obstacle that keeps away pathogen and foreign particle found in the environment from our body and thus serves as a first line of defense. The term skin microbiome collectively includes all such tiny organisms that used to present surrounding our ecosystem (Bates 2022). The good and healthy skin indicates the healthy functioning of microbiota. Various studies

K. Joshi · M. Panchasara · H. Goswami · A. K. Bishoyi
Department of Microbiology, Atmiya University, Rajkot, Gujarat, India

G. Sanghvi (✉)
Department of Microbiology, Marwadi University, Rajkot, Gujarat, India
e-mail: gaurav.sanghvi@marwadieducation.edu.in

I. Mani, V. Singh (eds.), *Multi-Omics Analysis of the Human Microbiome*,
https://doi.org/10.1007/978-981-97-1844-3_10

revealed link between various illness and microbiome of humans. Changes at genome, protein, and metabolites level are important to study about human health and diseases. Multi-omics data production and analysis at multiple levels has significant impact to understand disease biology with the advancement in sequencing technology. Clinical data analysis using multi-omics approach is more insightful to understand biological disorders. It appears that understanding complicated biology systematically and comprehensively may be achieved by the integration of multi-omics data that provide information on biomolecules from many levels (Roux et al. 2022). Shotgun metagenomics sequences whole DNA, which gives idea about genetic similarities and dissimilarities, same way 16S rRNA genes can be amplified and studies to know the taxonomic grouping for microbiome. Numerous analytical methods, like QIIME (quantitative insights into microbial ecology), have been created for the purpose of evaluating NGS data from focused 16S or metagenomics analysis (Skowron et al. 2021). Microorganisms usually found in the communities of specific niche in the environment where they get vital resources to grow and reproduce. Such microbial community plays a crucial role in hosts biological processes, like metabolism, immunity, and behavioral aspects.

After birth human skin is colonized by microorganisms. For neonates the level of microbial diversity is less, gradually it increase with the factors such as age, exposure with different environmental conditions, use of antibiotics, availability of oxygen, and other parameters (Skowron et al. 2021). However, healthy infant skin also exhibit the presence of certain phylum of bacterium such as Actinobacteria, Firmicutes, Proteobacteria, and Bacteroidetes. Majority of the skin is exposed with aerobic bacteria but some skin area such as hair follicle layer of a straight arm colonial can support anaerobic environment has been connected in the numerous host-associated conditions many of which is directly affect the human health. Some of the microbial community directly inhibit the pathogenic bacteria on the skin surface by producing the antimicrobial compound or by interacting with human system and activate the host innate immune response. Bacterial populations observed to be more diversified on human skin compared to fungal population. Among them, coagulase-negative staphylococci, particularly Staphylococcus epidermidis, anaerobic *Propionibacterium acnes*, *Corynebacterium*, *Micrococcus*, *Streptococcus*, and *Acinetobacter* are dominant species. However, skin microbiome also includes fungus, mostly of the *Ascomycota* and *Basidiomycota* families. The genus *Malassezia* is the most prevalent. Fungi diversity is highest on the feet. *Aspergillus*, *Rhodotorula Epicoccum*, and *Cryptococcus* are colonized on the feet. There is little information available regarding the abundance of virus population on human skin. This is due to the presence of a large number of microorganisms in the skin's microbiological environment. Viruses from *Papillomavirus* families (including alpha- and beta-virus, hepatitis-virus, and hepatitis-c virus) as well as family of *Polyomavirus* and *Circovirus* usually found in numerous demographic studies indicate viruses exhibiting a high degree of variation (Nash et al. 2018). The invasion, colonization, and infection by harmful microorganisms are prevented by microbiota itself which acts as a barrier (Ducarmon et al. 2019). Skin microorganism compared for a resource in the poly microbial community where they live and have developed the strategies

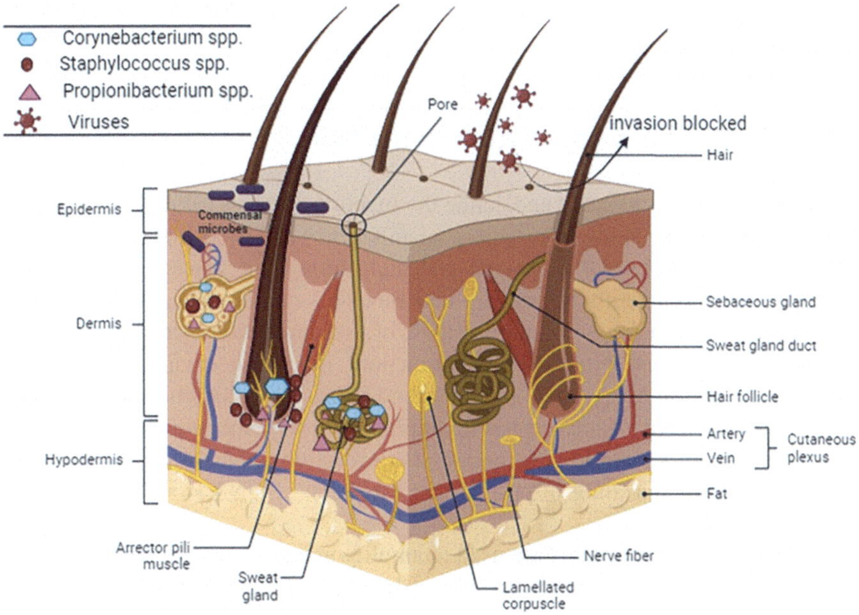

Fig. 10.1 Anatomy of skin cell

numerous a CoNS species, including *Staphylococcus hominis*, generate antibiotic that have distinctive chemical composition and high inhibitory effect against the main in pathogen staphylococcus aureus (Zhu et al. 2023). Beta *hemolytic streptococci, coryneform* bacteria, and *Staphylococcus aureus* are the primary pathogens responsible for the skin illness. When your systemic infection involves skin symptoms brought on by pathogens or talks, such as Malaysia, chicken pox, gonococcus, and staphylococcal scalded skin disease, these organisms frequently enter through a break in the skin, such as an insect bite. Due to their strong keratin affinity, dermatophyte fungi can enter the keratinized skin, hair, and nail tissue (Thye et al. 2022) (Fig. 10.1).

10.2 Skin Diseases and Microbes

Propionibacterium acne is the pathogen that causes acne vulgaris alters the circumstances that generate inflammatory response by increasing the quantity of sebum generated within a Pilosebaceous follicle. Staphyllococcus epidermis inhibits the growth of *Propionibacterium* Atopic dermatitis is caused by *S. aureus*, skin colonization, or infection by *Streptococcus*, and *Propionibacterium* (Alexander et al. 2020). Psoriasis disease is bacterial infection, all these show multiple microbial diversity. The process known as dysbiosis results in a damaged skin barrier and an unbalanced cutaneous microbiota contributes skin illness. However, overall pathogenicity of the skin microbiota is shaped by the balance between metagenomic components and the presence of disease-associated bacteria.

10.3 Multi-omics Approaches in the Skin

Genomics: its goal is to study and define the function of genes called genomics and metagenomics is the study of microorganism's genes and their function. With the advancement of molecular technique and high-throughput sequencing platforms microbial communities their structure and functions can be understood properly. The metagenomic study contributes isolation of vast number of microorganism having no homology, exhibits diversified functions in certain environments. DNA sequencing plays significant role to understand structure and functional differences of an organism. Gene of interest has to amplify prior than the sequencing with the help of technique known as polymerase chain reaction, which involved the steps like denaturation, annealing of primer, and extension of DNA strand using deoxy nucleotides.

Gene-specific PCR primers are used to amplify universal genes, such as the 16S rRNA gene found in bacteria, and sequence the target sections rather than the entire genomes. Following a database comparison, rRNA sequence similarity is used to construct operational taxonomic units. To comprehend the taxonomic diversity of a microbial community engaged in a certain metabolic pathway, non-universal, function-specific target genes can also be employed, such as nifH for nitrogen fixation. Although molecular markers have gained popularity as a method of analyzing taxonomic diversity, they do not isolate the function or abundance of the microbial community as a whole (Fadiji and Babalola 2020). Few techniques utilized for sequencing are Oligo Ligation Detection (SOLID), Illumina, Ion Torrent, HeliScope, Pacific Biosciences (PacBio), 454/Roche, and Oxford Nanopore.

10.3.1 Metagenomics Next-Generation Sequencing (mNGS)

In most studies, mNGS has sensitivity similar to specific PCR assays (Mitchell and Simner 2019). Two types of methods are widely used for metagenomics next generation sequencing. first one is ribosomal community profiling or 16s rRNA gene profiling and another is shotgun sequencing or whole-genome sequencing. 16s rRNA gene used for phylogenetic analysis is only used for bacteria, fungi, and other prokaryotes except for viruses because viruses do not have conserved regions, and shotgun sequencing is used for all prokaryotes analysis. Ribosomal community profiling of certain conserved regions (16s rRNA, 18s rRNA, ITS regions) are amplified with PCR primers and sequenced. These conserved regions have variable regions that allow for the identification of organisms. For shotgun metagenomics the quality and amount of data are generated based on the route to take, if high-quality datasets de novo sequencing is often used, and for lower-quality sets, marker

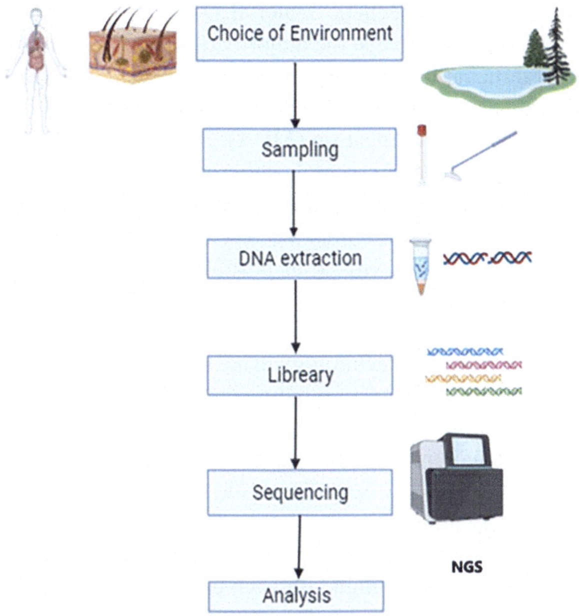

Fig. 10.2 Next-generation sequencing steps

gene analysis is preferred (Quince et al. 2017). There are three main steps for NGS (1) Preparation of library (2) Sequencing (3) Data analysis (Fig. 10.2).

For sample a significant amount of ambient environmental DNA was inserted into bacterial artificial chromosome or other plasmid and transformed into host *E.coli* cells, followed by sequencing via dideoxy nucleotide chain—termination sequencing. Further, the high-throughput sequencing systems were developed enables direct ambient DNA fragments sequencing which produces short-read datasets with tens of thousands to millions of reads. By finding related genes, biochemical pathways, and metabolic processes, this new method for microbial population determination allowed researchers to better comprehend a microbial community's taxonomic composition as well as its functional potential. The challenging part of metagenomics study is to isolate high quality of DNA from the environmental samples by eliminating the organic and inorganic pollutants (Quince et al. 2017).

Although this method yields a greater amount of unbiased DNA, it frequently leads to the simultaneous withdrawal of polyphenols and humic acid, which can interfere with subsequent enzymatic stages. As a result, it necessitates substantial purification, which may cause nucleic acid material to be lost. In addition to that various samples isolated from human stools, skin wash, clinical biopsies, and oral specimens are prone to be associated with host genetic material is also a major challenge to overcome. Such sample issues produced in datasets that are predominantly composed of human-originated reads must be eliminated before data analysis. Commercially available kits from the manufacturers like New England Biolabs (NEB), Molzym, and MoBio useful in extraction and purification of human DNA contaminated from microbial genetic material (Kelly 2013). The relatively poor DNA proportion recovered amongst the majority of samples isolated from host contributes another significant technological barrier to creating metagenomic libraries connected with humans.

Multiple displacement amplification (MDA), a random whole-genome amplification technique, has been used in several studies to boost DNA yield overall (Wang et al. 2022). However, this method is known to exhibit biases, such as preferentially amplifying low GC sequences. Illumina has recently developed a unique technique of library synthesis that uses in vitro transposition to build libraries from as little amount of starting material.

The intricacy of the information and the enormous volume of reads produced by the present order of sequencing technology make the processing of high-throughput metagenomic datasets a significant computing task. The recent rise in popularity of metagenomics has prompted the creation of a plethora of new tools that can handle massive data volumes and understand complex datasets to address the fundamental concerns about the structure and function of microbiological population (Lopatkin and Collins 2020). To accurately anticipate the functional and taxonomic makeup of a community of microorganisms, several stages must be taken, and a wide range of computational techniques are available for each one. A typical metagenomic analysis project entails, among other things, the following main steps:

(a) Screening raw data sets for contaminant from the environment, sequence artifacts, and poor grade reads.
(b) Categorized the filtered reads in different taxa to determine its evolutionary composition.
(c) All filtered reads are assembled to larger continuous to obtain reconstruction of the entire genome.
(d) Prediction of genes to assess functional potential of the community.

The machinery used for sequencing is called the NGS platform. There are several platforms on the market, and they vary in terms of technology and bulk sequencing methodology. Principal NGS platforms include:

10.3.2 Roche/454 (Pyro Sequencing)

It was introduced in 2005 and is a ground-breaking platform in NGS. This technique is based on the detection of pyrophosphate produced following the incorporation of a nucleotide in the newly synthesized DNA strand. The Platform can produce around one million reads per run and has a maximum read length of 1000 base pairs for amplicons and 600 base pairs for genomic DNA (Fig. 10.3).

Library preparation involves the extraction and purification of high molecular weight genomic DNA and its multiplexing. Followed by ligation onto tiny capture beads, each DNA fragment is then amplified in a water-oil emulsion complex called a micro-reactor, creating millions of clonally amplified DNA templates on each

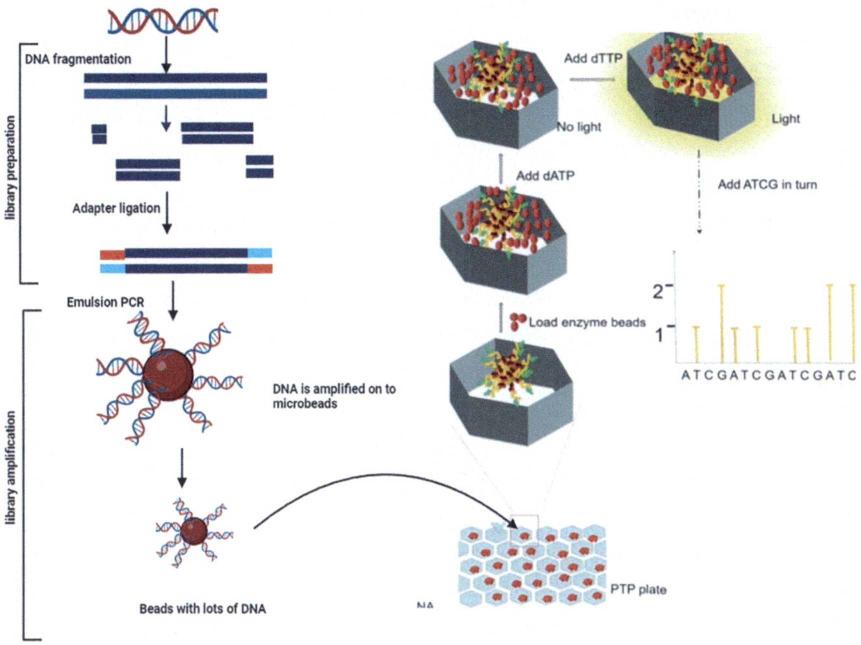

Fig. 10.3 Roche sequencing steps

bead. Each bead is placed onto a single Picotiter Plate, a plate well with a diameter of around 29 m after enrichment. Deoxynucleoside triphosphates (dNTPs) are consecutively flown over the immobilized templates in the order TACG during the sequencing-by-synthesis procedure, which results in the release of pyrophosphate (PPI) if a complementary base is integrated by DNA polymerase (Meera Krishna et al. 2019). The energy necessary for luciferase to oxidize luciferin is provided by the conversion of PPi to ATP in the presence of ATP sulfurylase and adenosine 5′ phosphosulfate (APS). This last reaction produces visible light that is measured by a charge-coupled device (CCD) camera and output as a flow gram in proportion to the number of bases included. Limitation of this technique is to sequence nucleotide sequence of certain length is difficult and data analysis is little complex.

10.3.3 Torrent Ion

The technology used by this platform is similar to Roche's. A hydrogen ion is produced as a by-product when a nucleotide is integrated into a DNA strand by a polymerase. Ion Torrent's Jon Personal Genome Machine (PGMTM) sequencer performs nucleotide incorporation in a highly parallel fashion using a high-density array of micro-machined wells (Foox et al. 2020) (Fig. 10.4).

Ion-sequencing devices for semiconductor ions may detect hydrogen ions. Similar to other semiconductor chips used in electrical equipment, these ion

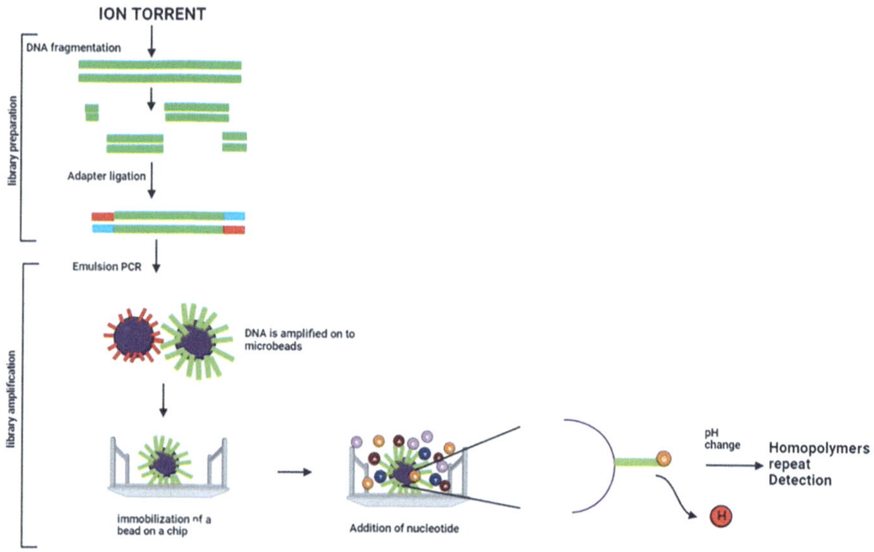

Fig. 10.4 Torrent Ion-sequencing steps

semiconductor chips are created and produced. Depending on the type of ion semiconductor sequencing chip utilized, the Ion Torrent PGM (Personal Genome Machine) produces a total data output of between 10 and 1000 MB. The Ion Proton, Ion Torrent's larger system, was introduced in September 2012. It makes use of bigger, higher-density chips, making it appropriate for whole- and exome-genome sequencing.

Ion Torrent produces reads with an average length of 200 bp that are needed to complete the assembly made by other technologies. Ion Torrent can be a good option in some situations, despite the read duration being significantly shorter than Roche-454 and PacBio because of cheaper costs. This technique's quick run time also makes it possible to run it numerous times at once to generate more data in a given amount of time (Tsuchiya et al. 2023).

10.3.4 Illumina's Sequencing Method

For quick and precise large-scale sequencing, Illumina's sequencing method makes use of clonal array construction and its reversible terminator technology. The sophisticated and adjustability sequencing system enables a broad array of applications in genomics, transcriptomics, and epigenomics (Fig. 10.5).

Sequencing templates are immobilized on a custom flow cell surface that is made to allow enzyme access while assuring high surface-bound template stability and minimal non-specific binding of fluorescently labeled nucleotides. Nearby (diameter of one micron or less), solid-phase amplification generates up to 1000 identical

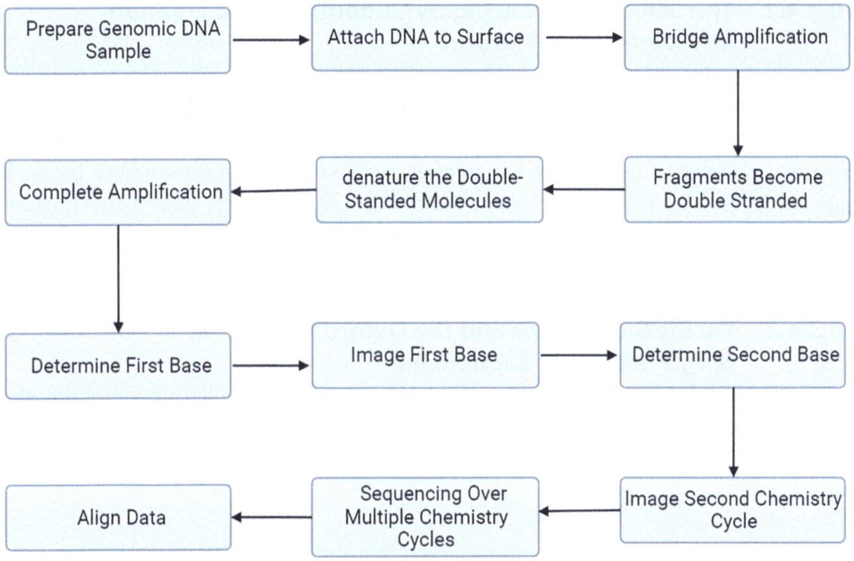

Fig. 10.5 Illumina sequencing steps

copies of each template molecule. This method produces densities of around ten million single-molecule clusters per square centimeter without the need for photolithography, mechanical spotting, or placing beads into wells.

The tens of millions of clusters on the flow cell surface are sequentially sequenced using the sequencing by synthesis (SBS) technique using four fluorescently labeled nucleotides (Knetsch et al. 2019). Each deoxy nucleotides are incorporated and photographed with the help of fluorescent dye for its identification followed by cleavage the enzyme for further incorporation of subsequent nucleotides. The labeled pair of nucleotide sequence serves as the closure loop for further polymerization process. In contrast to previous technologies, here signal intensity is assessed at the end of the cycle which significantly lowers raw error rates (Hu et al. 2021). The outcome is highly accurate base calling across the genome, including repeated sequence sections and inside homopolymers, which is made possible by precise base-by-base sequencing that removes sequence-context-specific mistakes (Hu et al. 2021).

10.3.4.1 Issues with Illumina Sequencing

The high necessity for sample loading control, which can lead to overlapping clusters and poor sequencing quality, is one of the key downsides of the Illumina/Solexa platform. As a result, the total error rate of this sequencing method is roughly 1%.

10.3.4.2 The SOLiD (Sequencing by Oligonucleotide Ligation and Detection) Platform

Under the heading of "Life Technology," applied biosystem debuted this technology in 2006. Similar to sanger sequencing, the SOLiD (sequencing by oligonucleotide ligation and detection) platform is based on the detection of fluorescence signals. However, different from sanger sequencing, which employs a fluorophore for each nucleotide, SOLiD sequencing uses a fluorophore for a specific pair of two nucleotides. Because of its expensive cost and short-read length, this sequencing approach is not very popular in the scientific community (El Bairi et al. 2020) (Fig. 10.6).

10.3.4.3 Pacific Biosciences and the Oxford Nanopore: Single-Molecule Sequencing

It is a third-generation sequencing and modern platform capable of overcoming the problems which occur in other sequencing methods, for example, short reads, reading errors or even bench work for sample preparation (Mikheenko et al. 2022).

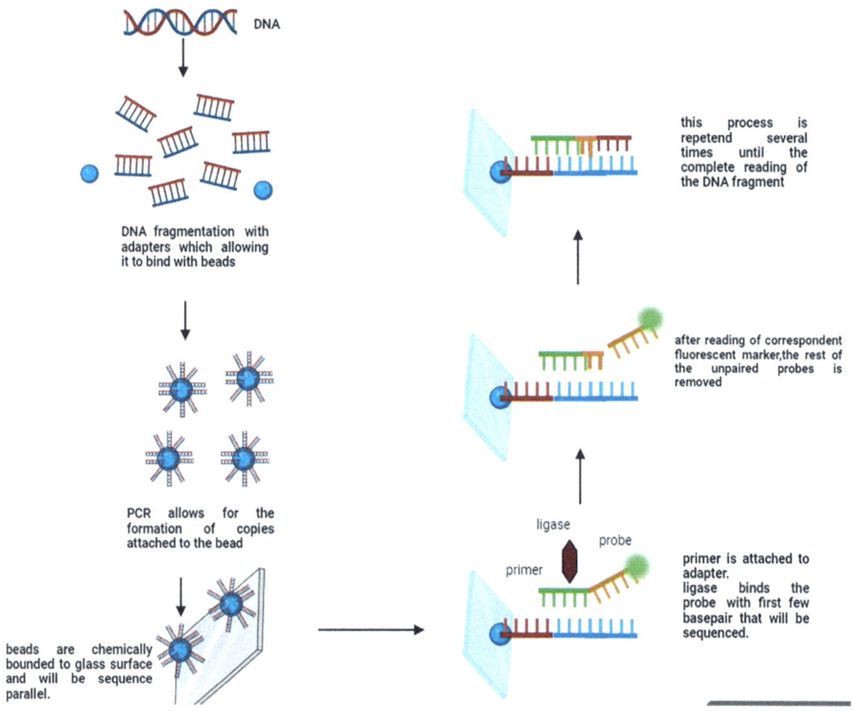

Fig. 10.6 Shotgun metagenomics approach

10.4 Approaches to Study Metagenomes

10.4.1 Shotgun Metagenomics

The majority of the genomes that are presented in an environmental sample can be fully sequenced using shotgun sequencing metagenomics. These result in the creation of profiling of community biodiversity that can be further linked to functional composition studies of both known and unidentified organism lineages that can be genera or taxa. Shotgun metagenomics has developed to address the issues of the presence of community of environmental organisms, their functional and microbial interaction with the environment, maintains healthy ecological niche (Zhang et al. 2021) (Fig. 10.7).

Entire genetic material is intended to be sequenced using shotgun metagenomics, which allows for the concurrent examination of the myco- and viro-biomes as well as the whole microbiome. Additionally, obtaining functional information is possible with sufficient depth of study which includes studying the evolution of resistance to antibiotics, finding new genes in specific settings, comparing the proportion of each of different microbial species, based on the goals of the study. According to Sui et al. (2020), enrichment procedures are often necessary when

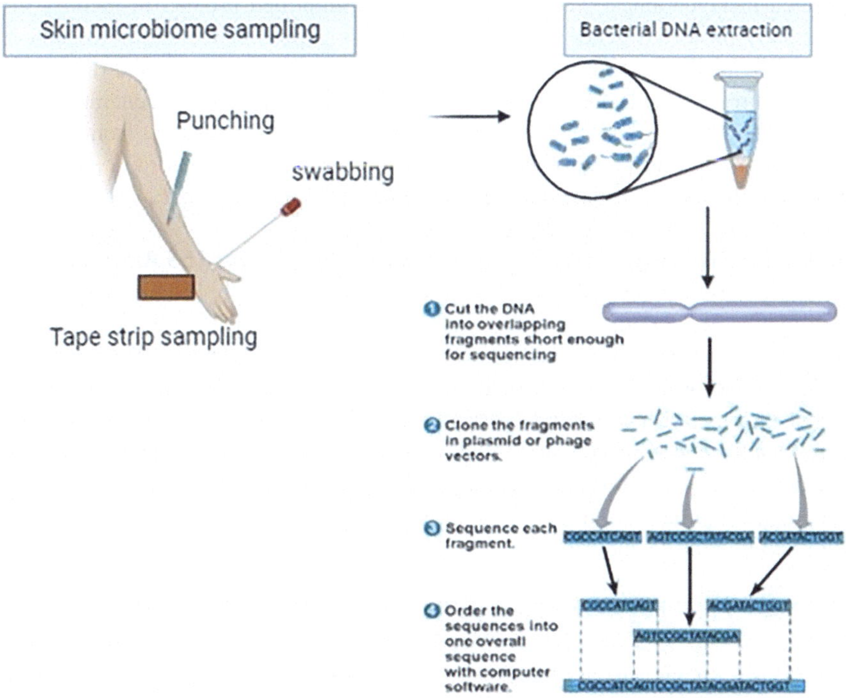

Fig. 10.7 Shotgun metagenomics approach

dealing with samples that have low microbial concentration because the majority of the genetic material originates from the host rather than the microbial cells as well as analyzing metagenomic data without enrichment is far expensive and computationally challenging. This technique involves extraction of DNA from eukaryotic cells, dividing them based its distinct features including their methylation pattern. However, such enrichment might induce bias into sequencing data on its own. In contrast, because the bacterial concentration of gut microbiome samples is high enough, enrichment is frequently unnecessary. Shotgun sequencing consumes far more compute and resources than a targeted technique. Analysis techniques are fast changing in response to data complexity, with innovative methodologies such as machine learning being used. Because of the large quantity of data obtained and the greater resolution, this is the preferable approach for dissecting high-complexity microbial communities or conducting comparative research. Some of the WGS issues may be rectified when third-generation sequencing technologies become more common. Pacific Biosciences' or Oxford Nanopore Sequencing, for example, do not use amplification technologies and instead use single-molecule sequencing (Mikheenko et al. 2022).

10.4.2 Targeted Metagenomics

Targeted metagenomics, which uses nucleotide amplified product to identify the universally protected gene, is a quicker and less expensive technique to achieve community profiling. Using the obtained results, the classification and dispersion can be correlated with the current present information (metadata) generated from the premises of the sampling being looked into. The several sequences from a single gene that are present in all of the species during targeted metagenomics can act as proxies for the taxa that they individually represent (Liu et al. 2022).

10.4.2.1 Profiling of the Bacterial Community Using 165 Amplicons

Most research involves examining the microbiota, particularly if there is any interest in the bacterial component. The gene that encodes the RNA for the ribosomal small subunit (SSU), commonly referred to as "16s," is by far the most widely used genomic area for research on bacterial diversification. With the remarkable exception of viruses, all living things possess the 16S rRNA gene, which accounts for more than 80% of all bacterial RNA. The gene is ideally suited for PCR amplification and sequencing since it contains both conserved and variable regions. The many gene fragments identified in the various organisms of a given DNA sample are amplified by PCR using primers.

10.4.2.2 Profiling of the Fungus Population Using ITS, LSU, and SSU Amplicon Sequencing

There is less unanimity over the gene of interest in surveys of fungal diversity. The Internal Transcribed Spacer (ITS), Large Subunit (LSU), and Small Subunit (SSU) sections are the three most often used ribosomal loci. It should be noted that "SSU" is the more general term for the gene known as "16s" in bacteria; however, it is usually referred to as "18s" in eukaryotes due to its higher molecular weight. ITS is the most efficient locus for delivering species level identifications (Rana et al. 2024).

10.4.2.3 Transcriptomics

RNA transcript sequencing is known as metatranscriptomics. The whole set of actively transcribed genes can be investigated in any environment using metatranscriptomic sequencing (Mukherjee and Reddy 2020). This method enables the capture of a moment in time of molecular activity occurring in the organisms present in a sample. Transcripts from both the host and the microbiome's members may be present, allowing researchers to analyze host–microbe interactions and learn not only how the microbiome functions on its own but also how the host may respond to a microbiome member.

The below steps are involved in metagenomics process, RNA isolation form microbiome with the help of metarancriptomics techniques, followed by RNA enrichment based of interest of RNA to be sequenced. RNA is fragmented to smaller pieces followed by cDNA synthesis. Ligation of adapters and 5′-3′ ends of the cDNA are repaired followed by the construction of cDNA metagenomic libraries. Similar to the metagenomics pipeline, the conventional bioinformatics pipeline for analyzing the data from a metatranscriptomic experiment is split into two strategies:

(a) Tagging sequence reads to genes and genomes of reference, and
(b) Assembling new transcriptome.

Skin microbiome study is very important using metatranscriptomic sequencing which tracks the involvement of microbiome in various illness and their response in treatment aspects. Researchers also found that transcriptional activity of skin microbiomes changes with types of illness. Moreover, few researchers have used metatranscriptomics to explore the skin microbiome, and these studies have typically focused on specific gene expression metrics [e.g., utilizing reverse transcription (RT)-quantitative PCR (qCR)]. Metatranscriptomics are used widely but is has few limitations like used widely there are certain limiting factors such as, sample collection and storage, to get sufficient quantity and quality of RNA from human microbiome, mRNA enrichment protocols using rRNA elimination, average lifespan of mRNA which affects detection of short term and quick reactions of environmental changes, insufficient transcriptome database, host RNA contamination and use of poly A-RNA selection kits for mRNA of prokaryotes.

10.5 Future Perspective

10.5.1 Developing Therapeutics and Diagnostics

Creating treatments for skin disorders is one of the long-term objectives of skin microbiome research, and this is why sequencing data is being utilized more and more. Ideally, a case-control study design is needed to develop treatments, where samples are taken from patients and healthy controls at skin areas exhibiting symptoms, in order to assess future results, ex vivo, in vivo, or in vitro preclinical models using bacteria or other additional relevant microbiological components should be investigated. If the bacteria of interest can mimic disease characteristics in animal models and/or skin systems, then more research may be done to discover how the illness is caused. The results of the above described procedures can then be used to develop and test a treatment further. A probiotic used in bacteriotherapy may be identified using a similar approach to identify good bacteria that are lacking in patients but present in controls. Through the application of bacterial consortia with beneficial traits, the ecology of skin with a particular medical condition can be modified through skin bacteriotherapy.

10.5.2 Multi-Kingdom Interactions

On the skin, this multi-kingdom interaction can occur across and within different species, subspecies, and strains of bacteria, viruses, fungi, archaea, and the host. Given the abundance of subject-specific species-, subspecies-, and strain-level diversity, it has been proposed that some multi-kingdom interactions may also be disease- and specific to the subject.

10.6 Conclusion

Identification of the skin microbiome is difficult to work with the samples of low biomass. Sequencing methods like metatranscriptomics is useful for taxonomic classification and the functional activity. Shotgun metagenomics and 16s rRNA are important tool for the idnetification of the normal, oppurtunistic and pathogenic flora associated with the skin microbiome.

References

Alexander H, Paller AS, Traidl-Hoffmann C, Beck LA, De Benedetto A, Dhar S et al (2020) The role of bacterial skin infections in atopic dermatitis: expert statement and review from the International Eczema Council Skin Infection Group. Br J Dermatol 182(6):1331–1342

Bates M (2022) The role of the skin microbiome in health and disease. IEEE Pulse 13(4):8–13

Ducarmon QR, Zwittink RD, Hornung BVH, Van Schaik W, Young VB, Kuijper EJ (2019) Gut microbiota and colonization resistance against bacterial enteric infection. Microbiol Mol Biol Rev 83(3):10–1128

El Bairi K, Azzam F, Amrani M (2020) The arrival of next-generation sequencing: an overview of current technologies. In: Illuminating colorectal cancer genomics by next-generation sequencing: a big chapter in the tale. Springer, Cham, pp 73–89

Fadiji AE, Babalola OO (2020) Metagenomics methods for the study of plant-associated microbial communities: a review. J Microbiol Methods 170:105860

Foox J, Tighe SW, Nicolet CM, Zook JM, Byrska-Bishop M, Clarke WE et al (2020) Multi-platform assessment of DNA sequencing performance using human and bacterial reference genomes in the ABRF next-generation sequencing study. bioRxiv, 2020-07

Hu T, Chitnis N, Monos D, Dinh A (2021) Next-generation sequencing technologies: an overview. Hum Immunol 82(11):801–811
Kelly J (2013) Metagenomic and genomic analysis of the skin microbiota. Doctoral dissertation, University of Liverpool
Knetsch CW, van der Veer EM, Henkel C, Taschner P (2019) DNA sequencing. In: Molecular diagnostics: part 1: technical backgrounds and quality aspects. Springer, pp 339–360
Liu S, Moon CD, Zheng N, Huws S, Zhao S, Wang J (2022) Opportunities and challenges of using metagenomic data to bring uncultured microbes into cultivation. Microbiome 10(1):76
Lopatkin AJ, Collins JJ (2020) Predictive biology: modelling, understanding and harnessing microbial complexity. Nat Rev Microbiol 18(9):507–520
Meera Krishna B, Khan MA, Khan ST (2019) Next-generation sequencing (NGS) platforms: an exciting era of genome sequence analysis. In: Microbial genomics in sustainable agroecosystems, vol 2. Springer, Singapore, pp 89–109
Mikheenko A, Prjibelski AD, Joglekar A, Tilgner HU (2022) Sequencing of individual barcoded cDNAs using Pacific Biosciences and Oxford Nanopore Technologies reveals platform-specific error patterns. Genome Res 32(4):726–737
Mitchell SL, Simner PJ (2019) Next-generation sequencing in clinical microbiology: are we there yet? Clin Lab Med 39(3):405–418
Mukherjee A, Reddy MS (2020) Metatranscriptomics: an approach for retrieving novel eukaryotic genes from polluted and related environments. 3 Biotech 10(2):71
Nash MV, Anesio AM, Barker G, Tranter M, Varliero G, Eloe-Fadrosh EA, Nielsen T, Turpin-Jelfs T, Benning LG, Sánchez-Baracaldo P (2018) Metagenomic insights into diazotrophic communities across Arctic glacier forefields. FEMS Microbiol Ecol 94(9):fiy114. https://doi.org/10.1093/femsec/fiy114. PMID: 29901729; PMCID: PMC6054269
Lopez-Ojeda W, Pandey A, Alhajj M, Oakley AM (2020) Anatomy, skin (integument)[Internet]. Treasure Island: StatPearls Publishing
Quince C, Walker AW, Simpson JT, Loman NJ, Segata N (2017) Shotgun metagenomics, from sampling to analysis. Nat Biotechnol 35(9):833–844
Rana S, Kumar M, Maurya DK, Singh SK (2024) Fungal metagenomics: an emerging approach to determine the imperceptible facets of fungal diversity. In: Applied mycology for agriculture and foods. Apple Academic Press, pp 19–36
Roux PF, Oddos T, Stamatas G (2022) Deciphering the role of skin surface microbiome in skin health: an integrative multiomics approach reveals three distinct metabolite–microbe clusters. J Invest Dermatol 142(2):469–479.e5. https://doi.org/10.1016/j.jid.2021.07.159. Epub 2021 Jul 31. PMID: 34343557
Skowron K, Bauza-Kaszewska J, Kraszewska Z, Wiktorczyk-Kapischke N, Grudlewska-Buda K, Kwiecinska-Piróg J, Wałecka-Zacharska E, Radtke L, Gospodarek-Komkowska E (2021) Human skin microbiome: impact of intrinsic and extrinsic factors on skin microbiota. Microorganisms 9:543. https://doi.org/10.3390/microorganisms9030543
Sui HY, Weil AA, Nuwagira E, Qadri F, Ryan ET, Mezzari MP et al (2020) Impact of DNA extraction method on variation in human and built environment microbial community and functional profiles assessed by shotgun metagenomics sequencing. Front Microbiol 11:953
Thye AYK, Bah YR, Law JWF, Tan LTH, He YW, Wong SH et al (2022) Gut–skin axis: unravelling the connection between the gut microbiome and psoriasis. Biomedicines 10(5):1037
Tsuchiya KD, Funke B, Hegde M, Santani A, Souers RJ, Szelinger S et al (2023) Implementation, evolution, and laboratory performance of methods-based proficiency testing for next-generation sequencing detection of germline sequence variants. Arch Pathol Lab Med 1:1–9
Wang X, Liu Y, Liu H, Pan W, Ren J, Zheng X et al (2022) Recent advances and application of whole genome amplification in molecular diagnosis and medicine. MedComm 3(1):e116
Zhang L, Chen F, Zeng Z, Xu M, Sun F, Yang L et al (2021) Advances in metagenomics and its application in environmental microorganisms. Front Microbiol 12:766364
Zhu Y, Yu X, Cheng G (2023) Human skin bacterial microbiota homeostasis: a delicate balance between health and disease. mLife 2(2):107–120

Multi-omics in Study of Oral Microbiome 11

Duy Ha Nguyen and Dinh-Toi Chu

11.1 Introduction

The method of studying how microbes impact host physiology has changed in the study of human microbiomes (Gilbert et al. 2018). The advent of independent culture techniques has improved the capacity to discover and categorize microbial species within the microbial community (Rappe and Giovannoni 2003), and next-generation sequencing and shotgun sequencing have become commonplace. It has become a widespread technique for ascertaining the composition and genetic constitution of the human microbiome (Bikel et al. 2015). Other multi-omics including genomics, metagenomics, proteomics, metaproteomics, and metabolomics help to establish mechanistic ideas about pathogenic microbial pathways involved in health and illness (Petruschke et al. 2020; Vernocchi et al. 2016). Exogenous foreign compounds are continually present in the oral microbial community (Baker and Edlund 2018). Such situations are defining microbial formation elements and their survival in this environment, while also producing a distinct interaction between bacteria and host. Pioneer bacteria in the oral cavity, like *Streptococcus salicylic (S. salicylic), Streptococcus gordonii (S. gordonii), Streptococcus sanguinis (S. sanguinis)*, and *Streptococcus mitis (S. mitis)* possess core traits that render them well adapted to this habitat because they can selectively bind filters with pre-dental tongue and buccal cells present and can outcompete other microbial species (Baker and Edlund 2018; Long and Swenson 1976). The underlying molecular pathways of

D. H. Nguyen
Vietnam Military Medical University, Hanoi, Vietnam

D.-T. Chu (✉)
Vietnam Military Medical University, Hanoi, Vietnam

Center for Biomedicine and Community Health, International School, Vietnam National University, Hanoi, Vietnam
e-mail: toicd@vnu.edu.vn

I. Mani, V. Singh (eds.), *Multi-Omics Analysis of the Human Microbiome*,
https://doi.org/10.1007/978-981-97-1844-3_11

various oral disorders remain unclear due to the intricacy of their development. Using multi-omics to better understand the pathophysiology of oral microbiome problems will enhance prevention, diagnosis, therapy, and repair techniques. The progress achieved in genomes, transcription, proteomics, metabolites, and the oral microbiome is reviewed in the following pages, and we explain how to combine data from various approaches using integromics and systems biology. Research utilizing multiple "omics" techniques clarifies possible pathogenic objectives, pathophysiological processes, and biomarkers for therapeutic involvement for the incidence and advancement of oral disease, and gives a significant new point regarding limiting its progression. Typically, each form of omics data gives a list of disease-related changes. These data may be valuable as disease process indicators as well as for providing insights into biological processes or processes that differ between disease and control groups. However, the study is confined to correlations, which mostly represent reactive rather than causative processes. Integrating various forms of omics data is frequently used to identify probable causative alterations to illness or therapeutic targets, which can subsequently be investigated in additional molecular research (Hasin et al. 2017).

11.2 Oral Microbiome

The oral cavity is found anteriorly above the face, beneath the nasal cavity. The lips, cheeks, palate, pharynx, and posterior (inner) pharynx link with the oral cavity. Through the slot, the gap opens out to the face. While in the back, most of the information flows through an organ known as the pharynx, which is surrounded by the oral cavity and the pharynx itself. Some frames add to the frame by creating a vacancy. The aperture frame and the temporal frame are coupled with the upper skeleton. Skeletal frameworks and isolated structures.

The human oral microbiome is defined as all bacteria identified on or in the human mouth cavity that end at the distal esophagus (Dewhirst et al. 2010). Humans have found oral microbiota owing to next-generation sequencing technology (NGS). Researchers uncovered approximately 700 species by examining 16S rRNA genes, of which roughly 280 (40%) are culture-identifiable and formally designated (Dewhirst et al. 2010; Yamashita and Takeshita 2017). As a result, the standard culture approach is limited in its ability to regulate all microorganisms in the oral cavity (Gao et al. 2018). The oral microflora is particularly rich and diverse because the oral cavity has a neutral pH, is rich in oxygen, and provides an abundant food supply for bacterial development. The HOMD database (Human Oral Microbial Database) contains 400 genomes and 1300 bacterial strains (Gao et al. 2018). Oral bacteria that have been sequenced account for roughly 26% of all locations in the human body (Griffen et al. 2011; Ahn et al. 2011). The human buccal cavity possesses a diverse microbial environment encompassing bacteria, infections, yeasts, and archaeans. Bacteria that make up a substantial portion of the oral microbiota are mostly from the phyla *Firmicutes*, *Fusobacteria*, *Proteobacteria*, and *Actinobacteria* (Costalonga and Herzberg 2014). The components alter depending on where they are in the mouth cavity. Microorganisms from the tongue biofilm are present in the salivary microbiome. The papillary surface of the tongue contains a microbiome dominated by anaerobic varieties such as the genera *Prevotella* and *Veillonella*, whereas the undersurface of the tongue is dominated by *Streptococcus* spp. and *Gemella* spp. (Mager et al. 2003). A number of genera have been characterized in the salivary microbiota, with the most frequent or native being *Prevotella, Streptococcus, Veillonella, Neisseria* spp., and *Haemophilus* spp. (Keijser et al. 2008). Transient or heterozygous genes have a frequency of less than 1% (Mager et al. 2003). The bacteria on the crown of the teeth differs somewhat from that within saliva, with the genera *Streptococcus*, *Veillonella* being the most predominant, then comes *Corynebacterium*, *Actinomyces*, *Fusobacterium*, and *Rothia* (Keijser et al. 2008). Because of differences in oxygen content, exposure to saliva, serum, and food plaque, the microbial makeup on the gums differs from that below the gums. The bacteria lives and grows in the ecological niches of the tooth. These bacteria are strongly connected to human oral disorders. The bacteria of healthy and cleft gingival sulcus has recently been determined using 16S rRNA gene sequencing of regions such as V1–V2, V4 (Simon-Soro et al. 2013; Griffen et al. 2012). The phylum *Proteobacteria*, particularly the genera *Gammaproteobacteria, Acinetobacter*, *Haemophilus,* and *Moraxella* is most frequent in healthy gingival grooves. The class *Bacilli*, which includes the genera *Streptococcus*, *Granulicatella*, and *Gemella* is also associated with health in the phylum *Firmicutes.* The genera are symbionts, and they return to the periodontal pocket at a significant rate following periodontal therapy (Yamanaka et al. 2012). Previously, *P. gingivalis*, *T. forsythia*, and *T. denticola* were thought to be the microbes most linked with deep periodontal disease using standard culture techniques (Socransky et al. 1998). Despite being encountered in modest figures in healthy individuals (Hinrichs et al. 1985; Kumar et al. 2003), these species are thought to be responsible for the disease's origin and development. Many more uncultured species have been discovered as science and technology have advanced, particularly with the 16S rRNA gene sequencing with the V1–2, V4–6, and V7–9 gene sections. The sequencing approach has resulted in the discovery of novel bacteria from the phylum *Spirochaetes*, including the genera *Treponema* and *Sinergistes* (Vartoukian et al. 2009), as well as the phylum *Bacteroidetes*, which includes the species, *Porphyromonas* spp., *Prevotella* spp., and *Tannerella* spp. *Fusobacteria* classes *Fusobacterium* spp., and *Leptotrichia* spp. are also strongly linked to the condition. As illness severity grew, as assessed by deeper vesicles, the *Negativicutes* genera *Seleomonas* and *Megasphera* were more numerous (Griffen et al. 2012; Kumar et al. 2005; Hutter et al. 2003) while *Clostridia*, including the genera *Filifactor, Lachnospiraceae*, and *Peptostreptococcus*, became more dominant. In deeper illness areas, *Erysipelotrichia* related genera *Erysipelothrix*, *Solobacterium*, and *Bulleidia (*Griffen et al. 2012*;* Hutter et al. 2003*)*. Author D'Agostino et al. did a study on 34 research publications, the findings of which revealed a variation in the richness of the bacterial flora inside the mouth cavity between children and infants (Table 11.1). For infants aged 12–24 months, prominent taxa such as *Strenotrophomas, Burkholderia*, and *Enterobacteriaceae* (D'Agostino et al. 2022).

Table 11.1 Oral microbiome in infants and children

Infants	Children	Authors/year
Proteobacteria	*Firmicutes*	
Fusobacterium	*Proteobacteria*	
Actynobacteria	*Actinobacillus*	
Bacteroidetes	*Bacteroidetes*	D'Agostino et al. (2022)
Firmicutes	*Fusobacterium*	
Synergistetes	*Streptococcus*	
Tenericutes	*Prevotella*	
Capnocytophaga	*Veillonella*	
Neisseria	*Neisseria*	
Sreptococcus	*Rothya*	
Kingella	*Haemophilus*	
Leptotrichia		

The oral microbiome is critical to microbial ecology and human health (Zarco et al. 2012). An imbalance in the microflora of the mouth contributes to dental irregularities like dental caries, periodontitis, mucosal diseases of the oral cavity, and systemic conditions including digestive and neurological disorders. Caries is a chronic bacterial infection that causes a persistent breakdown of tooth hard tissue and advances owing to a range of factors. Dental caries is a pervasive dilemma that can afflict individuals of all ages, from youngsters to senior citizens. The genera *Streptococcus, Porphyromonas*, and *Actinomyces* are significant causal organisms of caries and may be possible signals of dental caries in the primary teeth set (Ma et al. 2015). Wang et al. (2017) used a single-molecule real-time DNA sequencing technique to characterize the oral microbiota. The author proposed that *Prevotella, Lactobacillus, Dialister*, and *Filifactor* could be involved in the pathogenesis and progression of caries (Wang et al. 2017). Agnello et al. (2017) used next-generation sequencing to compare the dental plaque microbiota of children with severe caries versus children with no caries. The outcomes differed considerably across groups. *S. gordonii* and *S. sanguinis* in the SECC group, *Veillonella* HOT 780, and *Porphyromonas* HOT 284 in the non-caries group. *Streptococcus mutans* levels were extremely high in the SECC caries group (Agnello et al. 2017).

Periodontal disease occurs commonly in the human mouth and is classified into two types: gum disease and periodontitis. Because of their aggressiveness and close relationship with infections, the oral anaerobes, *P. gingivalis*, *T. denticola*, and *T. forsythia* have historically been thought to be the causal agents of periodontitis, afflicted mind (Socransky and Haffajee 2000). Specific 16S rRNA gene analyses have revealed that certain bacteria, like *Fusobacterium nucleatum*, are more populous in patients with periodontitis and less bountiful in the blue complex group except *A. actinomycetemcomitas,* with all LAP groups encompassing green and blue bacteria (Topcuoglu and Kulekci 2015).

Squamous cell carcinoma accounts for more than 90% of all malignant growths in the mouth cavity and surrounding areas (Gholizadeh et al. 2016). Several studies have used next-generation sequencing to compare microbiome changes between healthy persons, epithelial precursor lesions, and cancer patients. They discovered that the presence of *Slackia, Enterococcus, Peptostreptococcus, Parvimonas,* and *Bacillus* corresponded with categorization into two groups in epithelial precursor lesions and cancer patients (Lee et al. 2017). Author Yang et al., study examined the composition of the oral microbiota in individuals with oral squamous cell carcinoma (OSCC) and uncovered a probable association between the oral microbiome and mutagenic modifications in OSCC (Yang et al. 2018).

11.3 Multi-omics for Oral Microbiome

Omics-based research has offered a unique viewpoint on biological systems at the molecular level, with all levels engaging in distinct but interconnected biological processes. Multi-omics is also a useful tool for discovering and prioritizing biomarkers, which is a hot issue in clinical research (Sussulini et al. 2022). Many new disciplines of inquiry, sometimes referred to as "omics," have emerged as science and technology have advanced. Starting with genomics, the word "genomics" allows for the speedy and efficient elucidation of complete genomes and the study of all genes at once, as opposed to evaluating genes individually (Yadav 2007). Proteomics is a term used to describe protein analyses, (Aslam et al. 2017), metabolism (describing various types of small molecules from a variety of sample types, (Schrimpe-Rutledge et al. 2016), and epigenetics (of the impacts of chromatin architecture, including greater chromatin coiling and fixation to the nuclear framework (Wang and Chang 2018). It also gave rise to a slew of additional "omics" terminology, including lipidomics (Han and Gross 2005), metagenomics (Riesenfeld et al. 2004), and glycomics (Raman et al. 2005).

With the rapid advancement of next-generation sequencing, our cognizance of the significance and functionality of the microbiome is growing. Recent research has concentrated on the makeup of biomes, or "microbiota," which are groupings of tiny organisms in any particular place, such as a bodily site, specialized habitat, or the environment (Thursby and Juge 2017). Metagenomics, metatranscriptomics, metabolomics, and metaproteomics are some of the methods available for studying the microbiome. To alleviate these constraints, shotgun metagenomic sequencing provides an alternate way to study uncultured microbiota. DNA is reclaimed from all cells in a community en masse. Instead of a solitary genomic locale being targeted for amplification, the entire DNA is subsequently fragmented into minute segments that are sequenced independently. This leads to DNA sequences (reads) matching different genomic sites for the plethora of genomes present in the sample, including non-bacterial genomes (Sharpton 2014). This information may be utilized in the same way as amplification-based approaches are used to establish which taxa are present and their relative abundance (Xia et al. 2011), as well as to determine the coding sequence. Exploiting the functional potential of the microbial population through genetic engineering (Qin et al. 2010). Metatranscriptomic techniques apply the same analytical assumptions as shotgun metagenomics, but they only gather RNA generated by microbial cells, allowing evaluation of these organisms' expression activity (Bashiardes et al. 2016). Metabolic analyses primarily seek to determine the metabolites created by the microbiome and how these substances interact with both the bacteria and host metabolism (Zierer et al. 2018). Small molecules such as antibiotics, antibiotic bio-products, and host and/or bacterial metabolic intermediates are frequently quantified using these approaches. Mass spectrometry is often employed in metabolomics to recognize known metabolites. Metaproteomics, which uses mass spectrometry as well, is concerned with the identification and quantification of proteins present in the microbiome (Lai et al. 2019). Metaproteomics and metabolomics are both fast-growing microbiome research tools (Fig. 11.1).

11.3.1 Genomics

- Depict a DNA double helix to represent genomics.
- Indicate arrows pointing toward other omics techniques, symbolizing the flow of information.

11.3.2 Metagenomics

- Show a magnifying glass over the oral microbiota to represent metagenomics.
- Highlight, the collection of microbial DNA from an oral sample.
- Connect metagenomics with genomics to emphasize the extraction of individual microbial genomes.

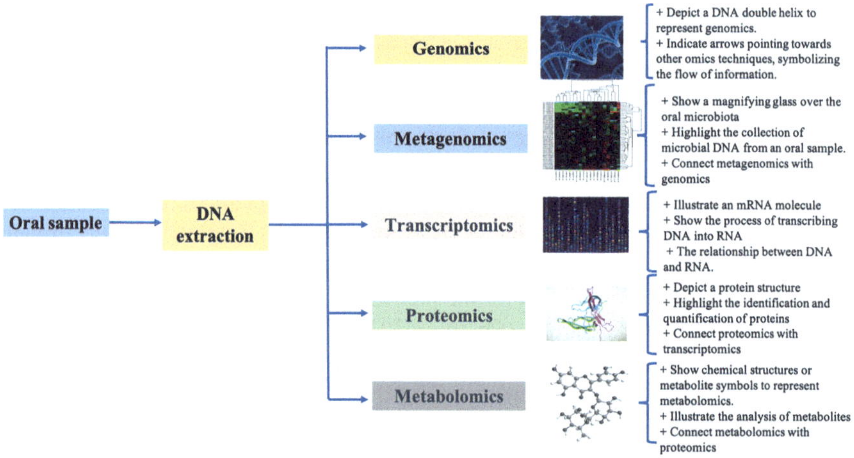

Fig. 11.1 A conceptual illustration of different "omics" methodologies used to examine oral microbial communities

11.3.3 Transcriptomics

- Illustrate an mRNA molecule to represent transcriptomics.
- Show the process of transcribing DNA into RNA within microbial cells.
- Connect transcriptomics with genomics to indicate the relationship between DNA and RNA.

11.3.4 Proteomics

- Depict a protein structure to represent proteomics.
- Highlight, the identification and quantification of proteins within the oral microbial community.
- Connect proteomics with transcriptomics to emphasize the translation of RNA into proteins.

11.3.5 Metabolomics

- Shows chemical structures or metabolite symbols to represent metabolomics.
- Illustrate the analysis of metabolites produced by the oral microbiota.
- Connect metabolomics with proteomics to indicate the downstream effects of protein activities on metabolite production.

Several studies have shown changes in plaque composition during advanced gingivitis (Huang et al. 2014; Teng et al. 2016), as well as the functional capacity of the oral microbiota during gingivitis development. Metagenomics and metatranscriptomic approaches (Kistler et al. 2013; Wang et al. 2013) have been used to characterize the findings. However, the molecular mechanisms enabling the onset and progression of gingivitis have not been precisely determined due to an absence of longitudinal perspective encompassing the microbiome, its metabolites, and the host immune response (Huang et al. 2014). The permanent and destructive gingivitis stage of periodontitis comes from persistent, untreated gingivitis, a different phylogenetic structure of the oral microbiome, as evidenced by 16S rRNA genes or metagenome sequences (Li et al. 2014). Periodontitis-associated microbial communities' phylogenetic traits have been widely examined. Using whole-genome shotgun sequencing, Dabdoub et colleagues examined 73 bacterial subsets from 25 people with widespread chronic periodontitis and 25 people who were periodontically healthy. Taxa and genes present in a minimum of 80% of individuals in each group were used to calculate the core metabolic network. The core microbiome encompasses 50% of the genes and species classified as healthy. Locations that are clinically healthy in patients with periodontitis are more aligned with diseased sites than with truly healthy sites (Dabdoub et al. 2016).

The typical morphological shape and performance of the oral epithelium are critical in maintaining oral homeostasis, with the microbiota and chronic stress playing important roles. According to a multi-omics investigation, the molecules in the tongue that changed between particular disease and specific disease rats were mainly concentrated in metabolism of mitochondria and immune responses. Integral analyses of transcription, proteomics, and metabolites were used to investigate oral homeostasis in germ-free and germ-free rat models. The findings indicate that the microbiome may impact the architectural remodeling of the oral epithelium, as shown in cellular pigmentation and attachment, and that the microbiota can govern homeostasis via respiratory metabolisms and immunological responses. Furthermore, prolonged stress may improve these regulatory systems (Long et al. 2022). Standard omic dataset analyses concentrate on community structure or function. Untangling the complicated web of relationships between constituent entities remains a major issue. Understanding the nature and dynamics of sickness in complex oral ecosystems requires multiple levels of research that incorporate taxonomic analysis (composition), prospective function (genomic), functional (metaproteomics), and encoded functions (metaproteomics and metabolomics) (Duran-Pinedo and Frias-Lopez 2015). Even so, the delicate interaction of host genomics and environmental factors complicates the research. All of these supplementary techniques necessitate coordination to pinpoint the culprits in this convoluted condition. We examine milestones in oral illness research, concentrating on contemporary single and integrated omics investigations, supplemented by bibliography analysis of networks to address the microbiome, mouth microorganisms, host variables, and their relationships (Moussa et al. 2022).

11.4 Conclusion

Despite indisputable breakthroughs in the use of multi-omics to investigate interactions between the microbiota of the mouth and the host, interpreting biological data remains problematic. It is still difficult to know which bacterial species create specific metabolites that may be involved in health and illness. In reality, detailed information on the source of the metabolites (whether microbial or host) as well as knowledge of potential multidirectional effects will be required for the study results to be relevant in establishing treatment strategies for a specific medical problem. To do this, powerful analytic frameworks capable of handling the large range of omic datasets in this study area are required. To compensate for the shortcomings of each method, researchers may need to integrate numerous omic methods in the future rather of depending on just one or two. Such an approach would be a significant step toward getting proper nutrition or medication, as well as aiding in the understanding of the precise mechanism through which the oral microbiota affects health and illness.

References

Agnello M, Marques J, Cen L, Mittermuller B, Huang A, Chaichanasakul Tran N, Shi W, He X, Schroth RJ (2017) Microbiome associated with severe caries in Canadian first nations children. J Dent Res 96(2):1378–1385

Ahn J et al (2011) Oral microbiome profiles: 16S rRNA pyrosequencing and microarray assay comparison. PLoS One 6(7):e22788

Aslam B et al (2017) Proteomics: technologies and their applications. J Chromatogr Sci 55(2):182–196

Baker JL, Edlund A (2018) Exploiting the oral microbiome to prevent tooth decay: has evolution already provided the best tools? Front Microbiol 9:3323

Bashiardes S, Zilberman-Schapira G, Elinav E (2016) Use of metatranscriptomics in microbiome research. Bioinform Biol Insights 10:19–25

Bikel S et al (2015) Combining metagenomics, metatranscriptomics and viromics to explore novel microbial interactions: towards a systems-level understanding of human microbiome. Comput Struct Biotechnol J 13:390–401

Costalonga M, Herzberg MC (2014) The oral microbiome and the immunobiology of periodontal disease and caries. Immunol Lett 162(2 Pt A):22–38

D'Agostino S et al (2022) Exploring oral microbiome in healthy infants and children: a systematic review. Int J Environ Res Public Health 19(18):11403

Dabdoub SM, Ganesan SM, Kumar PS (2016) Comparative metagenomics reveals taxonomically idiosyncratic yet functionally congruent communities in periodontitis. Sci Rep 6:38993

Dewhirst FE et al (2010) The human oral microbiome. J Bacteriol 192(19):5002–5017

Duran-Pinedo AE, Frias-Lopez J (2015) Beyond microbial community composition: functional activities of the oral microbiome in health and disease. Microbes Infect 17(7):505–516

Gao L et al (2018) Oral microbiomes: more and more importance in oral cavity and whole body. Protein Cell 9(5):488–500

Gholizadeh P et al (2016) Role of oral microbiome on oral cancers, a review. Biomed Pharmacother 84:552–558

Gilbert JA et al (2018) Current understanding of the human microbiome. Nat Med 24(4):392–400

Griffen AL et al (2011) CORE: a phylogenetically-curated 16S rDNA database of the CORE oral microbiome. PLoS One 6(4):e19051

Griffen AL et al (2012) Distinct and complex bacterial profiles in human periodontitis and health revealed by 16S pyrosequencing. ISME J 6(6):1176–1185

Han X, Gross RW (2005) Shotgun lipidomics multidimensional MS analysis of cellular lipidomes. Expert Rev Proteomics 2(2):253–264

Hasin Y, Seldin M, Lusis A (2017) Multi-omics approaches to disease. Genome Biol 18(1):83

Hinrichs JE, Wolff LF, Pihlstrom BL, Schaffer EM, Liljemark WF, Bandt CL (1985) Effects of scaling and root planning on subgingival microbial proportions standardized in terms of their naturally occurring distribution. J Periodontol 56(4):187–194

Huang S et al (2014) Predictive modeling of gingivitis severity and susceptibility via oral microbiota. ISME J 8(9):1768–1780

Hutter G et al (2003) Molecular analysis of bacteria in periodontitis: evaluation of clone libraries, novel phylotypes and putative pathogens. Microbiology 149(Pt 1):67–75

Keijser BJ, Zaura E, Huse SM, van der Vossen JM, Schuren FH, Montijn RC, ten Cate JM, Crielaard W (2008) Pyrosequencing analysis of the oral microflora of healthy adults. J Dent Res 87(11):1016–1020

Kistler JO et al (2013) Bacterial community development in experimental gingivitis. PLoS One 8(8):e71227

Kumar PS et al (2003) New bacterial species associated with chronic periodontitis. J Dent Res 82(5):338–344

Kumar PS et al (2005) Identification of candidate periodontal pathogens and beneficial species by quantitative 16S clonal analysis. J Clin Microbiol 43(8):3944–3955

Lai LA et al (2019) Metaproteomics study of the gut microbiome. Methods Mol Biol 1871:123–132
Lee WH et al (2017) Bacterial alterations in salivary microbiota and their association in oral cancer. Sci Rep 7(1):16540
Li Y et al (2014) Phylogenetic and functional gene structure shifts of the oral microbiomes in periodontitis patients. ISME J 8(9):1879–1891
Long SS, Swenson RM (1976) Determinants of the developing oral flora in normal newborns. Appl Environ Microbiol 32:494–497
Long H et al (2022) Multi-omics analysis reveals the effects of microbiota on oral homeostasis. Front Immunol 13:1005992
Ma C et al (2015) Comparison of oral microbial profiles between children with severe early childhood caries and caries-free children using the human oral microbe identification microarray. PLoS One 10(3):e0122075
Mager DL et al (2003) Distribution of selected bacterial species on intraoral surfaces. J Clin Periodontol 30:644–654
Moussa DG et al (2022) Current state and challenges of the global outcomes of dental caries research in the meta-omics era. Front Cell Infect Microbiol 12:887907
Petruschke H et al (2020) Enrichment and identification of small proteins in a simplified human gut microbiome. J Proteomics 213:103604
Qin J et al (2010) A human gut microbial gene catalogue established by metagenomic sequencing. Nature 464(7285):59–65
Raman R et al (2005) Glycomics: an integrated systems approach to structure-function relationships of glycans. Nat Methods 2(11):817–824
Rappe MS, Giovannoni SJ (2003) The uncultured microbial majority. Annu Rev Microbiol 57:369–394
Riesenfeld CS, Schloss PD, Handelsman J (2004) Metagenomics: genomic analysis of microbial communities. Annu Rev Genet 38:525–552
Schrimpe-Rutledge AC et al (2016) Untargeted metabolomics strategies-challenges and emerging directions. J Am Soc Mass Spectrom 27(12):1897–1905
Sharpton TJ (2014) An introduction to the analysis of shotgun metagenomic data. Front Plant Sci 5:209
Simon-Soro A et al (2013) Microbial geography of the oral cavity. J Dent Res 92(7):616–621
Socransky SS, Haffajee AD (2000) Periodontal microbial ecology. Periodontology 38:135–187
Socransky SS, Haffajee AD, Cugini MA, Smith C, Kent RL Jr (1998) Microbial complexes in subgingival plaque. J Clin Periodontol 25:134–144
Sussulini A, Xia J, Oresic M (2022) Editorial: multi-omics: trends and applications in clinical research. Front Mol Biosci 9:994239
Teng F et al (2016) Cetylpyridinium chloride mouth rinses alleviate experimental gingivitis by inhibiting dental plaque maturation. Int J Oral Sci 8(3):182–190
Thursby E, Juge N (2017) Introduction to the human gut microbiota. Biochem J 474(11):1823–1836
Topcuoglu N, Kulekci G (2015) 16S rRNA based microarray analysis of ten periodontal bacteria in patients with different forms of periodontitis. Anaerobe 35(Pt A):35–40
Vartoukian SR, Palmer RM, Wade WG (2009) Diversity and morphology of members of the phylum "synergistetes" in periodontal health and disease. Appl Environ Microbiol 75(11):3777–3786
Vernocchi P, Del Chierico F, Putignani L (2016) Gut microbiota profiling: metabolomics based approach to unravel compounds affecting human health. Front Microbiol 7:1144
Wang KC, Chang HY (2018) Epigenomics: technologies and applications. Circ Res 122(9):1191–1199
Wang J et al (2013) Metagenomic sequencing reveals microbiota and its functional potential associated with periodontal disease. Sci Rep 3:1843
Wang Y et al (2017) Profiling of Oral microbiota in early childhood caries using single-molecule real-time sequencing. Front Microbiol 8:2244
Xia LC et al (2011) Accurate genome relative abundance estimation based on shotgun metagenomic reads. PLoS One 6(12):e27992

Yadav SP (2007) The wholeness in suffix -omics, –omes, and the word om. J Biomol Tech 18(5):277
Yamanaka W et al (2012) Compositional stability of a salivary bacterial population against supragingival microbiota shift following periodontal therapy. PLoS One 7(8):e42806
Yamashita Y, Takeshita T (2017) The oral microbiome and human health. J Oral Sci 59(2):201–206
Yang SF et al (2018) Compositional and functional variations of oral microbiota associated with the mutational changes in oral cancer. Oral Oncol 77:1–8
Zarco MF, Vess TJ, Ginsburg GS (2012) The oral microbiome in health and disease and the potential impact on personalized dental medicine. Oral Dis 18(2):109–120
Zierer J et al (2018) The fecal metabolome as a functional readout of the gut microbiome. Nat Genet 50(6):790–795

Multi-omics in Study of Lung Microbiome 12

Dibakar Roy, Paramita Roy, and Sudipto Saha

12.1 Introduction

The lung microbiome (bacteria, fungi, and viruses) is pivotal in lung disease pathogenesis and progression (O'Dwyer et al. 2016; Natalini et al. 2023). Only a tiny proportion of microbes, especially bacteria, can be cultured in vitro. Therefore, relying solely on culture-based methods for diagnosing pulmonary diseases may not be accurate (Calderaro et al. 2022). The microbiome composition is studied through culture-independent techniques, including amplicon (16s/18s rRNA) sequencing and whole metagenome sequencing (WMS) (Carney et al. 2020). Unlike amplicon data, WMS data can be further analyzed to obtain the functional aspects of microbiomes, such as gene families and pathways, more precisely (Pérez-Cobas et al. 2020). Multi-omics techniques can characterize the microbial environment and assess its functional characteristics and connection to the host immune responses. The high-throughput analysis of microbial proteins, transcripts, and metabolites made possible by bioinformatics advancements has expanded our knowledge of the function of the human microbiome under various diseased conditions (Kwoji et al. 2023). The respiratory microbiome maintains lung homeostasis, and thus, it can be treated as a possible drug target for acute and chronic lung diseases (Chotirmall et al. 2022). This chapter describes the taxonomic profile, functional analyses, and multi-omics techniques that reveal the lung microbiome, host, and crosstalk between microbes–host, explicitly focusing on pulmonary diseases.

D. Roy · P. Roy · S. Saha (✉)
Department of Biological Sciences, Bose Institute, Kolkata, India
e-mail: ssaha4@jcbose.ac.in

I. Mani, V. Singh (eds.), *Multi-Omics Analysis of the Human Microbiome*,
https://doi.org/10.1007/978-981-97-1844-3_12

12.2 Pulmonary Diseases and Lung Microbiome

Pulmonary diseases are a leading cause of morbidity and mortality worldwide (Momtazmanesh et al. 2023). Pulmonary diseases can be characterized into different groups: obstructive diseases, such as asthma, bronchiectasis, and chronic obstructive pulmonary disease (COPD); restrictive diseases, namely idiopathic pulmonary disease and sarcoidosis; vascular diseases, including pulmonary hypertension pulmonary edema; infectious diseases, such as Corona Virus Disease of 2019 (COVID-19), cystic fibrosis, pneumonia, tuberculosis; and malignancies (lung cancer). As our knowledge of lung microbiome in pulmonary diseases has expanded, we have realized its significant role in developing lung diseases. Recent studies showed that the lung microbiota can cause or worsen diseases, increasing susceptibility (Belizário et al. 2023; Chen et al. 2023a).

The microbiome's size and makeup can vary based on the body part and are influenced by environmental and host variables (Maddi et al. 2019). Culture-independent studies have demonstrated that the lung microbiome differs from that in the oral, skin, and gut (Bassis et al. 2015). The biomass of the airway microbiome is much lesser than that of the gut, skin, and oral microbiome due to low nutrient availability. Thus, alteration or dysbiosis in the lung microbiome may result in disease conditions (Natalini et al. 2023). Table 12.1 summarizes the variations between the lung and other human microbiome locations.

The lung microbiome (bacteria, fungi, and viruses) can interact with the host immune cells, including epithelial cells and macrophages, through their metabolites (Huynh et al. 2023). These metabolites can cross the lung epithelial cell layer and interact with the cell surface receptors of host immune cells, such as G-protein coupled receptors (GPCRs), or enter the cells through passive diffusion (Ulven 2012; Kotlyarov 2022). These metabolites can excite the immune cells (macrophages), and as a result, these cells secrete several pro-inflammatory chemokines and cytokines, including CXCL1, CXCL8 and IL-1b, IL-6, TNF-a, which recruits neutrophils in the airways, which may, in turn, increase the inflammation in the lung, as shown in Fig. 12.1. Finally, these microbe-host interactions increase lung epithelial cell apoptosis.

Table 12.1 Comparison of lung microbiome with other human microbiome sites

Sl. no	Factors	Lung microbiome	Gut microbiome	Skin microbiome	Oral microbiome
1	Sampling	Difficult	Easy	Easy	Easy
2	Microbial mass	Low	High	High	High
3	Nutrient availability	Low	High	Low	High
4	Oxygen concentration	Healthy lung-low Diseased lung-high	Small intestine-high Large intestine-low	Moderate-high	Moderate-high
5	Deep sequencing	Required	Not required	Not required	Not required

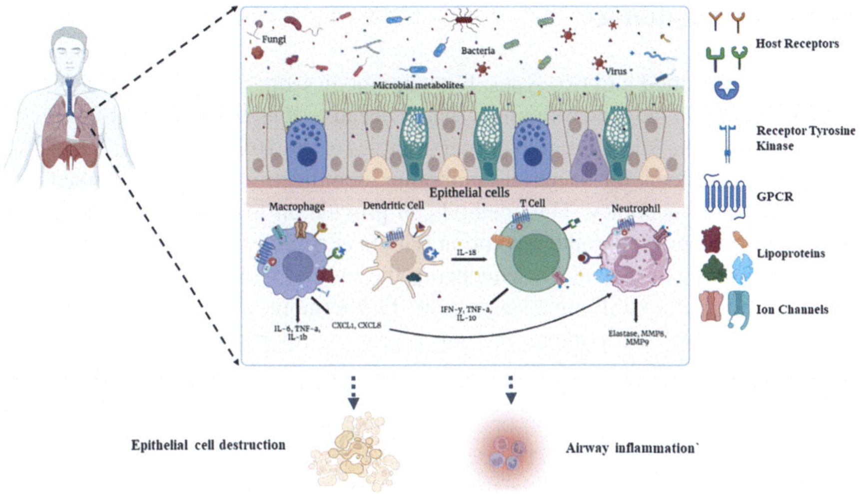

Fig. 12.1 Microbial metabolites and host immune cells crosstalk in the airways. The microbial metabolites cross the epithelial barrier and interact with the receptors of host immune cells, such as G-protein coupled receptors (GPCRs) and cytokine receptors on macrophages

12.3 Methods of Lung Microbiome Study

The significance of airway microbiome research in health and pulmonary diseases is well recognized (Scialò et al. 2023). Several methods are used for studying the taxonomic and functional aspects of lung microbiome in healthy and diseased states. These methods include sampling, detection methods, and bioinformatic approaches, including databases and software. Current sampling methods for culture-independent analysis involve several ways in both animal and human microbiome studies. The samples are bronchoalveolar lavage (BAL) fluid, sputum, bronchial brushings, exhaled breath condensate (EBC), tracheal aspirate, and lung tissue. BAL is intrusive, while sputum is non-intrusive (Chang et al. 2020). BAL is usually used for studying lung cancer, idiopathic pulmonary diseases, and COVID-19, while BAL and sputum can be utilized for asthma, COPD, and cystic fibrosis (Carney et al. 2020). Current detection methods reliant on culture are insufficient to detect all the microorganisms in a sample. On the other hand, culture-independent methods facilitate the detection of microorganisms present and can handle huge sample loads. The two most popular culture-independent approaches for identifying microbes are amplicon and WMS sequencing. In addition, various approaches like metatranscriptomics, metaproteomics, and metabolomics are utilized for the functional characterization of microbial products in lung microbiome research (Ghosh et al. 2021).

12.3.1 Metagenomics

This can be classified into two types. First, amplicon sequencing (16s/18s rRNA) is commonly used in lung microbiome research because of its technological simplicity and durability. The V1–V9 region (hypervariable) of the 16S rRNA gene is used for sequencing. It allows for bacterial taxonomic identification even with limited raw reads and higher accuracy. This method can provide up to genus-level taxonomic information but lacks in-depth resolution (Gupta et al. 2019; Kameoka et al. 2021). Second is WMS, that involves sequencing the entire DNA of all organisms in a sample, instead of amplifying a single gene. Furthermore, all functional genes may be revealed, as well as antibiotic resistance genes, virulence genes, and the organism's metabolism. Individual strains can also be identified using this method. But WMS gives higher coverage than amplicon (Durazzi et al. 2021; Purushothaman et al. 2022; Ren et al. 2019).

12.3.2 Metatranscriptomics

Metatranscriptomics provides information on the genes expressed by the microbiome. The total microbial RNA is sequenced to infer the functional repertoire of a community in different conditions (Aguiar-Pulido et al. 2016; Ojala et al. 2023).

12.3.3 Metaproteomics

Metaproteomics is an approach to investigate the presence and abundance of collective proteins in a microbiome. Mass spectrometry-based analyses of microbial proteins give direct insight into the molecular characteristics of microorganisms (Kleiner 2019; Salvato et al. 2021).

12.3.4 Metabolomics

Metabolomics is an approach to study the small molecules produced by the microbial community. The lung samples (sputum, BAL) containing the microbial byproducts are analyzed by mass spectrometry or nuclear magnetic resonance (NMR). Microbial metabolites are involved in various cellular processes (Bauermeister et al. 2022; Muhamadali et al. 2023).

In summary, metagenomics (amplicon and WMS) generates a taxonomic profile of the sample. Whereas metatranscriptomics and metaproteomics aid in obtaining a microbial functional aspect, and metabolomics help to identify which metabolites are secreted by the microbes into the lung environment.

12.4 Taxonomic Analysis of Lung Microbiome

12.4.1 Bioinformatics Tools for Taxonomic Analysis

The taxonomic analyses of the lung microbiota reveal the microbial composition in different health conditions. Several bioinformatics tools are available for discovering the taxonomic profile of lung microbiota in health and pulmonary diseases. There is some difference between the tools used for analyzing amplicon sequencing and WMS or shotgun metagenomics data.

Processing of amplicon sequencing data uses some specific tools that are widely used across studies. The sequencing data are processed through a standardized pipeline using QIIME/QIIME2 (Kuczynski et al. 2011; Estaki et al. 2020). Filtering and denoising of reads are done through DADA2, Trimmomatic, UCHIME, deblur, and FLASH (Callahan et al. 2016; Bolger et al. 2014; Edgar et al. 2011; Amir et al. 2017; Magoč and Salzberg 2011). The databases for assigning taxonomy to the sequencing reads are RDP classifier, SILVA, Greengenes, and EzBioCloud (Wang et al. 2007; Pruesse et al. 2007; DeSantis et al. 2006; Yoon et al. 2017).

For WMS or shotgun metagenomics data, reads are filtered using tools like Trimmomatic (Bolger et al. 2014). Human-derived sequences are removed using Bowtie2, and reads can be cleaned through MEGAHIT/metaSPAdes (Langmead and Salzberg 2012; Li et al. 2015; Nurk et al. 2017). The databases used for taxonomic assignment include the Standard database of Kraken2/Metaphlan2 (Wood et al. 2019; Truong et al. 2015).

Different kinds of plots such as krona, abundance plots (bar plots), bubble plots, tree structures (like radial trees, cladograms), and heatmaps are used to visualize the taxonomic profile. Linear discriminant analyses (LDA) allow identifying the differentially abundant microbes in different health and disease conditions. There are several R packages available for analyzing microbiome data. These packages include *phyloseq*, used for alpha and beta diversity and phylogenetic tree analyses. Another useful package is *MicrobiomeAnalyst* for alpha and beta diversity, differential analysis, and biomarker prediction. Package *microbiomemarker* is also used for LDA analysis, while *microeco* provides options for diversity analysis, microbiome composition, differential analysis, biomarker prediction, and correlation analysis (McMurdie and Holmes 2013; Chong et al. 2020; Liu et al. 2021; Cao et al. 2022).

12.4.2 Taxonomic Analysis of Lung Microbiome in Lung Diseases

Various studies are performed using the above tools for taxonomic analyses of the lung microbiome of pulmonary diseases like asthma, COPD, cystic fibrosis, lung cancer, and others. These studies revealed the microbial markers (the most abundant genus/species or differential markers) of diseases compared to healthy subjects, as given in Table 12.2.

Table 12.2 List of lung diseases that utilize amplicon or shotgun sequencing techniques to find the role of specific microbial markers in the lung microbiome

Disease	Sample	Sequencing type	Biomarkers
Asthma	Sputum	16s rRNA amplicon sequencing Shotgun metagenomics	*Haemophilus parainfluenzae, Haemophilus influenzae, Streptococci*, and *S. aureus* (Up) (Taylor et al. 2019)
		16s rRNA amplicon sequencing	Bacteria—*Streptococcus, Haemophilus* (Up); Fungi—*Candida* spp, *Malassezia* spp. (Up) (Al Bataineh et al. 2020)
			Streptococcus, Gemella, Veillonella, and *Megasphaera* (Up in severe asthma) (Wang et al. 2021)
			Capnocytophaga (Up); *Saccharimonas, Rothia, Gemella, Bulleidia*, and *Eubacterium_g10* (Down) (Kim et al. 2021)
COPD	Sputum	Targeted metagenomics Shotgun metagenomics	*Streptococcus, Haemophilus Granulicatella, Prevotella*, and *Veillonella* (Up) (Goolam Mahomed et al. 2021)
		16s rRNA amplicon sequencing	*Haemophilus* (Up); *Streptococcus* (Down) (Haldar et al. 2020)
			Actinomyces, Haemophilus, Rothia, Streptococcus, Neisseria, Leptotrichia, Veillonella, Granulicatella, Porphyromonas, and *Capnocytophaga* (Up) (Yang et al. 2021)
		16s rRNA amplicon sequencing Shotgun metagenomics	*Haemophilus parahaemolyticus* (Up in COPD smoker); *Ralstonia mannitolilytica* (Up in COPD exacerbation) (Wang et al. 2020a)
COVID-19	Sputum BAL	16s rRNA amplicon sequencing	*Acinetobacter, Klebsiella* and *Pseudomonas* (Up) (Xie et al. 2023)
	BAL		*Acinetobacter schindleri, Acinetobacter* spp., *Pseudomonas alcaligenes, Sphingobacterium* spp., *Clostridium hiranonis*, (Up) (Gaibani et al. 2021)
Cystic fibrosis	Sputum	16s rRNA amplicon sequencing Shotgun metagenomics	Bacteria—*Neisseria, Streptococcus, Prevotella, Rothia*, and *Haemophilus; Achromobacter xylosoxidans, Burkholderia cepacia complex, Pseudomonas aeruginosa, and Staphylococcus aureus* (Up); Fungi—*Aspergillus Fumigatus* (Up) (De Almeida et al. 2020)
		Whole genome sequencing	*Pseudomonas aeruginosa, Staphylococcus aureus, Prevotella melaninogenica*, and *Rothia mucilaginosa* (Up) (Bacci et al. 2020)

(continued)

Table 12.2 (continued)

Disease	Sample	Sequencing type	Biomarkers
Idiopathic pulmonary fibrosis	BAL	Shotgun metagenomics	*Acinetobacter, Neisseria, and Streptococcus.* (Up); *Bacteroides* (Down); *Microvirus, simplex virus*, and *varicella virus* (Up) (Tong et al. 2019)
Lung cancer	Bronchial washing fluid (BWF) sputum	16s rRNA amplicon sequencing	*Streptococcus, Veillonella* and *Rothia* (Up) (Huang et al. 2019)
	BAL		*TM7*, c: *TM7–3* and *Capnocytophaga* (Up) (Cheng et al. 2020)

12.4.2.1 Asthma

Studies of asthmatic amplicon sequencing data of sputum microbiome mainly showed an increased abundance of the genus *Streptococcus* and *Haemophilus* (Al Bataineh et al. 2020; Wang et al. 2021). Shotgun metagenomics data of asthmatic sputum microbiome showed a higher abundance of *Haemophilus influenzae* and *Staphylococcus aureus* (Taylor et al. 2019).

12.4.2.2 COPD

Studies of COPD amplicon sequencing data of sputum microbiome revealed that microbes like *Haemophilus*, *Veillonella*, and *Rothia* are the most abundant genus (Goolam Mahomed et al. 2021; Yang et al. 2021). Another study using shotgun metagenomics data of sputum microbiome showed *Haemophilus parahaemolyticus* and *Ralstonia mannitolilytica* as markers of COPD (Wang et al. 2020a).

12.4.2.3 COVID-19

A study using amplicon sequencing data showed *Acinetobacter*, *Klebsiella*, and *Pseudomonas* as the most abundant as well as differential markers of Sputum and BAL microbiome of COVID-19 (Xie et al. 2023).

12.4.2.4 Cystic Fibrosis

A study using both amplicon sequencing and WMS data showed genera *Rothia, Prevotella, Streptococcus*, *Neisseria*, and *Haemophilus* and species *Achromobacter xylosoxidans, Burkholderia cepacia* complex, *Pseudomonas aeruginosa, and Staphylococcus aureus* as abundant in Cystic Fibrosis sputum microbiome (De Almeida et al. 2020). Another study using WMS data showed *Pseudomonas aeruginosa, Staphylococcus aureus*, *Prevotella melaninogenica,* and *Rothia mucilaginosa* as the most abundant species in patients with cystic fibrosis (Bacci et al. 2020).

12.4.2.5 Idiopathic Pulmonary Fibrosis

Taxonomic analysis of shotgun metagenomics data showed that *Acinetobacter*, *Neisseria*, and *Streptococcus* are enriched, whereas *Bacteroides* are depleted in the BAL microbiome of idiopathic pulmonary fibrosis (Tong et al. 2019).

12.4.2.6 Lung Cancer

A study of amplicon data using sputum samples and bronchial washing fluid showed enriched *Streptococcus*, *Veillonella*, and *Rothia* in lung cancer patients (Huang et al. 2019). Another study of amplicon sequencing data showed *TM7, c: TM7–3,* and *Capnocytophaga* as abundant genera of the lung cancer BAL microbiome (Cheng et al. 2020).

12.5 Functional Analysis of Lung Microbiome

The by-products secreted by the airway microbiome are involved in various host biological processes, including regulating epithelial development, modulating the immune response, and facilitating the digestion of complex macromolecules (Hou et al. 2022). The alteration of the functional profile of the lung microbiome may be linked with pulmonary disease progression.

12.5.1 Bioinformatic Tools for Functional Analysis

There are two ways to study the functional profile of lung microbes: indirect approaches (amplicon and WMS) and direct approaches (metatranscriptomics, metaproteomics, and metabolomics).

12.5.1.1 Indirect Approaches

Indirect approaches for predicting the functional profile of microbiomes include amplicon and WMS data. Several publicly available tools, including the PICRUSt and Tax4Fun, use amplicon datasets to obtain the functional capabilities of microbiomes (Kim et al. 2021; Haldar et al. 2020; Gaibani et al. 2021; Hou et al. 2022). However, these approaches are only approximate estimates because they mostly rely on reference genomes and their annotations. Using a more conserved method that involves the core genes of each species, MicFunPred can predict the functional profiles from amplicon sequence data, reducing the number of false-positive outcomes (Mongad et al. 2021). Additionally, Piphilin uses the nearest neighbor matching method of OTUs/ASVs (Operational Taxonomic Unit/Amplicon Sequence Variants) based on the genomes from KEGG and BioCyc to predict the functional composition of the microbiome (Narayan et al. 2020; Kanehisa and Goto 2000; Karp et al. 2019). WMS or shotgun metagenomics data can more accurately reveal the functional repertoire of the microbiome than amplicon sequencing data. Publicly available tools, such as MetaGeneMark2, Glimmer-MG, MetaProdigal, FragGeneScan, and FragGeneScanRs, use homology-based search to predict genes from the WMS data (Gemayel et al. 2022; Kelley et al. 2012; Hyatt et al. 2012; Rho et al. 2010; Van Der Jeugt et al. 2022). HUMAnN is a pipeline that analyzes short DNA/RNA reads from metagenomics/metatranscriptomics sequencing data (Beghini et al. 2021). Through GO (gene ontology) annotations, MetaFunc, a microbiome analysis pipeline, can identify the functional contributions in a microbiome sample (Sulit et al. 2020). Using either a WMS or amplicon sequencing data, BugBase predicts microbial functional pathways and pathogenic potential within complex microbiome communities (Ward et al. 2017). MAAWf is a fully automated workflow platform for analyzing the composition and functions of microbiomes of WMS or amplicon sequencing data (Zhu et al. 2020).

12.5.1.2 Direct Approaches

Direct approaches include metatranscriptomics, metaproteomics, and metabolomics, which can provide quantitative measures of the lung microbiome's functional profile. Bioinformatic tools such as MetaTrans, COMAN, FMAP, SAMSA2, SqueezeMeta, IMP, and MOSCA are used to analyze the metatranscriptomics data (Martinez et al. 2016; Ni et al. 2016; Kim et al. 2016; Westreich et al. 2018; Tamames and Puente-Sánchez 2018; Narayanasamy et al. 2016; Sequeira et al. 2019). Taxonomic classifications can be performed using Kraken2, Metaphlan2, HMP Unified Metabolic Analysis Network (HUMANn), and Bracken (Wood et al. 2019; Truong et al. 2015; Beghini et al. 2021; Lu et al. 2017). Mass spectroscopy-based metaproteomics data can be analyzed and interpreted through web-based and open-source software tools such as MetaLab, iMetalab 1.0, Unipept 4.0, MetaGOmics, eggNOG-mapper, Meta4P, proPHAnE (http://www.prophane.de), MetaProteomeAnalyzer (Kanehisa and Goto 2000; Cheng et al. 2017; Liao et al. 2018; Gurdeep Singh et al. 2019; Cantalapiedra et al. 2021; Muth et al. 2015). MEtaGenome ANalyzer, or MEGAN, can analyze metaproteomics data through protein BLAST (Huson et al. 2007). The microbial metabolomics data (LC-MS based) can be analyzed through vendor-specific software such as progenesis QI (waters), mass hunter (Agilent), compound discoverer (thermos scientific), or by publicly available tools including XCMS, MS-DIAL, Mzmine 3, GNPS (Smith et al. 2006; Tsugawa et al. 2015; Schmid et al. 2023; Wang et al. 2016). Researchers generally use KEGG, Metlin, and HMDB databases to identify the metabolites (Kanehisa and Goto 2000; Smith et al. 2005; Wishart et al. 2007). Microbial metabolite database (MiMeDB) allows users to perform LC-MS-based searches for microbial metabolites (Wishart et al. 2023). There are few databases available to identify the metabolites originating from the microbiome, such as Web of Microbes (WoM), Gut Microbial Metabolite Association with Disease (GMMAD), and MetOrigin (Kosina et al. 2018; Wang et al. 2023; Yu et al. 2022).

12.5.2 Studies on Functional Analysis of Lung Microbiome in Lung Diseases

Here, we summarize the studies on the lung microbiota that used lower airway samples like BAL and sputum to predict microbial functional repertoire in various pulmonary diseases, as given in Table 12.3.

Table 12.3 List of lung diseases that use indirect/direct approaches to study functional aspects of lung microbiome

Disease	Sample	Tool(s)	Key predicted microbial functions
Asthma	Sputum	PICRUSt	Sucrose degradation, lactose and galactose degradation, Kdo transfer to lipid IVA, and glycogen biosynthesis, upregulated in severe asthmatics and peptidoglycan biosynthesis IV, pyruvate fermentation to propanoate I, and O-antigen building block biosynthesis, abundant in mild asthmatics (Wang et al. 2021)
			Glycan degradation pathways downregulated and lipopolysaccharide biosynthesis enriched in asthmatic children with exacerbation (Kim et al. 2021)
COPD	Sputum	PICRUSt2	Butyrate synthesis and nitrate reduction pathways downregulated significantly in patients with COPD (Wang et al. 2020a)
			Tetrapyrrole and heme biosynthesis-related pathways enriched in COPD patients, sucrose degradation, glycogen biosynthesis, and L-methionine biosynthesis pathways upregulated in patients with COPD frequent exacerbators (Kussek et al. 2022)
			Genes related to ABC transporters, bacterial motility, and lipopolysaccharide biosynthesis upregulated in patients with COPD (Haldar et al. 2020)
			L-tyrosine, L-arginine, L-ornithine, and androstenedione degradation pathways are enriched in acute exacerbation, and mycolate biosynthesis, ppGpp, and fatty acid super-pathway are depleted. Glucose oxidation, primordial inositol biosynthesis pathways, and L-tryptophan increased in the COPD stable group (Xue et al. 2023)
Cystic fibrosis	Sputum	HUMANn2	Phosphatidylcholine acyl editing, fatty acid elongation, palmitoleate biosynthesis, oleate biosynthesis, octanoyl biosynthesis, super-pathway of fatty acid biosynthesis initiation, gondoate biosynthesis, dodec-5-enoate biosynthesis, sulfate assimilation, and cysteine synthesis pathways, differentially abundant in follow-up samples (next clinical appointment) in comparison to treatment samples of cystic fibrosis patients (Felton et al. 2021)
		Tax4Fun	In CF with *S.* aur*eus* infection, antimicrobial resistance genes (K01467, K03327, K08218, K07481, K07485, K07489, K02456) are enriched, and in *P.* aerugi*nosa* group genes of secretion systems (K02459, K03195, K07344, K11891, K11896, K11937) are upregulated (Kussek et al. 2022)

Disease	Sample	Tool(s)	Key predicted microbial functions
Lung cancer	Sputum	PICRUSt2	Menaquinol and dimethyl-menaquinol biosynthesis, heme biosynthesis, purine and pyrimidine biosynthesis pathways are upregulated, and thiazole biosynthesis acidogenic fermentation pathways are downregulated in patients with NSCLC (Lu et al. 2021)
			Super-pathway of preQ0 and queuosine biosynthesis, L-aspartate and L-asparagine biosynthesis, abundant in the nonsmoking group with EGFR mutation and in EFGR negative group super-pathway of branched amino acids biosynthesis, L-isoleucine biosynthesis II, III, and L-isoleucine biosynthesis II, enriched (Huang et al. 2022)
		PICRUSt2 BugBase	In lung cancer patients with COPD, the super-pathway of L-threonine biosynthesis, the super-pathway of polyamine biosynthesis, L-glutamate and L-glutamine biosynthesis, the super-pathway of S-adenosyl-L-methionine biosynthesis, arginine, ornithine, and proline interconversion are significantly downregulated. Whereas ppGpp biosynthesis, the super-pathway of L-methionine biosynthesis, and stearate biosynthesis pathways are upregulated (He et al. 2022a)
	Lung tumor tissues, normal lung tissues, Para-tumor tissues, bronchial tissues	PICRUSt	Ansamycin biosynthesis, pantothenate and CoA synthesis, geraniol degradation, valine, leucine, isoleucine synthesis, bacterial chemotaxis, glutamate metabolism, and ketone metabolism are abundant in patients with NSCLC (Su et al. 2023)
Pneumonia	Endotracheal aspirates	Tax4Fun	Genes related to nitrogen metabolism, homologous recombination, porphyrin and chlorophyll metabolism, and mismatch repair upregulated in patients in mechanical ventilation with pneumonia (Woo et al. 2020)
	BAL	PICRUSt2	D-arginine and D-ornithine pathways are significantly increased in the CTD-ILD group, and glycerolipid metabolism androstenedione degradation upregulated in CAP patients (He et al. 2022b)
			Biosynthesis of unsaturated fatty acids, pyruvate metabolism, cytokine-receptor interactions, NOD-like receptor signaling pathways altered in post-HSCT pneumonia patients (He et al. 2022c)

(continued)

Table 12.3 (continued)

Disease	Sample	Tool(s)	Key predicted microbial functions
Tuberculosis	TB lesions, normal lung tissues	PICRUSt2	Chorismate metabolism, hexitol degradation, Menaquinone biosynthesis, penicillin resistance, upregulated in exacerbated TB and, D-galacturonate degradation, vanillate degradation, catechol degradation, heme b biosynthesis, and adenosyl-cobamide biosynthesis, upregulated in non-exacerbated TB (Du et al. 2022)
Recurrent wheezing and atopy	BAL	PICRUSt2	Cytoskeleton proteins, glutamatergic synapse, porphyrin, and chlorophyll metabolism pathways are altered between wheezing and non-wheezing patients (Tang et al. 2023)
Bronchiectasis	Sputum	HUMANn2	ppGpp biosynthesis and purine ribonucleosides degradation pathways are differentially abundant in bronchiectasis and COPD (Ghosh and Saha 2022)

12.5.2.1 Asthma

Microbial dysbiosis in the lungs associated with the asthma pathophysiology (Valverde-Molina and García-Marcos 2023). In individuals with severe asthma, specific microbial metabolic pathways are upregulated, such as L-lysine biosynthesis, L-glutamate and L-glutamine biosynthesis, L-histidine biosynthesis, lipid IVA biosynthesis, Kdo transfer to lipid IVA, and pyruvate fermentation to propanoate (Wang et al. 2021).

12.5.2.2 COPD

Recent reports suggest that the airway microbiome is involved in the disease progression and acute exacerbation in COPD (Pragman et al. 2012; Wilkinson et al. 2017; Wang et al. 2019). The change in lung microbiome functional repertoire with upregulation of tetrapyrrole and heme biosynthesis-related pathways associated with COPD pathogenesis (Dang et al. 2022). PICRUSt analysis revealed that COPD patients have increased lipopolysaccharide biosynthesis pathways, which induces pro-inflammatory responses (Haldar et al. 2020).

12.5.2.3 Cystic Fibrosis

Microbial infections are often associated with chronic inflammation in cystic fibrosis (Petrocheilou et al. 2022). In cystic fibrosis, long-chain fatty acid (LCFA) synthesis and fatty acid elongation pathways are abundant (Felton et al. 2021).

12.5.2.4 Lung Cancer

There is little evidence linking the airway microbiome to the pathophysiology of lung cancer. However, recent reports suggest that there may be a correlation between the lung microbiome and the pathogenesis of lung cancer (Dong et al. 2021). Patients with both COPD and lung cancer exhibit enhanced Gram-negative microbes, increasing potential pathogenicity, stress tolerance, and oxidative stress-related pathways (He et al. 2022a). In early-stage lung cancer patients, microbial pathways related to anhydro muro-peptides recycling, L-lysine biosynthesis II, and gondoate biosynthesis (anaerobic) are enriched. Whereas NAD salvage pathway I, phospho-pantothenate biosynthesis I, and incomplete reductive tricarboxylic acid (TCA) cycle pathways are abundant in the advanced-stage group (Huang et al. 2022).

12.5.2.5 Pneumonia

Alterations in the lung microbiota linked to microbial infections and disease progression in pneumonia. In community-acquired pneumonia (CAP), specific microbial pathways, such as glycerolipid metabolism and androstenedione degradation, are upregulated (He et al. 2022b).

12.5.2.6 Tuberculosis

The altered lung microbial functions in tuberculosis patients may be responsible for persistent *Mycobacterium tuberculosis* (MTB) infection. In tuberculosis (TB) exacerbation, microbial pathways responsible for chorismate metabolism and menaquinone synthesis are enriched. In TB non-exacerbation, the upregulation of heme b synthesis is linked to TB pathogenesis (Du et al. 2022).

12.5.2.7 Other Respiratory Diseases

PICRUSt2 analyses identified differences in predicted bacterial functions in patients with recurrent wheezing and atopy, including glutamatergic synapses, cytoskeleton proteins, porphyrin, and pathways associated with atopic wheezing (Tang et al. 2023). The pathways involved in the biosynthesis of ppGpp have been identified as differential markers for bronchiectasis (Ghosh and Saha 2022).

12.6 Multi-omics in Lung Microbiome

Multi-omics data integration provides more precise insights into molecular changes contributing to lung homeostasis, cellular processes, and disease than single-omics data. Researchers can more effectively link genotype to phenotype using integrative omics technologies, which also helps to identify novel biomarkers and therapeutic targets (Chu et al. 2021). Metagenomics, metatranscriptomics, metaproteomics, and metabolomics data can be integrated to study the functions of the lung microbiome in diseased conditions (Aguiar-Pulido et al. 2016). Combining various host omics datasets, genomics, epigenomics, transcriptomics, proteomics, and metabolomics, may accurately classify lung diseases and help identify host biomarkers, leading to better patient risk stratification (Chen et al. 2023b). Integrating multi-omics datasets from both the host and microbiome are used to investigate the microbe-host interactions in the lungs, as shown in Fig. 12.2. These interactions may involve either host DNA-microbial metabolite or host protein-microbial metabolite, which need to be identified for therapeutic targets.

12.6.1 Bioinformatics Tools for Integration of Multi-omics Data

Integrated omics approaches are used to investigate the role of lung microbiota in diseased conditions. Several publicly available bioinformatic tools can integrate multi-omics datasets of microbiome, host and microbe-host interactions in lungs. There are three different ways to examine multi-omics in the lungs.

The host factors in the lung can be studied using various bioinformatic tools. One is paintsOmics 3, a web-based interface that can merge host multi-omics data from transcriptomics, proteomics, and metabolomics datasets (Hernández-de-Diego et al. 2018). MOPED (Multi-omics Profiling Expression Database) is developed to integrate host genomic and protein expression data (Montague et al. 2014).

Several tools are available to integrate data from multiple omics of the lung microbiome. MicrobiomeAnalyst 2.0 can interpret associations between paired microbiome and metabolomic datasets (Lu et al. 2023). MIMOSA2 can integrate microbiome and metabolomics data based on reference databases on the metabolic capacities of microbial taxa (Noecker et al. 2022). Machine learning and neural

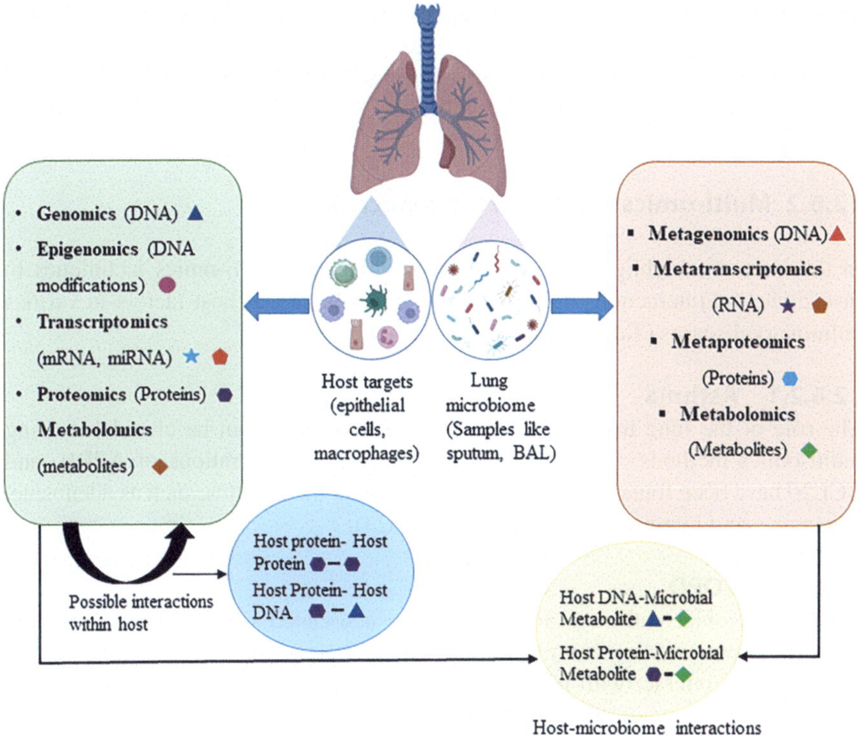

Fig. 12.2 Multi-omics approaches to study host targets (immune cells), lung microbiome, and their interactions. Multi-omics approaches for host factors are genomics, proteomics, and others that can be used to study immune cells like macrophages and epithelial cells, as well as interactions between their proteins and DNA. Multi-omics approaches to studying lung microbes are metagenomics, metatranscriptomics, metaproteomics, and metabolomics. Microbe-host interactions in the lungs, involving host DNA and proteins with microbial metabolites, are also studied

network-based tools mmvec and Mimenet are used to explore the microbiome and metabolome relationships (Morton et al. 2019; Reiman et al. 2021). Melonpann can predict the microbial metabolites from sequencing data based on pairwise correlations and regressions (Mallick et al. 2019). Metage2Metabo (M2M) uses graph-based metabolic analysis from annotated metagenomes to reconstruct metabolic networks (Belcour et al. 2020). Independent meta-omic pipeline (IMP) can analyze and merge paired metatranscriptomic data with metagenomics data (Narayanasamy et al. 2016).

Similarly, there are tools that can combine multi-omics data from both host and microbiome. The computational pipeline for the analysis of longitudinal multi-omics data (PALM) can integrate host expression data with the microbiome multi-omics data (Ruiz-Perez et al. 2021). gNOMO can analyze and combine metagenomics, metatranscriptomics, and metaproteomics datasets with host data (Muñoz-Benavent et al. 2020). OmicsNet is a web-based platform that integrates transcriptomics, proteomics, metabolomics, and lipidomics datasets from host () and microbial taxa (Zhou et al. 2022). *Mixomics*, an R package, is developed for multivariate analyses and integration of omics (metagenome, metabolome, host transcriptome, and proteome) datasets (Rohart et al. 2017).

12.6.2 Multi-omics Studies in Lung Microbiome

In this part, we highlighted research articles that use multi-omics techniques to investigate the interactions between airway microbiota and host factors in various pulmonary diseases (Table 12.4).

12.6.2.1 Asthma

The role of the lung microbiota in asthma pathogenesis can be elucidated using multi-omics methods. In pediatric asthma, higher concentrations of APRIL and CCL20 have been found to correlate with microbial metabolites, such as adenosine, guanosine, and hypoxanthine (Briard et al. 2023).

12.6.2.2 COPD

Altered microbial pathways and metabolites associated with COPD are studied using multi-omics. Microbial metabolites such as homocysteine, homocysteine, and palmitate can interact with host factors, such as genes, in COPD patients (Wang et al. 2020b). In neutrophilic COPD, *H. influenzae* secretes endotoxin that can induce IL-6TS in the lungs during neutrophil extracellular trap formation (NETosis) (Winslow et al. 2021). Yan et al. (2022) identified that altered tryptophan metabolism and reduced Indole-3-acetic acid (IAA) were associated with neutrophil-dominant COPD. Furthermore, in vivo and in vitro studies revealed that lung microbiota-secreted IAA improves emphysema and lung function (Yan et al. 2022). Liang et al. (2023) discovered that *Staphylococcus aureus* was linked to a decrease in lung function in individuals with COPD through multi-omics analysis. Homocysteine released from *S. aureus* triggers persistent neutrophilic inflammation in mice with emphysema, which was facilitated by the AKT1-S100A8/A9 pathway (Liang et al. 2023). According to Jude et al. (2023), microbial dysbiosis in individuals with HIV infection and COPD is linked to transcriptomic and epigenetic factors in the lung epithelium. Through multi-omics, they found correlations between microbes such as *Prevotella* and host genes ACVR1B, FUZ, FASTKD3, and host epigenetic factors CpG-PHLDB3 and CpG-FUZ (Jude et al. 2023).

12.6.2.3 COVID-19

Multi-omics techniques allowed researchers to understand the heterogeneity and complexity of COVID-19. Maras et al. (2021) found a correlation between SARS-CoV-2 infection and host proteins such as WARS (tryptophan-tRNA ligase) and MX1 (MX dynamin-like GTPase 1). Furthermore, this study also revealed increased respiratory pathogens such as *Klebsiella pneumoniae* and decreased *Lactobacillus*

Table 12.4 List of lung diseases that utilize multi-omics techniques to uncover the lung microbiome-host interaction

Disease	Sample	Omics type	Multi-omics integration method	Comments
Asthma	BAL	Metagenomics (M) Metabolomics (M) Cytokine assay (H)	Multiblock integration using Multi-Omics Factor Analysis (MOFA) in R ("MOFA2", version 1.8.0)	Cytokine APRIL positively correlated with microbial metabolites, α-pinene-oxide, hypoxanthine, guanosine, and adenosine, and cytokines CCL20, IL-8, and Matrix metalloproteinases 2 (MMP-2) negatively correlated with N-acetyl-ornithine, N6-acetyl-L-lysine, L-alanyl-L-proline, lysine, glycyl-L-leucine, L-prolyl-L-leucine (Pro-Leu), valyl-proline (Val-Pro), arginine, L-homoserine, xanthosine, methionine, and phenyl-pyruvic acid (Briard et al. 2023)

(continued)

Table 12.4 (continued)

Disease	Sample	Omics type	Multi-omics integration method	Comments
COPD	Sputum	Metagenomics (M) Metabolomics (M) Proteomics (H) Transcriptomics (H)	*Mediate* package in R, used to find significant mediation effects	Secretion of Indole-3-Acetic Acid (IAA) by the airway microbiota can mitigate lung function decline and neutrophilic inflammation in COPD by facilitating epithelial-macrophage crosstalk mediated by IL-22 (Yan et al. 2022)
		Metagenomics (M) Metabolomics (M) Transcriptomics (H)	Correlation-based analyses using HAllA (v0.8.17)	*Staphylococcus aureus* was associated with the decline of lung function by homocysteine-AKT1-S100A8/A9 crosstalk in COPD patients (Liang et al. 2023)
		Metagenomics (M) Metabolomics (M) Transcriptomics (H)	MIMOSA2 and correlation-based analyses	Microbiome-derived metabolites such as butyrate, urate, homocysteine, palmitate, and homocysteine strongly correlated with host genes of COPD patients (Wang et al. 2020b)
		Metagenomics (M) Proteomics (H) Transcriptomics (H)	Multivariate linear modeling and generalized linear mixed model (GLMM) were used for integration	Moraxella and Haemophilus were significantly linked to the host immune pathways and airway inflammation in COPD patients. Haemophilus correlated with neutrophilia, while Moraxella was linked to interferon signaling (Wang et al. 2019)
	Airway epithelial cell (AEC) brushings	Metagenomics (M) Methylomics (H) Transcriptomics (H)	Data integration analysis For biomarker discovery Using latent components (DIABLO) applied in the R package named mixOmics used	Patients suffering from both HIV and COPD had decreased as well as altered airway microbiome, and a decrease in genus *Prevotella* is linked to an imbalance in host gene expression such as FUZ, FASTKD3, and ACVR1B (Jude et al. 2023)
	Sputum, epithelial brushings, BAL	Metagenomics (M) Proteomics (H) Transcriptomics (H)	Correlation and hierarchical clustering-based analyses using Qlucore Omics Explorer 3	*Haemophilus* in the airways induced neutrophil extracellular trap formation (NETosis) and IL-6Ts in COPDs patients (Winslow et al. 2021)
	Bronchial brushings	Metagenomics (M) Transcriptomics (H)	Linear mixed models used to find the correlation	With increasing COPD severity, there was depletion of *Prevotella* and enriched *Moraxella* in the airways, that associated with upregulation of host pro-inflammatory genes and downregulation of epithelial defense genes (Ramsheh et al. 2021)

Disease	Sample	Omics type	Multi-omics integration method	Comments
COVID-19	Respiratory specimen	Metaproteomics (M) Metabolomics (M) Proteomics (H)	Correlation and hierarchal clustering-based analyses	Multi-omics data integration showed viral proteins significantly correlate with microbial and host metabolites and proteins (Maras et al. 2021)
	Background and supraglottic (buccal) samples, BAL	Metagenomics (M) Metatranscriptomics (M) Transcriptomics (H)	Correlation-based analyses	Viral and bacterial profiles associated with host transcriptomic factors and anti-SARS CoV-2 Spike IgG (Sulaiman et al. 2021)
Lung cancer	Lung tissue	Metagenomics (M) Transcriptomics (H)	No integration of multi-omics data; metagenomics and transcriptomics data were analyzed separately	In familial lung cancer, the microbiome's diversity was reduced, and the microbiome also varies according to age, tissue, blood type, gender, and smoking conditions RNA-seq data revealed pathways related to carcinogenesis, such as IL17, Ras, MAPK, and notch signaling, abundant in familial lung cancer patients (Chen et al. 2022)
	Background and supraglottic (buccal) samples, airway brushes	Metagenomics (M) Transcriptomics (H)	*Mmvec* package was used to inspect the co-occurrence probability, and networks were generated using the R package *ggnet* v 0.1.0	Lower airway dysbiosis, especially *Veillonella parvula,* was linked with upregulating host pathways such as PI3K, IL-17, ERK, and MAPK In a preclinical mice model mice, *V. parvula* treatment resulted in decreased survival, increased tumor burden, and upregulation of checkpoint inhibitor markers (Tsay et al. 2021)

Here, * (M) stands for Microbiome, and * (H) stands for Host *salivarius* in patients with COVID-19 (Maras et al. 2021). Through multi-omics, Sulaiman et al. (2021) identified microbial signatures associated with the COVID-19 outcome (Sulaiman et al. 2021).

12.6.2.4 Lung Cancer

The underlying mechanisms of microbiome and host interactions in lung cancer can be investigated through multi-omics. Tsay, Jun-Chieh et al. (2021) identified that microbial dysbiosis in lower airways was associated with disrupted host airway transcriptomes, such as upregulation of ERK, MAPK, IL-17, and PI3K pathways (Tsay et al. 2021). Chen et al. (2022) found that *Capnocytophaga* and *Fusobacterium* significantly increased familial lung cancer (FLC), while *Sphingomonas* and *Sphingopyxis* increased in FLC cancer tissue. The RNA-seq analysis revealed pathways linked to carcinogenesis, such as IL17, Ras, MAPK, and Notch, particularly in FLC (Chen et al. 2022).

12.7 Future Perspective

Current lung microbiome studies generally include taxonomic classification, functional analyses, and identification of microbial markers. However, several avenues can still be explored in the future. It is well-known that the airway microbiome actively participates in pulmonary diseases, and clinical exposures, such as antibiotics, can alter the lung microbiota (Segal et al. 2017; Nelson et al. 2020). According to a study, long-term erythromycin treatment reduced exacerbation rates and altered the microbiota in sputum by decreasing *P. aeruginosa* pathogenicity (Burr et al. 2016). This concludes the value of characterizing the airway microbiome as a potential therapeutic target (Chotirmall et al. 2022). Some novel strategies have been implemented for gastrointestinal diseases to manipulate the microbiome to restore ecological balance. The strategies include antibiotic modification and direct inoculation of microbes from healthy subjects (Gagliardi et al. 2018). The lung microbiome can also be manipulated to improve lung homeostasis by increasing the commensal microbes. The lung microbiome can regulate the inflammation in the brain. According to Hosang et al., perturbing the lung microbiota with neomycin altered the severity of a rat encephalomyelitis model. Thus, studying the lung microbiome can suggest a possible mechanism by which the lung microbiome regulates autoimmunity in the brain, establishing the "brain-lung axis" (Chen et al. 2023a; Hosang et al. 2022). Spatial mapping of microbial communities can help researchers to understand the lung microbiome's contribution to disease pathogenesis. Lung function tends to decline as we age, and the lung microbiome directly correlates with this decline (Bhattacharjee et al. 2022; Lee et al. 2019). Therefore, it is crucial to explore possible ways to restore the lung microbiome in elderly individuals to resemble that of a younger age, which could potentially restore their lung function.

Pulmonary diseases result from various levels of regulation and dysregulation of multiple factors, some of which are influenced by microbes. With the help of multi-omics techniques, the connection between the lung microbiome, host phenotype, and host endotype can be revealed. However, there are some challenges in investigating the lung microbiome: (1) study the influence of other microbiome data, such as gut on lung microbiome, the "gut–lung axis," (2) a standard pipeline for integration of omics datasets to uncover the microbe–host interface, (3) analysis of compositional data with quantitative data such as compositional data of lung microbiome with quantitative data of host proteomics, (4) appropriate statistical methods to incorporate several omics data, (5) the need for a well-characterized, large, and with clinical and geographic diversity cohort of WMS data, (6) A systems biology approach is necessary to create a multi-omics landscape of airway microbiota and its interactions with host factors. Finally, it may be worth exploring the aspects above in the near future as multi-omics studies of the lung microbiota can enable a more personalized approach to diagnosing and treating pulmonary diseases. Multi-omics can also aid in generating hypotheses for mechanistic study to comprehend the disease pathogenesis at the molecular level.

References

Aguiar-Pulido V, Huang W, Suarez-Ulloa V, Cickovski T, Mathee K, Narasimhan G (2016) Metagenomics, metatranscriptomics, and metabolomics approaches for microbiome analysis. Evol Bioinform Online 12:5–16. https://doi.org/10.4137/EBO.S36436

Al Bataineh MT, Hamoudi RA, Dash NR, Ramakrishnan RK, Almasalmeh MA, Sharif HA, Al-Hajjaj MS, Hamid Q (2020) Altered respiratory microbiota composition and functionality associated with asthma early in life. BMC Infect Dis 20:697. https://doi.org/10.1186/s12879-020-05427-3

Amir A, McDonald D, Navas-Molina JA, Kopylova E, Morton JT, Zech XZ, Kightley EP, Thompson LR, Hyde ER, Gonzalez A, Knight R (2017) Deblur rapidly resolves single-nucleotide community sequence patterns. mSystems 2:e00191–e00116. https://doi.org/10.1128/mSystems.00191-16

Bacci G, Taccetti G, Dolce D, Armanini F, Segata N, Di Cesare F, Lucidi V, Fiscarelli E, Morelli P, Casciaro R, Negroni A, Mengoni A, Bevivino A (2020) Untargeted metagenomic investigation of the airway microbiome of cystic fibrosis patients with moderate-severe lung disease. Microorganisms 8:1003. https://doi.org/10.3390/microorganisms8071003

Bassis CM, Erb-Downward JR, Dickson RP, Freeman CM, Schmidt TM, Young VB, Beck JM, Curtis JL, Huffnagle GB (2015) Analysis of the upper respiratory tract microbiotas as the source of the lung and gastric microbiotas in healthy individuals. MBio 6:e00037. https://doi.org/10.1128/mBio.00037-15

Bauermeister A, Mannochio-Russo H, Costa-Lotufo LV, Jarmusch AK, Dorrestein PC (2022) Mass spectrometry-based metabolomics in microbiome investigations. Nat Rev Microbiol 20:143–160. https://doi.org/10.1038/s41579-021-00621-9

Beghini F, McIver LJ, Blanco-Míguez A, Dubois L, Asnicar F, Maharjan S, Mailyan A, Manghi P, Scholz M, Thomas AM, Valles-Colomer M, Weingart G, Zhang Y, Zolfo M, Huttenhower C, Franzosa EA, Segata N (2021) Integrating taxonomic, functional, and strain-level profiling of diverse microbial communities with bioBakery 3. Elife 10:e65088. https://doi.org/10.7554/eLife.65088

Belcour A, Frioux C, Aite M, Bretaudeau A, Hildebrand F, Siegel A (2020) Metage2Metabo, microbiota-scale metabolic complementarity for the identification of key species. eLife 9:e61968. https://doi.org/10.7554/eLife.61968

Belizário J, Garay-Malpartida M, Faintuch J (2023) Lung microbiome and origins of the respiratory diseases. Curr Res Immunol 4:100065. https://doi.org/10.1016/j.crimmu.2023.100065

Bhattacharjee S, Saha B, Bhattacharyya P, Saha S (2022) LHSPred: a web based application for predicting lung health severity. Biomed Signal Process Control 77:103745. https://doi.org/10.1016/j.bspc.2022.103745

Bolger AM, Lohse M, Usadel B (2014) Trimmomatic: a flexible trimmer for Illumina sequence data. Bioinformatics 30:2114–2120. https://doi.org/10.1093/bioinformatics/btu170

Briard M, Guillon B, Venot E, Grauso M, Hennequet-Antier C, Bruneau A, Fenaille F, Castelli F, Thomas M, Lezmi G, Leite-de-Moraes M, Saint-Criq V, Adel-Patient K (2023) Cytokines and metabolites are the main descriptors of severe asthma in children when performing multi-omics analysis of bronchoaleveolar lavages, 13 June 2023, PREPRINT (Version 1) available at Research Square. https://doi.org/10.21203/rs.3.rs-3034067/v1

Burr LD, Rogers GB, Chen AC-H, Hamilton BR, Pool GF, Taylor SL, Venter D, Bowler SD, Biga S, McGuckin MA (2016) Macrolide treatment inhibits *Pseudomonas aeruginosa* quorum sensing in non-CF bronchiectasis: an analysis from the BLESS trial. Ann Am Thorac Soc 13:1697–1703. https://doi.org/10.1513/AnnalsATS.201601-044OC

Calderaro A, Buttrini M, Farina B, Montecchini S, De Conto F, Chezzi C (2022) Respiratory tract infections and laboratory diagnostic methods: a review with a focus on syndromic panel-based assays. Microorganisms 10:1856. https://doi.org/10.3390/microorganisms10091856

Callahan BJ, McMurdie PJ, Rosen MJ, Han AW, Johnson AJA, Holmes SP (2016) DADA2: high-resolution sample inference from Illumina amplicon data. Nat Methods 13:581–583. https://doi.org/10.1038/nmeth.3869

Cantalapiedra CP, Hernández-Plaza A, Letunic I, Bork P, Huerta-Cepas J (2021) eggNOG-mapper v2: functional annotation, orthology assignments, and domain prediction at the metagenomic scale. Mol Biol Evol 38:5825–5829. https://doi.org/10.1093/molbev/msab293

Cao Y, Dong Q, Wang D, Zhang P, Liu Y, Niu C (2022) microbiomeMarker: an R/Bioconductor package for microbiome marker identification and visualization. Bioinformatics 38:4027–4029. https://doi.org/10.1093/bioinformatics/btac438

Carney SM, Clemente JC, Cox MJ, Dickson RP, Huang YJ, Kitsios GD, Kloepfer KM, Leung JM, LeVan TD, Molyneaux PL, Moore BB, O'Dwyer DN, Segal LN, Garantziotis S (2020) Methods in lung microbiome research. Am J Respir Cell Mol Biol 62:283–299. https://doi.org/10.1165/rcmb.2019-0273TR

Chang D, Dela Cruz CS, Sharma L (2020) Challenges in understanding lung microbiome: it is NOT like the gut microbiome. Respirology 25:244–245. https://doi.org/10.1111/resp.13759

Chen Y, Huang Y, Ding X, Yang Z, He L, Ning M, Yang Z, He D, Yang L, Liu Z, Chen Y, Li G (2022) A multi-omics study of familial lung cancer: microbiome and host gene expression patterns. Front Immunol 13:827953. https://doi.org/10.3389/fimmu.2022.827953

Chen J, Li T, Ye C, Zhong J, Huang J-D, Ke Y, Sun H (2023a) The lung microbiome: a new frontier for lung and brain disease. Int J Mol Sci 24:2170. https://doi.org/10.3390/ijms24032170

Chen C, Wang J, Pan D, Wang X, Xu Y, Yan J, Wang L, Yang X, Yang M, Liu G-P (2023b) Applications of multi-omics analysis in human diseases. MedComm 4:e315. https://doi.org/10.1002/mco2.315

Cheng K, Ning Z, Zhang X, Li L, Liao B, Mayne J, Stintzi A, Figeys D (2017) MetaLab: an automated pipeline for metaproteomic data analysis. Microbiome 5:157. https://doi.org/10.1186/s40168-017-0375-2

Cheng C, Wang Z, Wang J, Ding C, Sun C, Liu P, Xu X, Liu Y, Chen B, Gu B (2020) Characterization of the lung microbiome and exploration of potential bacterial biomarkers for lung cancer. Transl Lung Cancer Res 9:693–704. https://doi.org/10.21037/tlcr-19-590

Chong J, Liu P, Zhou G, Xia J (2020) Using MicrobiomeAnalyst for comprehensive statistical, functional, and meta-analysis of microbiome data. Nat Protoc 15:799–821. https://doi.org/10.1038/s41596-019-0264-1

Chotirmall SH, Bogaert D, Chalmers JD, Cox MJ, Hansbro PM, Huang YJ, Molyneaux PL, O'Dwyer DN, Pragman AA, Rogers GB, Segal LN, Dickson RP (2022) Therapeutic targeting of the respiratory microbiome. Am J Respir Crit Care Med 206:535–544. https://doi.org/10.1164/rccm.202112-2704PP

Chu X, Zhang B, Koeken VACM, Gupta MK, Li Y (2021) Multi-omics approaches in immunological research. Front Immunol 12:668045. https://doi.org/10.3389/fimmu.2021.668045

Dang X, Kang Y, Wang X, Cao W, Li M, He Y, Pan X, Ye K, Xu D (2022) Frequent exacerbators of chronic obstructive pulmonary disease have distinguishable sputum microbiome signatures during clinical stability. Front Microbiol 13:1037037. https://doi.org/10.3389/fmicb.2022.1037037

De Almeida OGG, Capizzani CPDC, Tonani L, Grizante Barião PH, Da Cunha AF, De Martinis ECP, Torres LAGMM, Von Zeska Kress MR (2020) The lung microbiome of three Young Brazilian patients with cystic fibrosis colonized by fungi. Front Cell Infect Microbiol 10:598938. https://doi.org/10.3389/fcimb.2020.598938

DeSantis TZ, Hugenholtz P, Larsen N, Rojas M, Brodie EL, Keller K, Huber T, Dalevi D, Hu P, Andersen GL (2006) Greengenes, a chimera-checked 16S rRNA gene database and workbench compatible with ARB. Appl Environ Microbiol 72:5069–5072. https://doi.org/10.1128/AEM.03006-05

Dong Q, Chen ES, Zhao C, Jin C (2021) Host-microbiome interaction in lung cancer. Front Immunol 12:679829. https://doi.org/10.3389/fimmu.2021.679829

Du W, Zhao Y, Zhang L, Che J, Liu Z, Li K, Che N (2022) Bacteriomes in lesions of pulmonary tuberculosis and its association with status of mycobacterium tuberculosis excretion. BMC Microbiol 22:280. https://doi.org/10.1186/s12866-022-02698-5

Durazzi F, Sala C, Castellani G, Manfreda G, Remondini D, De Cesare A (2021) Comparison between 16S rRNA and shotgun sequencing data for the taxonomic characterization of the gut microbiota. Sci Rep 11:3030. https://doi.org/10.1038/s41598-021-82726-y

Edgar RC, Haas BJ, Clemente JC, Quince C, Knight R (2011) UCHIME improves sensitivity and speed of chimera detection. Bioinformatics 27:2194–2200. https://doi.org/10.1093/bioinformatics/btr381

Estaki M, Jiang L, Bokulich NA, McDonald D, González A, Kosciolek T, Martino C, Zhu Q, Birmingham A, Vázquez-Baeza Y, Dillon MR, Bolyen E, Caporaso JG, Knight R (2020) QIIME 2 enables comprehensive end-to-end analysis of diverse microbiome data and comparative studies with publicly available data. Curr Protoc Bioinformatics 70:e100. https://doi.org/10.1002/cpbi.100

Felton E, Burrell A, Chaney H, Sami I, Koumbourlis AC, Freishtat RJ, Crandall KA, Hahn A (2021) Inflammation in children with cystic fibrosis: contribution of bacterial production of long-chain fatty acids. Pediatr Res 90:99–108. https://doi.org/10.1038/s41390-021-01419-4

Gagliardi A, Totino V, Cacciotti F, Iebba V, Neroni B, Bonfiglio G, Trancassini M, Passariello C, Pantanella F, Schippa S (2018) Rebuilding the gut microbiota ecosystem. Int J Environ Res Public Health 15:1679. https://doi.org/10.3390/ijerph15081679

Gaibani P, Viciani E, Bartoletti M, Lewis RE, Tonetti T, Lombardo D, Castagnetti A, Bovo F, Horna CS, Ranieri M, Viale P, Re MC, Ambretti S (2021) The lower respiratory tract microbiome of critically ill patients with COVID-19. Sci Rep 11:10103. https://doi.org/10.1038/s41598-021-89516-6

Gemayel K, Lomsadze A, Borodovsky M (2022) MetaGeneMark-2: Improved Gene Prediction in Metagenomes. bioRxiv 2022.07.25.500264. https://doi.org/10.1101/2022.07.25.500264

Ghosh A, Saha S (2022) Meta-analysis of sputum microbiome studies identifies airway disease-specific taxonomic and functional signatures. J Med Microbiol 72. https://doi.org/10.1099/jmm.0.001617

Ghosh A, Firdous S, Saha S (2021) Bioinformatics for human microbiome. In: Singh V, Kumar A (eds) Advances in bioinformatics. Springer, Singapore, pp 333–350

Goolam Mahomed T, Peters RPH, Allam M, Ismail A, Mtshali S, Goolam Mahomed A, Ueckermann V, Kock MM, Ehlers MM (2021) Lung microbiome of stable and exacerbated COPD patients in Tshwane, South Africa. Sci Rep 11:19758. https://doi.org/10.1038/s41598-021-99127-w

Gupta S, Mortensen MS, Schjørring S, Trivedi U, Vestergaard G, Stokholm J, Bisgaard H, Krogfelt KA, Sørensen SJ (2019) Amplicon sequencing provides more accurate microbiome information in healthy children compared to culturing. Commun Biol 2:291. https://doi.org/10.1038/s42003-019-0540-1

Gurdeep Singh R, Tanca A, Palomba A, Van der Jeugt F, Verschaffelt P, Uzzau S, Martens L, Dawyndt P, Mesuere B (2019) Unipept 4.0: functional analysis of metaproteome data. J Proteome Res 18:606–615. https://doi.org/10.1021/acs.jproteome.8b00716

Haldar K, George L, Wang Z, Mistry V, Ramsheh MY, Free RC, John C, Reeve NF, Miller BE, Tal-Singer R, Webb AJ, Brookes AJ, Tobin MD, Singh D, Donaldson GC, Wedzicha JA, Brown JR, Barer MR, Brightling CE (2020) The sputum microbiome is distinct between COPD and health, independent of smoking history. Respir Res 21:183. https://doi.org/10.1186/s12931-020-01448-3

He J-Q, Chen Q, Wu S-J, Wang D-Q, Zhang S-Y, Zhang S-Z, Chen R-L, Wang J-F, Wang Z, Yu C-H (2022a) Potential implications of the lung microbiota in patients with chronic obstruction pulmonary disease and non-small cell lung cancer. Front Cell Infect Microbiol 12:937864. https://doi.org/10.3389/fcimb.2022.937864

He Y, Yu W, Ning P, Luo Q, Zhao L, Xie Y, Yu Y, Ma X, Chen L, Zheng Y, Gao Z (2022b) Shared and specific lung microbiota with metabolic profiles in bronchoalveolar lavage fluid between infectious and inflammatory respiratory diseases. J Inflamm Res 15:187–198. https://doi.org/10.2147/JIR.S342462

He Y, Li J, Yu W, Zheng Y, Yang D, Xu Y, Zhao L, Ma X, Gong P, Gao Z (2022c) Characteristics of lower respiratory tract microbiota in the patients with post-hematopoietic stem cell transplantation pneumonia. Front Cell Infect Microbiol 12:943317. https://doi.org/10.3389/fcimb.2022.943317

Hernández-de-Diego R, Tarazona S, Martínez-Mira C, Balzano-Nogueira L, Furió-Tarí P, Pappas GJ, Conesa A (2018) PaintOmics 3: a web resource for the pathway analysis and visualization of multi-omics data. Nucleic Acids Res 46:W503–W509. https://doi.org/10.1093/nar/gky466

Hosang L, Canals RC, van der Flier FJ, Hollensteiner J, Daniel R, Flügel A, Odoardi F (2022) The lung microbiome regulates brain autoimmunity. Nature 603:138–144. https://doi.org/10.1038/s41586-022-04427-4

Hou K, Wu Z-X, Chen X-Y, Wang J-Q, Zhang D, Xiao C, Zhu D, Koya JB, Wei L, Li J, Chen Z-S (2022) Microbiota in health and diseases. Signal Transduct Target Ther 7:135. https://doi.org/10.1038/s41392-022-00974-4

Huang D, Su X, Yuan M, Zhang S, He J, Deng Q, Qiu W, Dong H, Cai S (2019) The characterization of lung microbiome in lung cancer patients with different clinicopathology. Am J Cancer Res 9:2047–2063

Huang DH, He J, Su XF, Wen YN, Zhang SJ, Liu LY, Zhao H, Ye CP, Wu JH, Cai S, Dong H (2022) The airway microbiota of non-small cell lung cancer patients and its relationship to tumor stage and EGFR gene mutation. Thorac Cancer 13:858–869. https://doi.org/10.1111/1759-7714.14340

Huson DH, Auch AF, Qi J, Schuster SC (2007) MEGAN analysis of metagenomic data. Genome Res 17:377–386. https://doi.org/10.1101/gr.5969107

Huynh M, Crane MJ, Jamieson AM (2023) The lung, the niche, and the microbe: exploring the lung microbiome in cancer and immunity. Front Immunol 13:1094110. https://doi.org/10.3389/fimmu.2022.1094110

Hyatt D, LoCascio PF, Hauser LJ, Uberbacher EC (2012) Gene and translation initiation site prediction in metagenomic sequences. Bioinformatics 28:2223–2230. https://doi.org/10.1093/bioinformatics/bts429

Jude MS, Yang CX, Filho FSL, Hernandez Cordero AI, Yang J, Shaipanich T, Li X, Lin D, MacIsaac J, Kobor MS, Sinha S, Nislow C, Singh A, Lam W, Lam S, Guillemi S, Harris M, Montaner J, Ng RT, Carlsten C, Paul Man SF, Sin DD, Leung JM (2023) Microbial dysbiosis and the host airway epithelial response: insights into HIV-associated COPD using multi-omics profiling. Respir Res 24:124. https://doi.org/10.1186/s12931-023-02431-4

Kameoka S, Motooka D, Watanabe S, Kubo R, Jung N, Midorikawa Y, Shinozaki NO, Sawai Y, Takeda AK, Nakamura S (2021) Benchmark of 16S rRNA gene amplicon sequencing using Japanese gut microbiome data from the V1-V2 and V3-V4 primer sets. BMC Genomics 22:527. https://doi.org/10.1186/s12864-021-07746-4

Kanehisa M, Goto S (2000) KEGG: Kyoto encyclopedia of genes and genomes. Nucleic Acids Res 28:27–30. https://doi.org/10.1093/nar/28.1.27

Karp PD, Billington R, Caspi R, Fulcher CA, Latendresse M, Kothari A, Keseler IM, Krummenacker M, Midford PE, Ong Q, Ong WK, Paley SM, Subhraveti P (2019) The BioCyc collection of microbial genomes and metabolic pathways. Brief Bioinform 20:1085–1093. https://doi.org/10.1093/bib/bbx085

Kelley DR, Liu B, Delcher AL, Pop M, Salzberg SL (2012) Gene prediction with glimmer for metagenomic sequences augmented by classification and clustering. Nucleic Acids Res 40:e9–e9. https://doi.org/10.1093/nar/gkr1067

Kim J, Kim MS, Koh AY, Xie Y, Zhan X (2016) FMAP: functional mapping and analysis pipeline for metagenomics and metatranscriptomics studies. BMC Bioinformatics 17:420. https://doi.org/10.1186/s12859-016-1278-0

Kim YH, Jang H, Kim SY, Jung JH, Kim GE, Park MR, Hong JY, Kim MN, Kim EG, Kim MJ, Kim KW, Sohn MH (2021) Gram-negative microbiota is related to acute exacerbation in children with asthma. Clin Transl Allergy 11:e12069. https://doi.org/10.1002/clt2.12069

Kleiner M (2019) Metaproteomics: much more than measuring gene expression in microbial communities. mSystems 4:e00115–e00119. https://doi.org/10.1128/mSystems.00115-19

Kosina SM, Greiner AM, Lau RK, Jenkins S, Baran R, Bowen BP, Northen TR (2018) Web of microbes (WoM): a curated microbial exometabolomics database for linking chemistry and microbes. BMC Microbiol 18:115. https://doi.org/10.1186/s12866-018-1256-y

Kotlyarov S (2022) Role of short-chain fatty acids produced by gut microbiota in innate lung immunity and pathogenesis of the heterogeneous course of chronic obstructive pulmonary disease. Int J Mol Sci 23:4768. https://doi.org/10.3390/ijms23094768

Kuczynski J, Stombaugh J, Walters WA, González A, Caporaso JG, Knight R (2011) Using QIIME to analyze 16S rRNA gene sequences from microbial communities. Curr Protoc Bioinformatics 36:10.7.1. https://doi.org/10.1002/0471250953.bi1007s36

Kussek P, Mesa D, Vasconcelos TM, Rodrigues LS, Krul D, Ibanez H, Faoro H, Palmeiro JK, Dalla Costa LM (2022) Lower airway microbiota and decreasing lung function in young Brazilian cystic fibrosis patients with pulmonary staphylococcus and pseudomonas infection. PLoS One 17:e0273453. https://doi.org/10.1371/journal.pone.0273453

Kwoji ID, Aiyegoro OA, Okpeku M, Adeleke MA (2023) 'Multi-omics' data integration: applications in probiotics studies. NPJ Sci Food 7:25. https://doi.org/10.1038/s41538-023-00199-x

Langmead B, Salzberg SL (2012) Fast gapped-read alignment with bowtie 2. Nat Methods 9:357–359. https://doi.org/10.1038/nmeth.1923

Lee SY, Mac Aogáin M, Fam KD, Chia KL, Binte Mohamed Ali NA, Yap MMC, Yap EPH, Chotirmall SH, Lim CL (2019) Airway microbiome composition correlates with lung function and arterial stiffness in an age-dependent manner. PLoS One 14:e0225636. https://doi.org/10.1371/journal.pone.0225636

Li D, Liu C-M, Luo R, Sadakane K, Lam T-W (2015) MEGAHIT: an ultra-fast single-node solution for large and complex metagenomics assembly via succinct *de Bruijn* graph. Bioinformatics 31:1674–1676. https://doi.org/10.1093/bioinformatics/btv033

Liang W, Yang Y, Gong S, Wei M, Ma Y, Feng R, Gao J, Liu X, Tu F, Ma W, Yi X, Liang Z, Wang F, Wang L, Chen D, Shu W, Miller BE, Tal-Singer R, Donaldson GC, Wedzicha JA, Singh D, Wilkinson TMA, Brightling CE, Chen R, Zhong N, Wang Z (2023) Airway dysbiosis accelerates lung function decline in chronic obstructive pulmonary disease. Cell Host Microbe 31:1054–1070.e9. https://doi.org/10.1016/j.chom.2023.04.018

Liao B, Ning Z, Cheng K, Zhang X, Li L, Mayne J, Figeys D (2018) iMetaLab 1.0: a web platform for metaproteomics data analysis. Bioinformatics 34:3954–3956. https://doi.org/10.1093/bioinformatics/bty466

Liu C, Cui Y, Li X, Yao M (2021) *Microeco*: an R package for data mining in microbial community ecology. FEMS Microbiol Ecol 97:fiaa255. https://doi.org/10.1093/femsec/fiaa255

Lu J, Breitwieser FP, Thielen P, Salzberg SL (2017) Bracken: estimating species abundance in metagenomics data. PeerJ Comput Sci 3:e104. https://doi.org/10.7717/peerj-cs.104

Lu H, Gao NL, Tong F, Wang J, Li H, Zhang R, Ma H, Yang N, Zhang Y, Wang Y, Liang Z, Zeng H, Chen W-H, Dong X (2021) Alterations of the human lung and Gut microbiomes in non-small cell lung carcinomas and distant metastasis. Microbiol Spectr 9:e00802–e00821. https://doi.org/10.1128/Spectrum.00802-21

Lu Y, Zhou G, Ewald J, Pang Z, Shiri T, Xia J (2023) MicrobiomeAnalyst 2.0: comprehensive statistical, functional and integrative analysis of microbiome data. Nucleic Acids Res 51:W310–W318. https://doi.org/10.1093/nar/gkad407

Maddi A, Sabharwal A, Violante T, Manuballa S, Genco R, Patnaik S, Yendamuri S (2019) The microbiome and lung cancer. J Thorac Dis 11:280–291. https://doi.org/10.21037/jtd.2018.12.88

Magoč T, Salzberg SL (2011) FLASH: fast length adjustment of short reads to improve genome assemblies. Bioinformatics 27:2957–2963. https://doi.org/10.1093/bioinformatics/btr507

Mallick H, Franzosa EA, McIver LJ, Banerjee S, Sirota-Madi A, Kostic AD, Clish CB, Vlamakis H, Xavier RJ, Huttenhower C (2019) Predictive metabolomic profiling of microbial communities using amplicon or metagenomic sequences. Nat Commun 10:3136. https://doi.org/10.1038/s41467-019-10927-1

Maras JS, Sharma S, Bhat A, Rooge S, Aggrawal R, Gupta E, Sarin SK (2021) Multi-omics analysis of respiratory specimen characterizes baseline molecular determinants associated with SARS-CoV-2 outcome. iScience 24:102823. https://doi.org/10.1016/j.isci.2021.102823

Martinez X, Pozuelo M, Pascal V, Campos D, Gut I, Gut M, Azpiroz F, Guarner F, Manichanh C (2016) MetaTrans: an open-source pipeline for metatranscriptomics. Sci Rep 6:26447. https://doi.org/10.1038/srep26447

McMurdie PJ, Holmes S (2013) Phyloseq: an R package for reproducible interactive analysis and graphics of microbiome census data. PLoS One 8:e61217. https://doi.org/10.1371/journal.pone.0061217

Momtazmanesh S, Moghaddam SS, Ghamari S-H, Rad EM, Rezaei N, Shobeiri P, Aali A, Abbasi-Kangevari M, Abbasi-Kangevari Z, Abdelmasseh M, Abdoun M, Abdulah DM, Md Abdullah AY, Abedi A, Abolhassani H, Abrehdari-Tafreshi Z, Achappa B, Adane Adane DE, Adane TD, Addo IY, Adnan M, Sakilah Adnani QE, Ahmad S, Ahmadi A, Ahmadi K, Ahmed A, Ahmed A, Rashid TA, Al Hamad H, Alahdab F, Alemayehu A, Alif SM, Aljunid SM, Almustanyir S, Altirkawi KA, Alvis-Guzman N, Dehkordi JA, Amir-Behghadami M, Ancuceanu R, Andrei CL, Andrei T, Antony CM, Anyasodor AE, Arabloo J, Arulappan J, Ashraf T, Athari SS, Attia EF, Ayele MT, Azadnajafabad S, Babu AS, Bagherieh S, Baltatu OC, Banach M, Bardhan M, Barone-Adesi F, Barrow A, Basu S, Bayileyegn NS, Bensenor IM, Bhardwaj N, Bhardwaj P, Bhat AN, Bhattacharyya K, Bouaoud S, Braithwaite D, Brauer M, Butt MH, Butt ZA, Calina D, Cámera LA, Chanie GS, Charalampous P, Chattu VK, Chimed-Ochir O, Chu D-T, Cohen AJ, Cruz-Martins N, Dadras O, Darwesh AM, Das S, Debela SA, Delgado-Ortiz L, Dereje D, Dianatinasab M, Diao N, Diaz D, Digesa LE, Dirirsa G, Doku PN, Dongarwar D, Douiri A, Dsouza HL, Eini E, Ekholuenetale M, Ekundayo TC, Mustafa Elagali AE, Elhadi M, Enyew DB, Erkhembayar R, Etaee F, Fagbamigbe AF, Faro A, Fatehizadeh A, Fekadu G, Filip I, Fischer F, Foroutan M, Franklin RC, Gaal PA, Gaihre S, Gaipov A, Gebrehiwot M, Gerema U, Getachew ME, Getachew T, Ghafourifard M, Ghanbari R, Ghashghaee A, Gholami A, Gil AU, Golechha M, Goleij P, Golinelli D, Guadie HA, Gupta B, Gupta S, Gupta VB, Gupta VK, Hadei M, Halwani R, Hanif A, Hargono A, Harorani M, Hartono RK, Hasani H, Hashi A, Hay SI, Heidari M, Hellemons ME, Herteliu C, Holla R, Horita N, Hoseini M, Hosseinzadeh M, Huang J, Hussain S, Hwang B-F, Iavicoli I, Ibitoye SE, Ibrahim S, Ilesanmi OS, Ilic IM, Ilic MD, Immurana M, Ismail NE, Merin JL, Jakovljevic M, Jamshidi E, Janodia MD, Javaheri T, Jayapal SK, Jayaram S, Jha RP, Johnson O, Joo T, Joseph N, Jozwiak JJ, Kaambwa B, Kabir Z, Kalankesh LR, Kalhor R, Kandel H, Karanth SD, Karaye IM, Kassa BG, Kassie GM, Keikavoosi-Arani L, Keykhaei M, Khajuria H, Khan IA, MAB K, Khan YH, Khreis H, Kim MS, Kisa A, Kisa S, Knibbs LD, Kolkhir P, Komaki S, Kompani F, Koohestani HR, Koolivand A, Korzh O, Koyanagi A, Krishan K, Krohn KJ, Kumar N, Kumar N, Kurmi OP, Kuttikkattu A, La Vecchia C, Lám J, Lan Q, Lasrado S, Latief K, Lauriola P, Lee S, Lee YH, Legesse SM, Lenzi J, Li M-C, Lin R-T, Liu G, Liu W, Lo C-H, Lorenzovici L, Lu Y, Mahalingam S, Mahmoudi E, Mahotra NB, Malekpour M-R, Malik AA, Mallhi TH, Malta DC, Mansouri B,

Mathews E, Maulud SQ, Mechili EA, Nasab EM, Menezes RG, Mengistu DA, Mentis A-F, Meshkat M, Mestrovic T, AC MGNDS, Mirrakhimov EM, Misganaw A, Mithra P, Moghadasi J, Mohammadi E, Mohammadi M, Mohammadshahi M, Mohammed S, Mohan S, Moka N, Monasta L, Moni MA, Moniruzzaman M, Montazeri F, Moradi M, Mostafavi E, Mpundu-Kaambwa C, Murillo-Zamora E, CJL M, Nair TS, Nangia V, Swamy SN, Narayana AI, Natto ZS, Nayak BP, Negash WW, Nena E, Kandel SN, Niazi RK, Nogueira De Sá AT, Nowroozi A, Nzoputam CI, Nzoputam OJ, Oancea B, Obaidur RM, Odukoya OO, Okati-Aliabad H, Okekunle AP, Okonji OC, Olagunju AT, Bali AO, Ostojic SM, MP A, Padron-Monedero A, Padubidri JR, Pahlevan Fallahy MT, Palicz T, Pana A, Park E-K, Patel J, Paudel R, Paudel U, Pedersini P, Pereira M, Pereira RB, Petcu I-R, Pirestani M, Postma MJ, Prashant A, Rabiee M, Radfar A, Rafiei S, Rahim F, Ur Rahman MH, Rahman M, Rahman MA, Rahmani AM, Rahmani S, Rahmanian V, Rajput P, Rana J, Rao CR, Rao SJ, Rashedi S, Rashidi M-M, Ratan ZA, Rawaf DL, Rawaf S, Rawal L, Rawassizadeh R, Razeghinia MS, Mohamed Redwan EM, Rezaei M, Rezaei N, Rezaei N, Rezaeian M, Rodrigues M, Buendia Rodriguez JA, Roever L, Rojas-Rueda D, Rudd KE, AMA S, Sabour S, Saddik B, Sadeghi E, Sadeghi M, Saeed U, Sahebazzamani M, Sahebkar A, Sahoo H, Sajid MR, Sakhamuri S, Salehi S, Samy AM, Santric-Milicevic MM, Sao Jose BP, Sathian B, Satpathy M, Saya GK, Senthilkumaran S, Seylani A, Shahabi S, Shaikh MA, Shanawaz M, Shannawaz M, Sheikhi RA, Shekhar S, Sibhat MM, Simpson CR, Singh JA, Singh P, Singh S, Skryabin VY, Skryabina AA, Soltani-Zangbar MS, Song S, Soyiri IN, Steiropoulos P, Stockfelt L, Sun J, Takahashi K, Talaat IM, Tan K-K, Tat NY, Tat VY, Taye BT, Thangaraju P, Thapar R, Thienemann F, Tiyuri A, Ngoc Tran MT, Tripathy JP, Car LT, Tusa BS, Ullah I, Ullah S, Vacante M, Valdez PR, Valizadeh R, JFM VB, Vasankari TJ, Vaziri S, Violante FS, Vo B, Wang N, Wei MY, Westerman R, Wickramasinghe ND, Xu S, Xu X, Yadav L, Yismaw Y, Yon DK, Yonemoto N, Yu C, Yu Y, Yunusa I, Zahir M, Zangiabadian M, Zareshahrabadi Z, Zarrintan A, Zastrozhin MS, Zegeye ZB, Zhang Y, Naghavi M, Larijani B, Farzadfar F (2023) Global burden of chronic respiratory diseases and risk factors, 1990–2019: an update from the global burden of disease study 2019. eClinicalMedicine 59:101936. https://doi.org/10.1016/j.eclinm.2023.101936

Mongad DS, Chavan NS, Narwade NP, Dixit K, Shouche YS, Dhotre DP (2021) MicFunPred: a conserved approach to predict functional profiles from 16S rRNA gene sequence data. Genomics 113:3635–3643. https://doi.org/10.1016/j.ygeno.2021.08.016

Montague E, Stanberry L, Higdon R, Janko I, Lee E, Anderson N, Choiniere J, Stewart E, Yandl G, Broomall W, Kolker N, Kolker E (2014) MOPED 2.5—an integrated multi-omics resource: multi-omics profiling expression database now includes transcriptomics data. OMICS J Integr Biol 18:335–343. https://doi.org/10.1089/omi.2014.0061

Morton JT, Aksenov AA, Nothias LF, Foulds JR, Quinn RA, Badri MH, Swenson TL, Van Goethem MW, Northen TR, Vazquez-Baeza Y, Wang M, Bokulich NA, Watters A, Song SJ, Bonneau R, Dorrestein PC, Knight R (2019) Learning representations of microbe–metabolite interactions. Nat Methods 16:1306–1314. https://doi.org/10.1038/s41592-019-0616-3

Muhamadali H, Winder CL, Dunn WB, Goodacre R (2023) Unlocking the secrets of the microbiome: exploring the dynamic microbial interplay with humans through metabolomics and their manipulation for synthetic biology applications. Biochem J 480:891–908. https://doi.org/10.1042/BCJ20210534

Muñoz-Benavent M, Hartkopf F, Van Den Bossche T, Piro VC, García-Ferris C, Latorre A, Renard BY, Muth T (2020) gNOMO: a multi-omics pipeline for integrated host and microbiome analysis of non-model organisms. NAR Genom Bioinformatics 2:lqaa058. https://doi.org/10.1093/nargab/lqaa058

Muth T, Behne A, Heyer R, Kohrs F, Benndorf D, Hoffmann M, Lehtevä M, Reichl U, Martens L, Rapp E (2015) The MetaProteomeAnalyzer: a powerful open-source software suite for metaproteomics data analysis and interpretation. J Proteome Res 14:1557–1565. https://doi.org/10.1021/pr501246w

Narayan NR, Weinmaier T, Laserna-Mendieta EJ, Claesson MJ, Shanahan F, Dabbagh K, Iwai S, DeSantis TZ (2020) Piphillin predicts metagenomic composition and dynamics from

DADA2-corrected 16S rDNA sequences. BMC Genomics 21:56. https://doi.org/10.1186/s12864-019-6427-1
Narayanasamy S, Jarosz Y, Muller EEL, Heintz-Buschart A, Herold M, Kaysen A, Laczny CC, Pinel N, May P, Wilmes P (2016) IMP: a pipeline for reproducible reference-independent integrated metagenomic and metatranscriptomic analyses. Genome Biol 17:260. https://doi.org/10.1186/s13059-016-1116-8
Natalini JG, Singh S, Segal LN (2023) The dynamic lung microbiome in health and disease. Nat Rev Microbiol 21:222–235. https://doi.org/10.1038/s41579-022-00821-x
Nelson MT, Wolter DJ, Eng A, Weiss EJ, Vo AT, Brittnacher MJ, Hayden HS, Ravishankar S, Bautista G, Ratjen A, Blackledge M, McNamara S, Nay L, Majors C, Miller SI, Borenstein E, Simon RH, LiPuma JJ, Hoffman LR (2020) Maintenance tobramycin primarily affects untargeted bacteria in the CF sputum microbiome. Thorax 75:780–790. https://doi.org/10.1136/thoraxjnl-2019-214187
Ni Y, Li J, Panagiotou G (2016) COMAN: a web server for comprehensive metatranscriptomics analysis. BMC Genomics 17:622. https://doi.org/10.1186/s12864-016-2964-z
Noecker C, Eng A, Muller E, Borenstein E (2022) MIMOSA2: a metabolic network-based tool for inferring mechanism-supported relationships in microbiome-metabolome data. Bioinformatics 38:1615–1623. https://doi.org/10.1093/bioinformatics/btac003
Nurk S, Meleshko D, Korobeynikov A, Pevzner PA (2017) metaSPAdes: a new versatile metagenomic assembler. Genome Res 27:824–834. https://doi.org/10.1101/gr.213959.116
O'Dwyer DN, Dickson RP, Moore BB (2016) The lung microbiome, immunity, and the pathogenesis of chronic lung disease. J Immunol 196:4839–4847. https://doi.org/10.4049/jimmunol.1600279
Ojala T, Kankuri E, Kankainen M (2023) Understanding human health through metatranscriptomics. Trends Mol Med 29:376–389. https://doi.org/10.1016/j.molmed.2023.02.002
Pérez-Cobas AE, Gomez-Valero L, Buchrieser C (2020) Metagenomic approaches in microbial ecology: an update on whole-genome and marker gene sequencing analyses. Microb Genom 6:mgen000409. https://doi.org/10.1099/mgen.0.000409
Petrocheilou A, Moudaki A, Kaditis AG (2022) Inflammation and infection in cystic fibrosis: update for the clinician. Children 9:1898. https://doi.org/10.3390/children9121898
Pragman AA, Kim HB, Reilly CS, Wendt C, Isaacson RE (2012) The lung microbiome in moderate and severe chronic obstructive pulmonary disease. PLoS One 7:e47305. https://doi.org/10.1371/journal.pone.0047305
Pruesse E, Quast C, Knittel K, Fuchs BM, Ludwig W, Peplies J, Glockner FO (2007) SILVA: a comprehensive online resource for quality checked and aligned ribosomal RNA sequence data compatible with ARB. Nucleic Acids Res 35:7188–7196. https://doi.org/10.1093/nar/gkm864
Purushothaman S, Meola M, Egli A (2022) Combination of whole genome sequencing and metagenomics for microbiological diagnostics. Int J Mol Sci 23:9834. https://doi.org/10.3390/ijms23179834
Ramsheh MY, Haldar K, Esteve-Codina A, Purser LF, Richardson M, Müller-Quernheim J, Greulich T, Nowinski A, Barta I, Stendardo M, Boschetto P, Korzybski D, Prasse A, Parr DG, Hohlfeld JM, Döme B, Welte T, Heath S, Gut I, Morrissey JA, Ziegler-Heitbrock L, Barer MR, Singh D, Brightling CE (2021) Lung microbiome composition and bronchial epithelial gene expression in patients with COPD versus healthy individuals: a bacterial 16S rRNA gene sequencing and host transcriptomic analysis. Lancet Microb 2:e300–e310. https://doi.org/10.1016/S2666-5247(21)00035-5
Reiman D, Layden BT, Dai Y (2021) MiMeNet: exploring microbiome-metabolome relationships using neural networks. PLoS Comput Biol 17:e1009021. https://doi.org/10.1371/journal.pcbi.1009021
Ren Y, Su H, She Y, Dai C, Xie D, Narrandes S, Huang S, Chen C, Xu W (2019) Whole genome sequencing revealed microbiome in lung adenocarcinomas presented as ground-glass nodules. Transl Lung Cancer Res 8:235–246. https://doi.org/10.21037/tlcr.2019.06.11
Rho M, Tang H, Ye Y (2010) FragGeneScan: predicting genes in short and error-prone reads. Nucleic Acids Res 38:e191–e191. https://doi.org/10.1093/nar/gkq747

Rohart F, Gautier B, Singh A, Lê Cao K-A (2017) mixOmics: an R package for 'omics' feature selection and multiple data integration. PLoS Comput Biol 13:e1005752. https://doi.org/10.1371/journal.pcbi.1005752

Ruiz-Perez D, Lugo-Martinez J, Bourguignon N, Mathee K, Lerner B, Bar-Joseph Z, Narasimhan G (2021) Dynamic Bayesian networks for integrating multi-omics time series microbiome data. mSystems 6:e01105–e01120. https://doi.org/10.1128/mSystems.01105-20

Salvato F, Hettich RL, Kleiner M (2021) Five key aspects of metaproteomics as a tool to understand functional interactions in host-associated microbiomes. PLoS Pathog 17:e1009245. https://doi.org/10.1371/journal.ppat.1009245

Schmid R, Heuckeroth S, Korf A, Smirnov A, Myers O, Dyrlund TS, Bushuiev R, Murray KJ, Hoffmann N, Lu M, Sarvepalli A, Zhang Z, Fleischauer M, Dührkop K, Wesner M, Hoogstra SJ, Rudt E, Mokshyna O, Brungs C, Ponomarov K, Mutabdžija L, Damiani T, Pudney CJ, Earll M, Helmer PO, Fallon TR, Schulze T, Rivas-Ubach A, Bilbao A, Richter H, Nothias L-F, Wang M, Orešič M, Weng J-K, Böcker S, Jeibmann A, Hayen H, Karst U, Dorrestein PC, Petras D, Du X, Pluskal T (2023) Integrative analysis of multimodal mass spectrometry data in MZmine 3. Nat Biotechnol 41:447–449. https://doi.org/10.1038/s41587-023-01690-2

Scialò F, Vitale M, D'Agnano V, Mariniello DF, Perrotta F, Castaldo A, Campbell SFM, Pastore L, Cazzola M, Bianco A (2023) Lung microbiome as a treatable trait in chronic respiratory disorders. Lung 201:455. https://doi.org/10.1007/s00408-023-00645-3

Segal LN, Clemente JC, Wu BG, Wikoff WR, Gao Z, Li Y, Ko JP, Rom WN, Blaser MJ, Weiden MD (2017) Randomised, double-blind, placebo-controlled trial with azithromycin selects for anti-inflammatory microbial metabolites in the emphysematous lung. Thorax 72:13–22. https://doi.org/10.1136/thoraxjnl-2016-208599

Sequeira JC, Rocha M, Madalena Alves M, Salvador AF (2019) MOSCA: an automated pipeline for integrated metagenomics and metatranscriptomics data analysis. In: Fdez-Riverola F, Mohamad MS, Rocha M, De Paz JF, González P (eds) Practical applications of computational biology and bioinformatics, 12th international conference. Springer International, Cham, pp 183–191

Smith CA, O'Maille G, Want EJ, Qin C, Trauger SA, Brandon TR, Custodio DE, Abagyan R, Siuzdak G (2005) METLIN: a metabolite mass spectral database. Ther Drug Monit 27:747–751. https://doi.org/10.1097/01.ftd.0000179845.53213.39

Smith CA, Want EJ, O'Maille G, Abagyan R, Siuzdak G (2006) XCMS: processing mass spectrometry data for metabolite profiling using nonlinear peak alignment, matching, and identification. Anal Chem 78:779–787. https://doi.org/10.1021/ac051437y

Su K, Gao Y, He J (2023) A comparison of the microbiome composition in lower respiratory tract at different sites in early lung cancer patients. Transl Lung Cancer Res 12:1264–1275. https://doi.org/10.21037/tlcr-23-231

Sulaiman I, Chung M, Angel L, Tsay J-CJ WBG, Yeung ST, Krolikowski K, Li Y, Duerr R, Schluger R, Thannickal SA, Koide A, Rafeq S, Barnett C, Postelnicu R, Wang C, Banakis S, Perez-Perez L, Jour G, Shen G, Meyn P, Carpenito J, Liu X, Ji K, Collazo D, Labarbiera A, Amoroso N, Brosnahan S, Mukherjee V, Kaufman D, Bakker J, Lubinsky A, Pradhan D, Sterman DH, Weiden M, Hegu A, Evans L, Uyeki TM, Clemente JC, De Wit E, Schmidt AM, Shopsin B, Desvignes L, Wang C, Li H, Zhang B, Forst CV, Koide S, Stapleford KA, Khanna KM, Ghedin E, Segal LN (2021) Microbial signatures in the lower airways of mechanically ventilated COVID19 patients associated with poor clinical outcome. Infectious diseases (except HIV/AIDS). Nat Microbiol 6:1245–1258

Sulit AK, Kolisnik T, Frizelle FA, Purcell R, Schmeier S (2020) MetaFunc: taxonomic and functional analyses of high throughput sequencing for microbiomes. Gut Microbiome. 4:e4. https://doi.org/10.1017/gmb.2022.12

Tamames J, Puente-Sánchez F (2018) SqueezeMeta, a highly portable, fully automatic metagenomic analysis pipeline. Front Microbiol 9:3349. https://doi.org/10.3389/fmicb.2018.03349

Tang W, Zhang L, Ai T, Xia W, Xie C, Fan Y, Chen S, Chen Z, Yao J, Peng Y (2023) A pilot study exploring the association of bronchial bacterial microbiota and recurrent wheezing in infants with atopy. Front Cell Infect Microbiol 13:1013809. https://doi.org/10.3389/fcimb.2023.1013809

Taylor SL, Leong LEX, Mobegi FM, Choo JM, Wesselingh S, Yang IA, Upham JW, Reynolds PN, Hodge S, James AL, Jenkins C, Peters MJ, Baraket M, Marks GB, Gibson PG, Rogers GB, Simpson JL (2019) Long-term azithromycin reduces *Haemophilus influenzae* and increases antibiotic resistance in severe asthma. Am J Respir Crit Care Med 200:309–317. https://doi.org/10.1164/rccm.201809-1739OC

Tong X, Su F, Xu X, Xu H, Yang T, Xu Q, Dai H, Huang K, Zou L, Zhang W, Pei S, Xiao F, Li Y, Wang C (2019) Alterations to the lung microbiome in idiopathic pulmonary fibrosis patients. Front Cell Infect Microbiol 9:149. https://doi.org/10.3389/fcimb.2019.00149

Truong DT, Franzosa EA, Tickle TL, Scholz M, Weingart G, Pasolli E, Tett A, Huttenhower C, Segata N (2015) MetaPhlAn2 for enhanced metagenomic taxonomic profiling. Nat Methods 12:902–903. https://doi.org/10.1038/nmeth.3589

Tsay J-CJ, Wu BG, Sulaiman I, Gershner K, Schluger R, Li Y, Yie T-A, Meyn P, Olsen E, Perez L, Franca B, Carpenito J, Iizumi T, El-Ashmawy M, Badri M, Morton JT, Shen N, He L, Michaud G, Rafeq S, Bessich JL, Smith RL, Sauthoff H, Felner K, Pillai R, Zavitsanou A-M, Koralov SB, Mezzano V, Loomis CA, Moreira AL, Moore W, Tsirigos A, Heguy A, Rom WN, Sterman DH, Pass HI, Clemente JC, Li H, Bonneau R, Wong K-K, Papagiannakopoulos T, Segal LN (2021) Lower airway dysbiosis affects lung cancer progression. Cancer Discov 11:293–307. https://doi.org/10.1158/2159-8290.CD-20-0263

Tsugawa H, Cajka T, Kind T, Ma Y, Higgins B, Ikeda K, Kanazawa M, VanderGheynst J, Fiehn O, Arita M (2015) MS-DIAL: data-independent MS/MS deconvolution for comprehensive metabolome analysis. Nat Methods 12:523–526. https://doi.org/10.1038/nmeth.3393

Ulven T (2012) Short-chain free fatty acid receptors FFA2/GPR43 and FFA3/GPR41 as new potential therapeutic targets. Front Endocrinol 3:111. https://doi.org/10.3389/fendo.2012.00111

Valverde-Molina J, García-Marcos L (2023) Microbiome and asthma: microbial dysbiosis and the origins, phenotypes, persistence, and severity of asthma. Nutrients 15:486. https://doi.org/10.3390/nu15030486

Van Der Jeugt F, Dawyndt P, Mesuere B (2022) FragGeneScanRs: faster gene prediction for short reads. BMC Bioinformatics 23:198. https://doi.org/10.1186/s12859-022-04736-5

Wang Q, Garrity GM, Tiedje JM, Cole JR (2007) Naïve Bayesian classifier for rapid assignment of rRNA sequences into the new bacterial taxonomy. Appl Environ Microbiol 73:5261–5267. https://doi.org/10.1128/AEM.00062-07

Wang M, Carver JJ, Phelan VV, Sanchez LM, Garg N, Peng Y, Nguyen DD, Watrous J, Kapono CA, Luzzatto-Knaan T, Porto C, Bouslimani A, Melnik AV, Meehan MJ, Liu W-T, Crüsemann M, Boudreau PD, Esquenazi E, Sandoval-Calderón M, Kersten RD, Pace LA, Quinn RA, Duncan KR, Hsu C-C, Floros DJ, Gavilan RG, Kleigrewe K, Northen T, Dutton RJ, Parrot D, Carlson EE, Aigle B, Michelsen CF, Jelsbak L, Sohlenkamp C, Pevzner P, Edlund A, McLean J, Piel J, Murphy BT, Gerwick L, Liaw C-C, Yang Y-L, Humpf H-U, Maansson M, Keyzers RA, Sims AC, Johnson AR, Sidebottom AM, Sedio BE, Klitgaard A, Larson CB, Torres-Mendoza D, Gonzalez DJ, Silva DB, Marques LM, Demarque DP, Pociute E, O'Neill EC, Briand E, EJN H, Granatosky EA, Glukhov E, Ryffel F, Houson H, Mohimani H, Kharbush JJ, Zeng Y, Vorholt JA, Kurita KL, Charusanti P, McPhail KL, Nielsen KF, Vuong L, Elfeki M, Traxler MF, Engene N, Koyama N, Vining OB, Baric R, Silva RR, Mascuch SJ, Tomasi S, Jenkins S, Macherla V, Hoffman T, Agarwal V, Williams PG, Dai J, Neupane R, Gurr J, AMC R, Lamsa A, Zhang C, Dorrestein K, Duggan BM, Almaliti J, Allard P-M, Phapale P, Nothias L-F, Alexandrov T, Litaudon M, Wolfender J-L, Kyle JE, Metz TO, Peryea T, Nguyen D-T, Van Leer D, Shinn P, Jadhav A, Müller R, Waters KM, Shi W, Liu X, Zhang L, Knight R, Jensen PR, Palsson BO, Pogliano K, Linington RG, Gutiérrez M, Lopes NP, Gerwick WH, Moore BS, Dorrestein PC, Bandeira N (2016) Sharing and community curation of mass spectrometry data with global natural products social molecular networking. Nat Biotechnol 34:828–837. https://doi.org/10.1038/nbt.3597

Wang Z, Maschera B, Lea S, Kolsum U, Michalovich D, Van Horn S, Traini C, Brown JR, Hessel EM, Singh D (2019) Airway host-microbiome interactions in chronic obstructive pulmonary disease. Respir Res 20:113. https://doi.org/10.1186/s12931-019-1085-z

Wang Z, Liu H, Wang F, Yang Y, Wang X, Chen B, Stampfli MR, Zhou H, Shu W, Brightling CE, Liang Z, Chen R (2020a) A refined view of airway microbiome in chronic obstructive pulmonary disease at species and strain-levels. Front Microbiol 11:1758. https://doi.org/10.3389/fmicb.2020.01758

Wang Z, Yang Y, Yan Z, Liu H, Chen B, Liang Z, Wang F, Miller BE, Tal-Singer R, Yi X, Li J, Stampfli MR, Zhou H, Brightling CE, Brown JR, Wu M, Chen R, Shu W (2020b) Multi-omic meta-analysis identifies functional signatures of airway microbiome in chronic obstructive pulmonary disease. ISME J 14:2748–2765. https://doi.org/10.1038/s41396-020-0727-y

Wang Q, Guo A, Sheng M, Zhou H (2021) The changes of respiratory microbiome between mild and severe asthma patients. Microbiol Immunol 65:204–213. https://doi.org/10.1111/1348-0421.12881

Wang C-Y, Kuang X, Wang Q-Q, Zhang G-Q, Cheng Z-S, Deng Z-X, Guo F-B (2023) GMMAD: a comprehensive database of human gut microbial metabolite associations with diseases. BMC Genomics 24:482. https://doi.org/10.1186/s12864-023-09599-5

Ward T, Larson J, Meulemans J, Hillmann B, Lynch J, Sidiropoulos D, Spear JR, Caporaso G, Blekhman R, Knight R, Fink R, Knights D (2017) BugBase predicts organism-level microbiome phenotypes. bioRxiv:133462. https://doi.org/10.1101/133462

Westreich ST, Treiber ML, Mills DA, Korf I, Lemay DG (2018) SAMSA2: a standalone metatranscriptome analysis pipeline. BMC Bioinformatics 19:175. https://doi.org/10.1186/s12859-018-2189-z

Wilkinson TMA, Aris E, Bourne S, Clarke SC, Peeters M, Pascal TG, Schoonbroodt S, Tuck AC, Kim V, Ostridge K, Staples KJ, Williams N, Williams A, Wootton S, Devaster J-M (2017) A prospective, observational cohort study of the seasonal dynamics of airway pathogens in the aetiology of exacerbations in COPD. Thorax 72:919–927. https://doi.org/10.1136/thoraxjnl-2016-209023

Winslow S, Odqvist L, Diver S, Riise R, Abdillahi S, Wingren C, Lindmark H, Wellner A, Lundin S, Yrlid L, Ax E, Djukanovic R, Sridhar S, Higham A, Singh D, Southworth T, Brightling CE, Olsson HK, Jevnikar Z (2021) Multi-omics links IL-6 trans-signalling with neutrophil extracellular trap formation and *Haemophilus* infection in COPD. Eur Respir J 58:2003312. https://doi.org/10.1183/13993003.03312-2020

Wishart DS, Tzur D, Knox C, Eisner R, Guo AC, Young N, Cheng D, Jewell K, Arndt D, Sawhney S, Fung C, Nikolai L, Lewis M, Coutouly M-A, Forsythe I, Tang P, Shrivastava S, Jeroncic K, Stothard P, Amegbey G, Block D, Hau DD, Wagner J, Miniaci J, Clements M, Gebremedhin M, Guo N, Zhang Y, Duggan GE, Macinnis GD, Weljie AM, Dowlatabadi R, Bamforth F, Clive D, Greiner R, Li L, Marrie T, Sykes BD, Vogel HJ, Querengesser L (2007) HMDB: the human metabolome database. Nucleic Acids Res 35:D521–D526. https://doi.org/10.1093/nar/gkl923

Wishart DS, Oler E, Peters H, Guo A, Girod S, Han S, Saha S, Lui VW, LeVatte M, Gautam V, Kaddurah-Daouk R, Karu N (2023) MiMeDB: the human microbial metabolome database. Nucleic Acids Res 51:D611–D620. https://doi.org/10.1093/nar/gkac868

Woo S, Park S-Y, Kim Y, Jeon JP, Lee JJ, Hong JY (2020) The dynamics of respiratory microbiota during mechanical ventilation in patients with pneumonia. J Clin Med 9:638. https://doi.org/10.3390/jcm9030638

Wood DE, Lu J, Langmead B (2019) Improved metagenomic analysis with kraken 2. Genome Biol 20:257. https://doi.org/10.1186/s13059-019-1891-0

Xie L, Chen L, Li X, Zhou J, Tian H, Zhao J, Li Z, Li Y (2023) Analysis of lung microbiome in COVID-19 patients during time of hospitalization. Pathogens 12:944. https://doi.org/10.3390/pathogens12070944

Xue Q, Xie Y, He Y, Yu Y, Fang G, Yu W, Wu J, Li J, Zhao L, Deng X, Li R, Wang F, Zheng Y, Gao Z (2023) Lung microbiome and cytokine profiles in different disease states of COPD: a cohort study. Sci Rep 13:5715. https://doi.org/10.1038/s41598-023-32901-0

Yan Z, Chen B, Yang Y, Yi X, Wei M, Ecklu-Mensah G, Buschmann MM, Liu H, Gao J, Liang W, Liu X, Yang J, Ma W, Liang Z, Wang F, Chen D, Wang L, Shi W, Stampfli MR, Li P, Gong S, Chen X, Shu W, El-Omar EM, Gilbert JA, Blaser MJ, Zhou H, Chen R, Wang Z (2022) Multi-omics analyses of airway host–microbe interactions in chronic obstructive pulmonary disease identify potential therapeutic interventions. Nat Microbiol 7:1361–1375. https://doi.org/10.1038/s41564-022-01196-8

Yang C-Y, Li S-W, Chin C-Y, Hsu C-W, Lee C-C, Yeh Y-M, Wu K-A (2021) Association of exacerbation phenotype with the sputum microbiome in chronic obstructive pulmonary disease patients during the clinically stable state. J Transl Med 19:121. https://doi.org/10.1186/s12967-021-02788-4

Yoon S-H, Ha S-M, Kwon S, Lim J, Kim Y, Seo H, Chun J (2017) Introducing EzBioCloud: a taxonomically united database of 16S rRNA gene sequences and whole-genome assemblies. Int J Syst Evol Microbiol 67:1613–1617. https://doi.org/10.1099/ijsem.0.001755

Yu G, Xu C, Zhang D, Ju F, Ni Y (2022) MetOrigin: discriminating the origins of microbial metabolites for integrative analysis of the gut microbiome and metabolome. iMeta 1:e10. https://doi.org/10.1002/imt2.10

Zhou G, Pang Z, Lu Y, Ewald J, Xia J (2022) OmicsNet 2.0: a web-based platform for multi-omics integration and network visual analytics. Nucleic Acids Res 50:W527–W533. https://doi.org/10.1093/nar/gkac376

Zhu S, Sun T, Zhu C, Qing T, Jiang Y, Ding R, Su H, Sun Y, Xu X, Xu K, Suo C, Yuan Z, Zhang T, Zhao G, Ye W, Jin L, Chen X (2020) MAAWf: An Integrated and Visual Tool for Microbiome Data Analyses, 11 February 2020, PREPRINT (Version 1) available at Research Square. https://doi.org/10.21203/rs.2.23215/v1

Multi-omics in Viral Microbiome

13

Maha AlKhazindar, Waled Morsy El-Senousy,
and Yasmein Abuhadema

13.1 Introduction

The human virome refers to the complete collection of viruses that reside in the human body either as independent particles or integrated into the genomes of human cells and other microorganisms. The human virome is a significant constituent of the human microbiome, which encompasses all microorganisms living in and on humans. The human virome could impact human health and disease in numerous ways, including modifying the immune system, altering the bacterial composition, and triggering or preventing infections. Therefore, studying the human virome is essential for comprehending the intricate interactions between humans and microbes. However, studying the human virome is a challenging task, as viruses are diverse, dynamic, and difficult to detect. Human virome research has progressed through three phases: culture-based, sequence-based, and metatranscriptomic. Each phase carries its own set of advantages and limitations and has been fundamental in shaping our understanding of the human virome. The first phase of human virome research relied on culturing viruses in cell lines or animal models. This approach was valuable for identifying and characterizing viruses that cause recognizable diseases, such as poliovirus, hepatitis B virus, and human immunodeficiency virus (HIV). However, this approach was limited by the availability and suitability of host cells for virus propagation, and by the bias toward pathogenic viruses. Many viruses that inhabit asymptomatic humans could not be cultured or detected by this method. The second phase of human virome research emerged with the emergence of advanced sequencing technologies known as next-generation sequencing (NGS).

M. AlKhazindar (✉) · Y. Abuhadema
Botany and Microbiology Department, Faculty of Science, Cairo University, Cairo, Egypt
e-mail: malkhazi@aucegypt.edu; myoussef@sci.cu.edu.eg

W. M. El-Senousy
Water Pollution Research Department, National Research Centre (NRC), Cairo, Egypt

I. Mani, V. Singh (eds.), *Multi-Omics Analysis of the Human Microbiome*,
https://doi.org/10.1007/978-981-97-1844-3_13

NGS enabled the direct sequencing of viral nucleic acids from various biological samples, such as blood, saliva, feces, and tissues. This approach was able to detect novel and unknown viruses that could not be cultured or recognized by conventional methods. However, this approach also faced several challenges, such as removing host and bacterial DNA contamination, assembling viral genomes from short reads, annotating viral genes and functions, and distinguishing between active and latent infections. The third phase of human virome research is still ongoing and involves multi-omics analysis. Multi-omics refers to the integration of different types of omics data, including genomics, transcriptomics, proteomics, and metabolomics. Multi-omics has a significant role in the identification of the human virome, which entails the detection and characterization of viruses that inhabit the human body. Combining different omics data can be useful to obtain a more comprehensive view of viral diversity, activity, and interaction with the host and other microbes. For instance, genomics and transcriptomics can be used to sequence viral DNA and RNA from a variety of biological samples, revealing novel and unknown viruses that conventional methods cannot detect. However, this approach faces challenges such as removing host and bacterial DNA contamination, assembling viral genomes from short reads, annotating viral genes and functions, and distinguishing between active and latent infections. Transcriptomics and proteomics can also gain an understanding of the interactions between the human host and the virome such as how virus infection affects host gene expression and immune responses, and how host factors regulate viral replication and latency. Furthermore, proteomics and metabolomics can be used to analyze the metabolites produced by both the host and viruses in different tissues, revealing the roles and impacts of the human virome on human health and disease, such as modulating the immune system, altering the bacterial composition, and causing or preventing infections. Multi-omics can also play a crucial role in identifying potential biomarkers and therapeutic targets for various human diseases related to viral infections.

In conclusion, the techniques used to study the human virome have evolved over time, from culture-based to sequence-based to meta-omics methods. Each technique has its own strengths and weaknesses and has contributed to different aspects of human virome knowledge. By using multiple omics techniques and disciplines, we can gain a comprehensive understanding of the human virome, its complex interplay with the human body and the microbiome, and its role in both health and disease.

13.2 Human Virus Microbiome: An Overview

The human virome, a collection of viruses coexisting within the human body, represents a complex and diverse entity with significant ecological implications. This virome encompasses a spectrum of viral inhabitants, ranging from free virions capable of infecting eukaryotic cells, archaea, and bacteria (bacteriophages) to viral elements integrated into host chromosomes such as human endogenous retroviruses, prophages proviruses or episomes. The virome is an ever-changing entity that may exert a substantial influence on human health and disease. Some viruses can cause acute or chronic infections, while others can modulate the immune system or interact with the bacterial microbiome. It can be found in different regions of the body, such as the skin, oral cavity, respiratory system, gastrointestinal system, urogenital tract, and bloodstream (Lysholm et al. 2012). Understanding the virome and how it functions is essential for maintaining good health and preventing disease.

The makeup of viral communities in the human body is remarkably complex and affected by various variables such as host genetics, age, geography, and environmental exposures. Viral entities that coexist within the human host, intricately interact with both microbial communities and the host itself. The virome constitutes a highly diverse and complex assemblage. Bacteriophages represent the major portion of the human virome. They exert a profound influence on bacterial populations by modifying bacterial diversity and driving evolutionary dynamics. They have the potential to enhance human health, such as modulating the immune system, preventing pathogen colonization, and producing antimicrobial substances. They are especially plentiful at body sites where they encounter high densities of their bacterial hosts, for example, in areas like the gastrointestinal system or the oropharynx (Liang and Bushman 2021). Bacteriophages are abundant in body regions with high bacterial densities, such as the gastrointestinal tract and oropharynx. The interactions between viruses and their host bacteria involve mechanisms such as predation, symbiosis, and horizontal gene transfer. This dynamic ecological interplay shapes bacterial populations and potentially impacts host health (Bordigoni et al. 2019).

More than 140,000 distinct viruses thrive within the human intestinal ecosystem. These viruses include eukaryotic viruses that largely infect human cells, bacteriophages, archaeal viruses, and plant viruses originating from food and the environment (Rashed et al. 2022; Eifan et al. 2023). Intriguingly, the presence of these viruses in the gut environment can elicit an immune response, with potential implications for the onset of conditions affecting the intestines, like inflammatory bowel disease (Lecuit and Eloit 2017). Endogenous retroviruses (ER) are eukaryotic viruses that make up about 8% of human DNA. ER integrate into the human genome which for certain reasons, can be reactivated, giving rise to viral protein expression or particles. They showcase their multifaceted roles in human biology, such as regulating gene expression, influencing immune responses, and contributing to placental development. Such integration can also have significant evolutionary implications if the virus infiltrates the germ line and is transmitted across generations. Additionally, some persistent viruses, such as human papillomaviruses (HPV), adenoviruses, polyomaviruses, or herpesviruses, can induce cell transformation and cancer development (Fig. 13.1).

Various virus families capture attention due to their intriguing characteristics, each contributing unique insights into the intricate connections between humans and microbes. One such family is the anelloviruses, which are small, circular DNA viruses found widely in human blood and tissues. Anelloviruses are not typically linked to causing diseases; however, they tend to increase in individuals with weakened immune systems, hinting at their potential as indicators of the host's immune status. They can also interact with other viruses, potentially altering their ability to

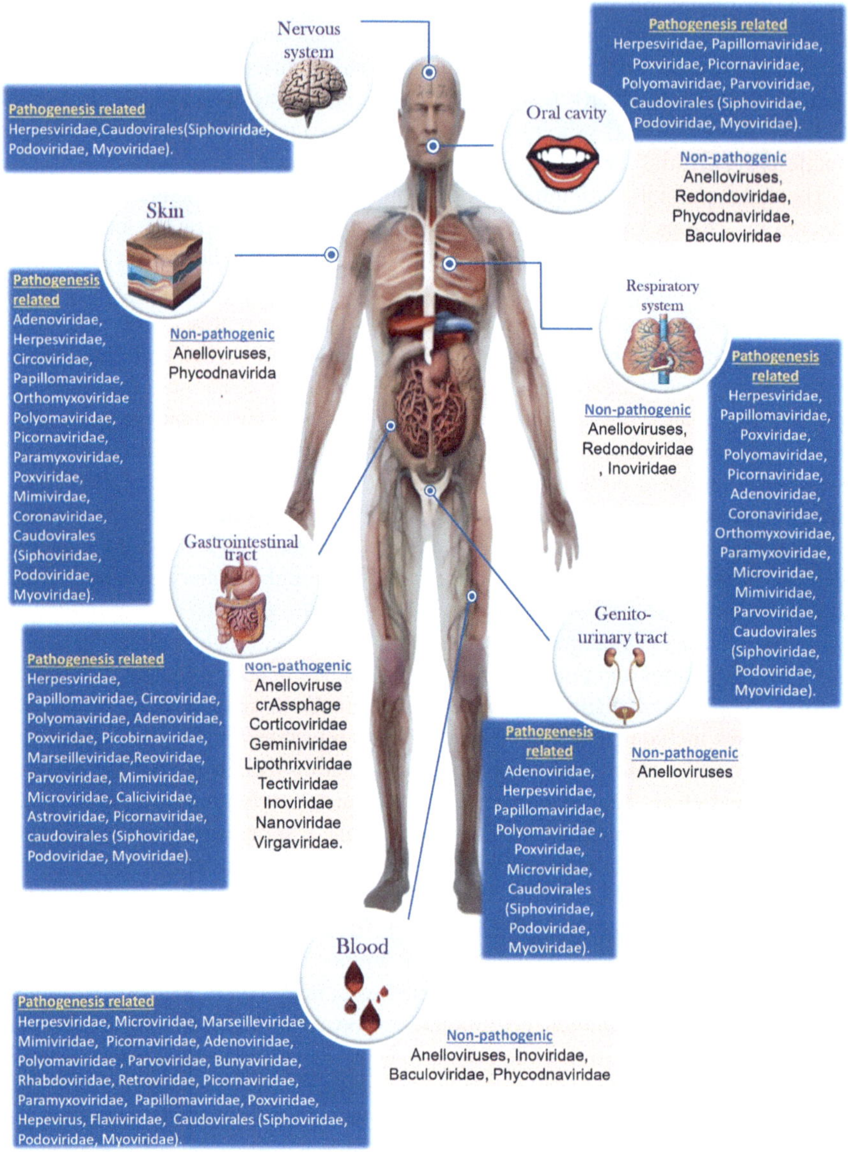

Fig. 13.1 Spatial distribution of the human virome in various anatomical regions. A graphical representation of the human body highlighting the presence of viral communities (virome) in different anatomical regions. The virome composition varies across body sites, with distinct viral populations inhabiting the gut, skin, respiratory tract, and other locations. This figure provides a visual overview of the virome's diversity and distribution within the human body

replicate or cause infection. Another group of anelloviruses, known as torque teno viruses, gain prominence due to their substantial prevalence and diversity in the human population. While their exact role in diseases like hepatitis, respiratory infections, and cancer remains uncertain, their significance persists. Torque teno viruses might not only contribute to adverse effects but could also have beneficial impacts. These impacts range from enhancing responses to vaccines to providing defense against autoimmune conditions and even prompting antiviral immune reactions (Cebriá-Mendoza et al. 2021).

Among the human diverse virome, archaeal viruses stand out. These viruses are found across different regions of the human body, such as the oral cavity, skin, and gastrointestinal system. Notably, archaeal viruses constitute approximately 10% of the anaerobic community residing within the human gut. Importantly, no link between these viruses and human diseases has been established. Instead, an intriguing possibility arises: archaeal viruses might play a favorable role in human well-being. Studies propose that these viruses potentially contribute to the regulation of both the microbiome and the immune system, ultimately promoting a beneficial effect on human health (Wirth and Young 2020).

A prominent feature within the human virome is the emergence of virus-like particles (VLPs). These entities bear structural similarities to viruses but diverge by lacking genetic material and the capability to provoke infections. VLPs are encountered across various bodily systems, with a notable predominance observed primarily within the gastrointestinal tract. Furthermore, their prevalence in bodily regions such as the oropharynx and maxillary sinus flush is also significant. Conversely, they exhibit comparatively lower quantities in blood, dental plaque, urine, and skin. The importance of VLPs goes beyond their direct involvement in causing diseases. Rather, their presence serves as a potential marker for other influential factors, such as microbial diversity, immune responses, and environmental exposures (Bordigoni et al. 2019).

The Human Microbiome Project (HMP) has significantly advanced our understanding of the human microbiome (Turnbaugh et al. 2007) and dsDNA viruses using metagenomic sequences (Wylie et al. 2014). The study of the human virome poses several challenges due to various factors: the absence of common viral markers, the heterogeneous nature of virome components, the constraints posed by low biomass samples, potential contamination from host DNA, the lack of standardized computational tools for virome analysis, and the exponential growth of the virome database. Nevertheless, despite these challenges, limited research suggests that viruses can have either positive or negative effects on human health. For example, it has been observed that the specific expansion of Caudovirales (bacteriophages) in Crohn's disease is associated with decreased bacterial diversity, suggesting that the virome may play a role in intestinal inflammation and bacterial dysbiosis (Norman et al. 2015). Conversely, emerging evidence suggests that the hepatitis G virus (HGV) may have a protective effect against HIV-associated diseases (Lanteri et al. 2015).

As research progresses, the diverse viral components within the human virome provide insights into the intricate relationships that shape human health and the balanced existence of microorganisms. However, there are likely many more viruses that remain unknown in terms of identification and function. Food and environmental factors can influence changes in the human natural virome. For instance, the gut virome, which represents the viral part of the gut microbiome, can be impacted by dietary choices that change the availability of nutrients, the pH, and interactions with other microorganisms (Li 2017; Gregory et al. 2020). Environmental factors such as eating habits, differences in ethnic groups, blood transfusions, and medical treatments can also impact the diversity and makeup of the human virome by introducing new viruses or modifying existing ones (AboNahas et al. 2022).

Comprehending the dynamic relationship of the human virome, microbiome, and the body's immune system is crucial for human health. This complex interrelationship holds implications for various diseases and health conditions. This is where the concept of multi-omics comes into play. Applying multi-omics approaches to virome research uncovers deeper insights into how viruses interact with our bodies, how they influence the microbiome, and how they impact health.

13.3 Multi-omics Technologies and Applications: Methods and Strategies for Studying Virus Microbiomes

The study of the human virome, encompassing the intricate interactions between viruses and their microbial hosts, has gained significant attention due to its implications for human health, ecosystem dynamics, and disease emergence. Traditional single-omic approaches often fall short in capturing the complexity of these systems. Multi-omics technologies, including metagenomics, metatranscriptomics, metaproteomics, and metabolomics, have emerged as powerful tools, enabling to study the profile of various biomolecules, such as nucleic acids, proteins, and metabolites thereby enhancing our understanding of the roles of human virome in diverse environments (Table 13.1).

Metagenomics involves the sequencing of DNA directly extracted from environmental specimens, providing insights into the genetic diversity and potential functions of viruses and their microbial hosts. By extracting and sequencing viral DNA, researchers can identify viral genomes, characterize their taxonomic and functional profiles, and predict viral-host interactions (Edwards and Rohwer 2005). Advances in sequencing technologies, such as high-throughput sequencing and long-read sequencing, have enabled comprehensive exploration of virus microbiomes. Metagenomics has enabled researchers to identify novel viruses and explore their genetic diversity within the gut (Daliri et al. 2021). Minot et al. (2012) used metagenomic sequencing to discover an extensive repertoire of previously unknown bacteriophages in the human gut, shedding light on the diversity of viruses present and the possible functions they might have in modulating bacterial populations.

The human intestine contains more than 1000 diverse microbial species, which provide crucial protection against harmful microbes, regulate metabolic processes, modulate immunity and are even considered to operate as an organ with endocrine functions. Traditional culture-based methods fall short when it comes to identifying

Table 13.1 Comparison of different multi-omics technologies and tools for studying virus microbiomes

Omics	Definition	Data type	Analysis method	Enrichment method
Metagenomics	Analysis of the entire DNA community from a given sample	DNA sequences	Shotgun sequencing, targeted sequencing, single-virus genomics	Filtration, centrifugation, nuclease treatment, immunocapture
Metatranscriptomics	Analysis of the entire RNA community from a given sample	RNA sequences	RNA-sequencing, targeted RNA-sequencing, single-cell RNA-sequencing	Poly(A) selection, ribosomal RNA depletion, immunocapture
Metaproteomics	Analysis of the entire protein community from a given sample	Protein masses and peptides	Mass spectrometry, tandem mass spectrometry, cross-linking mass spectrometry	Affinity purification, co-immunoprecipitation, proximity labeling
Metabolomics	Analysis of the entire metabolite community from a given sample	Metabolite masses and fragments	Nuclear magnetic resonance, mass spectrometry, tandem mass spectrometry	Liquid chromatography, gas chromatography, capillary electrophoresis

these complex microbial populations. Advanced molecular techniques, such as metagenomic sequencing, are highly recommended for investigating the diversity and dysbiosis of the intestinal microbiome and its connections to health and disease. Metagenomic functional analysis has the capability to unearth new functional genes, microbial pathways, antibiotic resistance genes, and changes in the functional balance of the intestinal microbiome. It also facilitates the exploration of interactions and the evolutionary dynamics between the microbiota and the host. Nevertheless, there are specific limitations to this method. Metatranscriptomics, metaproteomics, and metabolomics are supplementary techniques that can be employed to enhance our understanding of the microbiota in the human intestinal tract (Wang et al. 2015).

A significant milestone in the realm of metagenomics is the detection of crAssphage, which ranks as the most prevalent virus accompanying humans. This virus is prominently detected in around half of the viromes within the human gut, often comprising up to 90% of the viral sequences. Research has indicated that crAssphage can be transmitted vertically to infant and may also spread through fecal transplants, which could potentially lead to the spread among individuals (Koonin and Yutin 2020).

Different bacterial species belonging to the Bacteroidetes phylum have been suggested as potential primary hosts for crAssphage. The presence of such virus on various surfaces and hands during norovirus outbreaks underscores their value as markers for detecting human fecal contamination in a variety of environmental settings beyond just sewage-contaminated water. Detecting crAssphage on environmental surfaces could yield a more comprehensive assessment of the risk of human norovirus exposure in public spaces. Furthermore, it could aid in the identification of frequently touched surfaces that are frequently contaminated with fecal matter, making them critical areas for implementing heightened cleaning practices (Park et al. 2020; Lee et al. 2023).

Metatranscriptomics is the study of all RNA transcripts in a given sample, which can reveal the expression and activity of both host and viral genes. This method can offer a better understanding of the functional activities and interactions between the human host and the virome, such as how virus infection affects host gene expression and immune responses, and how host factors regulate viral replication and latency (Rajagopala et al. 2021). Metatranscriptomics can also identify new viral species and variants by comparing RNA sequences to existing databases (Terrón-Camero et al. 2022). Similarly, Kumata et al. (2020) performed a metatranscriptomic analysis of various human tissues from healthy individuals and detected 39 viral species, correlating their presence with the host gene expression and metabolite levels in each tissue. The study showed that some viruses were associated with altered levels of metabolites involved in immune responses, such as interferons, cytokines, and chemokines. It was also found that some viruses were associated with metabolites related to tissue-specific functions, such as digestion, respiration, and reproduction.

Moreover, Vazquez-Castellanos et al. (2018) employed metagenomics and metatranscriptomics to track the dynamics of the gut virome in HIV-infected individuals. This approach not only identified changes in viral diversity but also revealed potential interactions between the HIV virus and other gut-residing viruses. Combining these multi-omics datasets gained insights into the intricate interplay between viral populations, which could influence disease progression and treatment strategies.

Metaproteomics involves the identification and determination of proteins within a microbial community. This technique is essential for examining the host-pathogen interactions within virus microbiomes. This is particularly relevant in understanding how viruses manipulate host cellular machinery to enhance their replication and evade the immune system. For instance, during an influenza infection, the virus remodels the host cell's proteome to facilitate its own replication. By applying metaproteomics, the changes in protein expression within the infected host cells can be monitored. A study by Bradel-Tretheway et al. (2011) utilized metaproteomics to reveal the host cell responses to influenza virus infection, shedding light on the complex interplay between viral and host proteins during infection.

Integrating metaproteomics with other omics data helps researchers understand how microbial communities respond to viral infections and environmental changes. In addition, new metaproteomic tools can now measure the biomass of individual species to determine the structure of the community, the sources of carbon used by each member, and how each member takes in labeled substrates while in their natural environment (Kleiner 2019).

Metabolomics explores the metabolic profiles of biological systems, from cells and tissues to larger organisms. It shows the biochemical changes and interactions that occur in response to various stimuli such as diseases, drugs, or environmental factors. Metabolomics can provide valuable perspectives on the roles and impacts of the human virome on human health and disease thus, revealing how viruses manipulate host metabolism and influence ecosystem processes (Cai et al. 2023). Recent studies have shed light on the complex and dynamic interactions between viruses and their human hosts within different tiers of biological structure. Li et al. (2022) examined the correlations involving the gut virome, microbiome, and host metabolome in patients with irritable bowel syndrome (IBS). The study revealed that changed viruses in both the IBS and control groups exhibited concurrent fluctuations alongside modified bacteria and metabolites. Additionally, it pinpointed specific metabolites, including short-chain fatty acids, bile acids, and amino acids, that exhibited associations with distinct viral taxa. These metabolites could reflect the interplay between the gut virome and various aspects of the host's physiology, including the immune system, intestinal barrier function, and gut–brain axis. (Li et al. 2022). This demonstrates the potential of multi-omics analysis of the human virome to reveal crucial insights into viral infections and their effects on human health. Such analysis can also help identify potential biomarkers and therapeutic targets for various human diseases related to viral infections.

Researchers have employed multi-omics approaches to uncover the significant role of the Epstein–Barr virus (EBV) in the development of Natural Killer/T Cell Lymphomas (NK/T lymphomas) and Nasopharyngeal Carcinoma (NPC). Specifically, they found that a protein called EBNA-1 derived from NPC tumors loses its ability to maintain latency, leading to the activation of the EBV lytic cycle

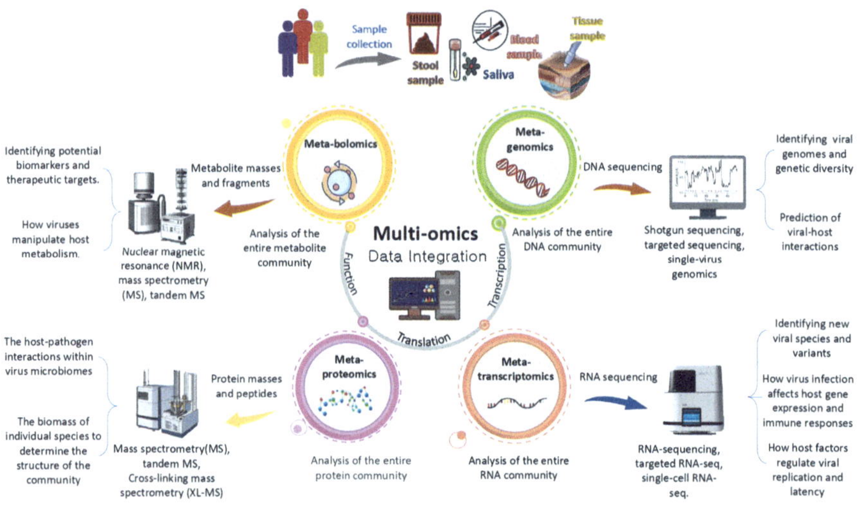

Fig. 13.2 Integration of multi-omics approaches for comprehensive analysis of virus microbiomes. A graphical illustration showing the specific roles of each omics technology in the examination of virus microbiomes. This diagram exemplifies how their synergistic integration contributes to a comprehensive understanding of these intricate ecosystems

at an epigenetic level. Methylation modifications were also observed in gastric epithelial cells after EBV infection, causing abnormalities in the activity of multiple host genes in EBV-positive gastric malignancies, both in vitro and in vivo. The involvement of epigenetic changes in EBV-related carcinogenesis suggests that epigenetic inhibitors could be a promising approach for treating EBV-associated malignancies, as shown in an animal study using a mouse model of NPC (Dheekollu et al. 2017).

Multi-omics technologies have revolutionized the study of virus microbiomes. However, the real potential of multi-omics is realized when data from different omics approaches are integrated (Fig. 13.2). This integration allows for the creation of extensive models of virus microbiomes that connect taxonomic information, functional activities, gene expression, and metabolic interactions. These analyses help reveal the intricate interactions between viruses and their microbial hosts, ultimately uncovering the complexity of such relationships.

13.4 Multi-omics Analysis of Virome and Microbiota Balance

The human body houses a complex ecosystem of bacteria within the gastrointestinal tract, collectively shaping health and disease outcomes. Among these microorganisms, those conferring health benefits are commonly referred to as "beneficial bacteria." However, when confronted with viral infections, the delicate balance within this microbial community is disrupted, leading to a decline in beneficial bacterial populations and an upsurge in potentially harmful bacteria.

These intricate dynamics have been the subject of investigation using multi-omics approaches, including metagenomics, metatranscriptomics, metaproteomics, and metabolomics, which collectively provide a comprehensive view of the interplay between viruses, the microbiota, and host responses.

As an example, within the microbiota associated with HIV, the Firmicutes phylum displays the most pronounced transcriptional activity. Conversely, the expression of anti-inflammatory processes, encompassing the biosynthesis of short-chain fatty acids and the production of indole, experiences a marked reduction. These microbial communities consequently foster an environment characterized by inflammation. Notably, Wang et al. (2015) observed disruptions in intestinal microbiota following respiratory influenza viral infection. In this scenario, there is a substantial increase in the population of enterobacteria, with a significant decrease in segmented filamentous bacteria and lactic acid bacteria. In a parallel context, Rotavirus, responsible for enteritis, provokes a decline in Lactobacillus populations and a concurrent rise in Escherichia and Streptococcus in calves.

While the intestinal microbiota possesses various mechanisms to resist viral intrusions, any disruption caused by external factors during treatment can have severe consequences. Previous studies suggested that the intestinal microbiota may get disrupted by external forces during treatment, resulting in severe consequences. The potential disruption of the intestinal microbiota and its consequences can be better understood through multi-omics approaches. Lv et al. (2021) demonstarted that a combination of antiviral medications, probiotics, and antibiotics used to treat the H7N9 virus led to improvements in microbial diversity and the richness of beneficial bacteria in the intestines of H7N9 patients.

In the context of microbiota and virome balance, metagenomics methods have led to the discovery of numerous novel HPV genomes in individuals with a deficiency in the DOCK-8 protein, which plays a crucial role in T cell movement through tissues rich in collagen. Additionally, a vaginal microbiota rich in Lactobacillus has been associated with protection against HPV infections and diminishing the risk of cancer-causing viruses (Tamarelle et al. 2019).

Viruses residing in a commensal state on the skin offer cancer protection by modulating the host's immune system. The absence of CD8(+) T cellular immunity against commensal HPVs in the skin microbiota increases the risk of skin cancer, contrary to prior beliefs. Studies in mice have highlighted the significance of robust T cell responses in defending against chemically and UV-induced cancer (Kitsou et al. 2021).

In summary, multi-omics studies have shed light on the mechanisms of tumorigenesis related to oncoviruses, highlighting the role of epigenetic changes, the potential of epigenetic inhibitors in treatment, and the influence of microbiota and virome balance in cancer development. Additionally, skin microbiota and T cell immunity play crucial roles in protecting against skin cancer.

13.5 Multi-omics Challenges and Solutions: Issues and Improvements for Multi-omics Approaches

The use of multi-omics techniques has greatly enhanced our understanding of viral infections, shedding light on viral diversity, host responses, and the underlying molecular mechanisms. However, each individual multi-omics approach presents distinct advantages and limitations, which must be carefully considered when investigating viral infections. Each of these individual multi-omics methods offers distinct advantages and brings forth specific considerations, highlighting their potential benefits and limitations (Table 13.2).

Genomics enables tracing the evolutionary trajectory of viruses, facilitating the identification of new strains, and tracking transmission routes (Saravanan et al. 2022). However, during viral infections, gene expression may shift dynamically and may not always be accurately captured by genomic analyses. Although genomics showcases potential genetic changes, translating these alterations into functional shifts within the host-virus interaction network might not always be straightforward (Tam et al. 2019).

Other omics approaches, such as transcriptomics, proteomics, and metabolomics, hold significant promise in complementing genomics to provide a more comprehensive understanding of viral infections. Transcriptomics, for instance, can capture the dynamic changes in gene expression during viral infections, shedding light on the host responses to viral invasion (Sonawane et al. 2019). This can be particularly valuable in elucidating the molecular mechanisms underlying pathogenesis and immune responses. However, transcriptomic data interpretation can be complex due to alternative splicing, post-transcriptional modifications, and temporal variations in gene expression (Barton et al. 2017).

On the other hand, proteomics can bridge the gap between genomic changes and functional alterations by directly measuring the proteins involved in host-virus interactions (Wang et al. 2019). Nonetheless, proteomic analyses can be challenging due to the vast diversity of proteins involved and the need for advanced instrumentation. Moreover, metabolomics, which focuses on small molecules within cells and tissues, can reveal metabolic changes induced by viral infections. Understanding these alterations can uncover potential therapeutic targets and metabolic pathways critical for viral replication. Nevertheless, metabolomic data interpretation can be intricate, requiring sophisticated techniques for compound identification and pathway analysis (Byers et al. 2019).

Although each omics approach provides valuable information, the synergistic use of genomics, transcriptomics, proteomics, and metabolomics offers a more complete understanding of viral infections, ultimately advancing our understanding of host-virus interactions and the development of targeted therapeutic interventions.

A notable illustration of the effectiveness of multi-omics approaches is evident in the investigation of SARS-CoV-2, the virus responsible for the COVID-19 pandemic. Through the integration of diverse omics techniques, researchers have

Table 13.2 Advantages and disadvantages of multi-omics

Omics	Advantages	Disadvantages
Metagenomics	– Provides information on the identity and abundance of viruses in a microbiome – Provides information on the genomic features and potential functions of viruses – Enables the discovery of novel viruses and their genes	– Cannot capture the functional activity or expression of viral genes – Cannot reveal the interactions and pathways between viruses and their hosts or other microbes – Requires high sequencing depth and computational resources for assembly and annotation
Metatranscriptomics	– Provides information on the expression levels of viral genes – Identifies active or latent infections – Reveals the transcriptional regulation and dynamics of virus microbiomes	– Cannot provide information on the identity and abundance of viruses in a microbiome – Cannot identify viral proteins and their interactions – Requires high sequencing depth and computational resources for alignment and quantification
Metaproteomics	– Provides information on the identity and abundance of viral proteins and their interactions with host or microbial proteins – Reveals the functional activity and expression of viral genes – Reveals the post-translational modifications and dynamics of viral proteins	– Cannot provide information on the genomic features and potential functions of viruses in a microbiome – Cannot identify viral metabolites and their effects – Requires high mass spectrometry sensitivity and resolution, as well as comprehensive reference databases for identification and quantification
Metabolomics	– Provides information on the identity and abundance of viral metabolites and their effects on host or microbial metabolism – Reveals the metabolic regulation and dynamics of virus microbiomes	– Cannot provide information on the identity and abundance of viruses in a microbiome – Cannot identify viral genes and their expression – Requires high mass spectrometry sensitivity and resolution, as well as comprehensive reference databases for identification and quantification

acquired a deeper understanding of the virus mechanisms, resulting in more comprehensive insights than what individual methods could provide.

Genomic analyses have been pivotal in understanding SARS-CoV-2's genetic makeup. This includes tracking mutations and identifying key variants (Saravanan et al. 2022). However, with the outbreak of COVID-19, there has been an unprecedented influx of genetic data from the SARS-CoV-2 virus, aimed at understanding its origins, evolution, and clinical implications (Wu et al. 2020). However, the rapid mutation rate and intricate network of infections have introduced complexities in accurately interpreting this data.

The surge in the SARS-CoV-2 virus mutations often introduces noise into datasets, potentially compromising the accuracy of analyses. Moreover, limited sampling during the beginning of the pandemic has hindered a comprehensive understanding of the virus dynamics. Traditional phylogenomic methods, though effective, might struggle to account for noisy or sparse data, leading to unreliable results (Deng et al. 2020).

In response to challenges in understanding SARS-CoV2 evolution, the VERSO (Viral Evolution ReconStructiOn) framework has been introduced. VERSO offers a detailed view of viral evolution by analyzing raw sequencing data, producing accurate phylogenetic trees, and detecting complex evolutionary relationships. It stands out by incorporating minor variant profiles, identifying hidden infection pathways, and providing insights into mutational hotspots, enhancing our understanding of viral evolution (Ramazzotti et al. 2021).

Transcriptomics sheds light on how hosts respond to viral infections by profiling changes in gene expression. Nonetheless, transcriptomics data may lack the contextual understanding of actual proteins and metabolites responsible for functional shifts. It might not fully encompass post-transcriptional regulatory mechanisms, such as those driven by microRNAs (Wyler et al. 2021). However, a significant constraint in transcriptomic data emerges due to the biological consequences of exclusively measuring mRNA. Due to RNA silencing or post-transcriptional gene silencing, mRNA does not invariably terminate in protein synthesis. As a result, predicting the complete biological effects of DNA through transcriptome analysis can be a challenging task (Karahalil 2016). Therefore, introducing proteomics has revealed shifts in protein expression and post-translational modifications induced by SARS-CoV-2 (Messner et al. 2020), thus facing some challenges in sample preparation for bottom-up proteomics analysis stem from contamination risks, detergent interference, protein depletion techniques, handling-related contaminants, limitations of separation techniques, sample variability, and the complexity of multistep processes (Dupree et al. 2020). Metabolomics has uncovered altered metabolic pathways due to SARS-CoV-2 infections (Valdés et al. 2022). However, conducting a metabolomic investigation demands the meticulous observation of various genetic and environmental factors to minimize confounding variables and enhance information retrieval. Another constraint within metabolomics, particularly evident in liquid chromatography (LC)-MS-based metabolomics, involves the challenge of accurately attributing identity to biomarkers (Johnson and Gonzalez 2012).

Moreover, a notable application of multi-omics integration in SARS-CoV-2 research is the study of cytokine storms. Transcriptomics identified overexpressed immune-related genes (Blanco-Melo et al. 2020), proteomics confirmed elevated protein levels (Messner et al. 2020), and metabolomics revealed disrupted metabolic pathways (Lee et al. 2020). This multi-omics tactic enabled a comprehension of the molecular basis of cytokine storms.

Integrating multi-omics techniques, including genomics, transcriptomics, proteomics, and metabolomics, has revolutionized our understanding of the human virus microbiome. These approaches enable the comprehensive investigation of viral diversity, host interactions, and the underlying biological mechanisms. Nevertheless, the integration of multi-omics data from viral microbiomes introduces a distinctive set of challenges.

13.6 The Power of Multi-omics Integration for Precision Medicine and Beyond

In recent years, the integration of multi-omics approaches has significantly reshaped virus microbiome research. This multidisciplinary strategy has enabled novel discoveries and transformative applications, fundamentally altering the landscape of virus microbiome research. The integration of multiple omics datasets has paved the way for personalized medicine regarding viral infections. Clinicians can tailor treatment plans and interventions by examining an individual's distinct omics profile, leading to improved outcomes for patients. This approach represents a significant change in managing the virus microbiome, shifting from general strategies to precision medicine.

Precision medicine has shown promising results in managing viral infections. For example, mRNA vaccines are being used to elicit an immune response against the spike protein of the SARS-CoV-2 virus. These vaccines can be customized to different variants of the virus and have been proven to provide high levels of protection against severe disease and death (Enyue et al. 2022). Furthermore, accustomed methods such as PCR or serology can diagnose viral infections, but they can only target a limited number of viruses at a time. These methods are not suitable for identifying new or unexpected viruses that may cause a certain disease, or rare viruses that may exist in a specific region. Therefore, metagenomics has proven successful in diagnosing viruses causing various types of infections like encephalitis, pneumonia, fever/sepsis, or eye infections thus ensuring proper treatment (d'Humières et al. 2021).

Except for HIV and viral hepatitis, antiviral therapies are still quite scarce. As a result, the use of comprehensive metagenomic sequencing to diagnose viral infections frequently does not result in immediate treatment options. Nonetheless, in some cases, metagenomics has played a critical role in identifying targeted treatments. For instance, Murkey et al. (2017) reported that a lung transplant recipient suffering from hepatitis E virus-related meningoencephalitis was diagnosed using metagenomics and successfully treated with ribavirin.

Multi-omics has also played a crucial role in advancing the treatment of hepatitis, a group of liver infections primarily caused by five distinct viruses. These viruses are referred to as hepatitis A (HAV), hepatitis B (HBV), hepatitis C (HCV), hepatitis D (HDV), and hepatitis E (HEV). Clinical precision uses a comprehensive approach that encompasses various omics disciplines to know the molecular functions within biological cells. Multi-omics strategies have allowed scientists to investigate the origins of these hepatitis viruses, identify biomarkers to aid in diagnosis and prognosis, and discover potential therapeutic targets to combat hepatitis and its associated ailments (Xiang et al. 2022).

The use of multi-omics has been essential in improving our understanding of SARS-CoV-2 which causes COVID-19, providing us with a valuable understanding of its etiology, immunology, and pathology (Bai et al. 2022). By utilizing omics technologies, researchers have been able to swiftly gather multidimensional data, leading to a comprehensive understanding of the virus. The combination of findings from both multi-omic and traditional approaches has strengthened the validity of omics results, ultimately contributing to the development of effective vaccines and therapies for this pandemic (Wang et al. 2021).

Moreover, multi-omics research has significantly improved predictive modeling. By utilizing the vast amount of omics data available, scientists can create models to predict viral outbreaks, providing valuable insights into how diseases progress and transmit. This allows for timely and effective interventions, which is crucial in managing viral infections on a larger scale. In light of the COVID-19 pandemic, machine learning (ML) has been pivotal in diagnosing and predicting the severity of the disease. Liu et al. (2023) employed different ML models to diagnose COVID-19 and forecast the disease severity. The findings indicated that integrating multi-omics data resulted in better model performance than using single omics datasets. Furthermore, omics-based models outperformed models based on images or physiological features, highlighting the significance of omics data for future development and improving healthcare resource allocation and system resilience.

In the biotechnological realm, multi-omics studies have led to the identification of novel biomarkers and drug targets. These discoveries have driven the development of innovative diagnostic tools and therapeutic agents, offering promising avenues for improving the diagnosis and treatment of viral infections. Such advancements have the potential to revolutionize virus microbiome management. They have helped uncover disease mechanisms, predict treatment responses, and identify novel therapeutic targets, particularly in the context of immunology, cancer research, and infectious diseases like COVID-19 (Zielinski et al. 2021).

In conclusion, multi-omics approaches have marked the onset of a new era in virus microbiome research, offering unparalleled opportunities for understanding, diagnosing, and managing viral infections. By analyzing diverse omics datasets, researchers can better understand the dynamics of viral spread, helping design more effective public health strategies and interventions to control outbreaks. However, combining multiple omics datasets is not an easy task. Researchers must deal with challenges related to data integration, storage, and analysis. Thus, creating reliable computational tools and workflows is crucial to exploit the full potential of multi-omics data.

13.7 Conclusion: Multi-omics and Virus Microbiomes

The study of the human virome and its interactions with microbial hosts has become crucial for understanding human health, disease emergence, and ecosystem dynamics. Traditional single-omic approaches are limited in capturing the complexity of these systems. Multi-omics technologies have revolutionized the study of virus microbiomes by providing comprehensive insights into viral diversity, host interactions, and underlying molecular mechanisms. The integration of genomics, transcriptomics, proteomics, and metabolomics enables a more complete understanding of host–virus interactions. These approaches have practical applications in precision medicine, allowing tailored treatments based on an individual's omics profile. In the context of viral outbreaks like COVID-19, multi-omics research has been instrumental in diagnosis, vaccine development, and predictive modeling. Furthermore, multi-omics studies have identified biomarkers, drug targets, and innovative diagnostic tools, offering promising avenues for managing viral infections and improving healthcare. However, the integration of multi-omics data presents challenges, and the development of reliable computational tools is crucial to fully harness their potential in virus microbiome research.

References

AboNahas HH, El-Maradny YA, Gezaf SA, Abo Nouh FA, AboNahas YH, Landa-Acuña D, Abdel-Azeem AM et al (2022) Virome and microbiome interaction and their implication in the human gut health. In: Microbiome-gut-brain axis: implications on health. Springer, Singapore, pp 97–138

Bai GH, Lin SC, Hsu YH, Chen SY (2022) The human virome: viral metagenomics, relations with human diseases, and therapeutic applications. Viruses 14(2):278

Barton AJ, Hill J, Pollard AJ, Blohmke CJ (2017) Transcriptomics in human challenge models. Front Immunol 8:1839

Blanco-Melo D, Nilsson-Payant BE, Liu WC, Uhl S, Hoagland D, Møller R, Albrecht RA et al (2020) Imbalanced host response to SARS-CoV-2 drives development of COVID-19. Cell 181(5):1036–1045

Bordigoni A, Halary S, Desnues C (2019) The human virome. In: Encyclopedia of virology, 4th edn, pp 552–558

Bradel-Tretheway BG, Mattiacio JL, Krasnoselsky A, Stevenson C, Purdy D, Dewhurst S, Katze MG (2011) Comprehensive proteomic analysis of influenza virus polymerase complex reveals a novel association with mitochondrial proteins and RNA polymerase accessory factors. J Virol 85(17):8569–8581

Byers NM, Fleshman AC, Perera R, Molins CR (2019) Metabolomic insights into human arboviral infections: dengue, chikungunya, and zika viruses. Viruses 11(3):225

Cai L, Weinbauer MG, Xie L, Zhang R (2023) The smallest in the deepest: the enigmatic role of viruses in the deep biosphere. Natl Sci Rev 10(4):nwad009

Cebriá-Mendoza M, Arbona C, Larrea L, Díaz W, Arnau V, Peña C, Cuevas JM et al (2021) Deep viral blood metagenomics reveals extensive anellovirus diversity in healthy humans. Sci Rep 11(1):6921

d'Humières C, Salmona M, Dellière S, Leo S, Rodriguez C, Angebault C, Ruppé E et al (2021) The potential role of clinical metagenomics in infectious diseases: therapeutic perspectives. Drugs 81(13):1453–1466

Daliri EBM, Ofosu FK, Chelliah R, Lee BH, Oh DH (2021) Challenges and perspective in integrated multi-omics in gut microbiota studies. Biomol Ther 11(2):300

Deng X, Gu W, Federman S, Du Plessis L, Pybus OG, Faria NR, Chiu CY et al (2020) Genomic surveillance reveals multiple introductions of SARS-CoV-2 into northern California. Science 369(6503):582–587

Dheekollu J, Malecka K, Wiedmer A, Delecluse HJ, Chiang AK, Altieri DC, Lieberman PM et al (2017) Carcinoma-risk variant of EBNA1 deregulates Epstein-Barr virus episomal latency. Oncotarget 8(5):7248

Dupree EJ, Jayathirtha M, Yorkey H, Mihasan M, Petre BA, Darie CC (2020) A critical review of bottom-up proteomics: the good, the bad, and the future of this field. Proteomes 8(3):14

Edwards RA, Rohwer F (2005) Viral metagenomics. Nat Rev Microbiol 3(6):504–510

Eifan S, Maniah K, Nour I, Hanif A, Yassin MT, Al-Ashkar I, Abid I (2023) Pepper mild mottle virus as a potential indicator of fecal contamination in influents of wastewater treatment plants in Riyadh, Saudi Arabia. Microorganisms 11(4):1038

Enyue F, Liu X, Li M, Zelun Z, Song L, Baiyu Z, Li Y et al (2022) Advances in COVID-19 mRNA vaccine development. Signal Transduct Target Ther 7(1):94

Gregory AC, Zablocki O, Zayed AA, Howell A, Bolduc B, Sullivan MB (2020) The gut virome database reveals age-dependent patterns of virome diversity in the human gut. Cell Host Microbe 28(5):724–740

Johnson CH, Gonzalez FJ (2012) Challenges and opportunities of metabolomics. J Cell Physiol 227(8):2975–2981

Karahalil B (2016) Overview of systems biology and omics technologies. Curr Med Chem 23(37):4221–4230

Kitsou K, Iliopoulou M, Spoulou V, Lagiou P, Magiorkinis G (2021) Viral causality of human cancer and potential roles of human endogenous retroviruses in the multi-omics era: an evolutionary epidemiology review. Front Oncol 11:687631

Kleiner M (2019) Metaproteomics: much more than measuring gene expression in microbial communities. mSystems 4(3):10–1128

Koonin EV, Yutin N (2020) The crAss-like phage group: how metagenomics reshaped the human virome. Trends Microbiol 28(5):349–359

Kumata R, Ito J, Takahashi K, Suzuki T, Sato K (2020) A tissue level atlas of the healthy human virome. BMC Biol 18:1–15

Lanteri MC, Vahidnia F, Tan S, Stapleton JT, Norris PJ, Heitman J, Stapleton JT et al (2015) Downregulation of cytokines and chemokines by GB virus C after transmission via blood transfusion in HIV-positive blood recipients. J Infect Dis 211(10):1585–1596

Lecuit M, Eloit M (2017) The viruses of the gut microbiota. In: The microbiota in gastrointestinal pathophysiology. Academic Press, Boston, MA, pp 179–183

Lee W, Ahn JH, Park HH, Kim HN, Kim H, Yoo Y, Seo YK et al (2020) COVID-19-activated SREBP2 disturbs cholesterol biosynthesis and leads to cytokine storm. Signal Transduct Target Ther 5(1):186

Lee SY, Yang J, Lee JH (2023) Improvement of crAssphage detection/quantification method and its extensive application for food safety. Front Microbiol 14:1185788

Li AM (2017) Ecological determinants of health: food and environment on human health. Environ Sci Pollut Res 24(10):9002–9015

Li M, Wang C, Guo Q, Xu C, Xie Z, Tan J, Zhu H et al (2022) More positive or more negative? Metagenomic analysis reveals roles of virome in human disease-related gut microbiome. Front Cell Infect Microbiol 12:846063

Liang G, Bushman FD (2021) The human virome: assembly, composition and host interactions. Nat Rev Microbiol 19(8):514–527

Liu X, Hasan MR, Ahmed KA, Hossain MZ (2023) Machine learning to analyse omic-data for COVID-19 diagnosis and prognosis. BMC Bioinform 24(1):7

Lv Z, Xiong D, Shi J, Long M, Chen Z (2021) The interaction between viruses and intestinal microbiota: a review. Curr Microbiol 78:3597–3608

Lysholm F, Wetterbom A, Lindau C, Darban H, Bjerkner A, Fahlander K, Andersson B et al (2012) Characterization of the viral microbiome in patients with severe lower respiratory tract infections, using metagenomic sequencing. PLoS One 7(2):e30875

Messner CB, Demichev V, Wendisch D, Michalick L, White M, Freiwald A, Ralser M et al (2020) Ultra-high-throughput clinical proteomics reveals classifiers of COVID-19 infection. Cell Syst 11(1):11–24

Minot S, Grunberg S, Wu GD, Lewis JD, Bushman FD (2012) Hypervariable loci in the human gut virome. Proc Natl Acad Sci 109(10):3962–3966

Murkey JA, Chew KW, Carlson M, Shannon CL, Sirohi D, Sample HA, Chiu CY et al (2017) Hepatitis E virus–associated meningoencephalitis in a lung transplant recipient diagnosed by clinical metagenomic sequencing. Open Forum Infect Dis 4(3):121

Norman JM, Handley SA, Baldridge MT, Droit L, Liu CY, Keller BC, Virgin HW et al (2015) Disease-specific alterations in the enteric virome in inflammatory bowel disease. Cell 160(3):447–460

Park GW, Ng TFF, Freeland AL, Marconi VC, Boom JA, Staat MA, Vinjé J et al (2020) CrAssphage as a novel tool to detect human fecal contamination on environmental surfaces and hands. Emerg Infect Dis 26(8):1731

Rajagopala SV, Bakhoum NG, Pakala SB, Shilts MH, Rosas-Salazar C, Mai A, Boone HH, McHenry R, Yooseph S, Halasa N, Das SR (2021) Metatranscriptomics to characterize respiratory virome, microbiome, and host response directly from clinical samples. Cell Rep Methods 1(6):100091

Ramazzotti D, Angaroni F, Maspero D, Gambacorti-Passerini C, Antoniotti M, Graudenzi A, Piazza R (2021) VERSO: a comprehensive framework for the inference of robust phylogenies and the quantification of intra-host genomic diversity of viral samples. Patterns 2(3):100212

Rashed MK, El-Senousy WM, Sayed ETAE, AlKhazindar M (2022) Infectious pepper mild mottle virus and human adenoviruses as viral indices in sewage and water samples. Food Environ Virol 14(3):246–257

Saravanan KA, Panigrahi M, Kumar H, Rajawat D, Nayak SS, Bhushan B, Dutt T (2022) Role of genomics in combating COVID-19 pandemic. Gene 823:146387

Sonawane AR, Tian L, Chu CY, Qiu X, Wang L, Holden-Wiltse J, Liu YY et al (2019) Microbiome-transcriptome interactions related to severity of respiratory syncytial virus infection. Sci Rep 9(1):13824

Tam V, Patel N, Turcotte M, Bossé Y, Paré G, Meyre D (2019) Benefits and limitations of genome-wide association studies. Nat Rev Genet 20(8):467–484

Tamarelle J, Thiébaut AC, De Barbeyrac B, Bebear C, Ravel J, Delarocque-Astagneau E (2019) The vaginal microbiota and its association with human papillomavirus, Chlamydia trachomatis, Neisseria gonorrhoeae and Mycoplasma genitalium infections: a systematic review and meta-analysis. Clin Microbiol Infect 25(1):35–47

Terrón-Camero LC, Gordillo-González F, Salas-Espejo E, Andrés-León E (2022) Comparison of metagenomics and metatranscriptomics tools: a guide to making the right choice. Genes 13(12):2280

Turnbaugh PJ, Ley RE, Hamady M, Fraser-Liggett CM, Knight R, Gordon JI (2007) The human microbiome project. Nature 449(7164):804–810

Valdés A, Moreno LO, Rello SR, Orduña A, Bernardo D, Cifuentes A (2022) Metabolomics study of COVID-19 patients in four different clinical stages. Sci Rep 12(1):1650

Vazquez-Castellanos JF, Serrano-Villar S, Jimenez-Hernandez N, Soto del Rio MD, Gayo S, Rojo D, Gosalbes MJ et al (2018) Interplay between gut microbiota metabolism and inflammation in HIV infection. ISME J 12(8):1964–1976

Wang WL, Xu SY, Ren ZG, Tao L, Jiang JW, Zheng SS (2015) Application of metagenomics in the human gut microbiome. World J Gastroenterol 21(3):803

Wang Q, Peng WX, Wang L, Ye L (2019) Toward multiomics-based next-generation diagnostics for precision medicine. Pers Med 16(2):157–170

Wang X, Xu G, Liu X, Liu Y, Zhang S, Zhang Z (2021) Multiomics: unraveling the panoramic landscapes of SARS-CoV-2 infection. Cell Mol Immunol 18(10):2313–2324

Wirth J, Young M (2020) The intriguing world of archaeal viruses. PLoS Pathog 16(8):e1008574

Wu F, Zhao S, Yu B, Chen YM, Wang W, Song ZG, Zhang YZ et al (2020) A new coronavirus associated with human respiratory disease in China. Nature 579(7798):265–269

Wyler E, Mösbauer K, Franke V, Diag A, Gottula LT, Arsiè R, Landthaler M et al (2021) Transcriptomic profiling of SARS-CoV-2 infected human cell lines identifies HSP90 as target for COVID-19 therapy. IScience 24(3):102151

Wylie KM, Mihindukulasuriya KA, Zhou Y, Sodergren E, Storch GA, Weinstock GM (2014) Metagenomic analysis of double-stranded DNA viruses in healthy adults. BMC Biol 12(1):1–10

Xiang Z, Li J, Lu D, Wei X, Xu X (2022) Advances in multi-omics research on viral hepatitis. Front Microbiol 13:987324

Zielinski JM, Luke JJ, Guglietta S, Krieg C (2021) High throughput multi-omics approaches for clinical trial evaluation and drug discovery. Front Immunol 12:590742

14 Multi-omics and Its Clinical Application

Ramneet Kaur, Dibita Mandal, Vedika Potdar, Mrudula Surendra Gosavi, and Ajay Kumar

14.1 Introduction

At the beginning of the human genome project, the advancement of novel technologies has made it possible to obtain a large number of molecular measurements within tissues or cells. These technologies can be employed for the purpose of examination or analysis of the underlying operations of living systems at a resolution that has never been possible. This field of study, which involves measuring biomolecules in a high-throughput way, is called "omics." Omics studies aim to describe the entirety of the collection of biomolecules in a cell, tissue, or organism. This includes identifying molecules, describing their structure and function, and quantifying their abundance. By understanding the whole collection of molecules in a biological system, researchers can gain insights into how the system works and responds to changes. Omics technologies allow scientists to study the entire set of genes (genomics), RNA transcripts (transcriptomics), proteins (proteomics), and metabolites (metabolomics) in biological systems.

R. Kaur
Department of Life Sciences, RIMT University, Mandi Gobindgarh, Punjab, India

D. Mandal
Department of Biosciences, SBST, Vellore Institute of Technology, Vellore, Tamil Nadu, India

V. Potdar
PES's Modern College of Arts, Science and Commerce, Shivajinagar, Savitribai Phule Pune University, Pune, Maharashtra, India

M. S. Gosavi
Sunandan Divatia School of Sciences, NMIMS, Mumbai, Maharashtra, India

A. Kumar (✉)
Department of Biotechnology, Faculty of Engineering and Technology, Rama University, Kanpur, Uttar Pradesh, India

I. Mani, V. Singh (eds.), *Multi-Omics Analysis of the Human Microbiome*,
https://doi.org/10.1007/978-981-97-1844-3_14

Omics tools have revolutionized our understanding of biological systems by enabling comprehensive analysis of various molecular components. Genomics, transcriptomics, proteomics, and metabolomics are key disciplines within the Omics field, providing insights into the genetic, transcriptional, protein, and metabolic profiles, respectively. Integrating data from multiple Omics disciplines, known as multi-omics, holds immense potential in advancing our understanding of complex biological phenomena. Research studies have shown the evidence of relevance and benefits of multi-omics in clinical applications. For instance, research conducted by Johnson and Hennessy (2019) scrutinized the integration of genomics, transcriptomics, and proteomics in lung cancer research. The authors revealed the metabolic dysregulation associated with lung cancer through the integration of diverse omics datasets. This integrative approach provided a deeper understanding of the disease and identified potential targets for therapeutic intervention.

Integrating genomics data to understand cancer biology and develop personalized treatment strategies. The study showcased how multi-omics can uncover molecular signatures and pathways associated with different cancer subtypes, leading to more precise diagnostics and targeted therapies. Moreover, a report by Wang et al. (2022) showed the utility of multi-omics integration for biomarker discovery in cancer. The authors integrated multiple omics data to identify robust biomarkers with high sensitivity and specificity. This integrated approach has the potential to enhance early detection, risk assessment, and medical outlook evaluation in cancer patients.

Present day therapeutic treatment is fundamentally based on alleviating the disease after its occurrence instead of adopting a preventive approach that averts the disease and restricts it is from progressing to a chronic stage. This in turn would prevent the high cost treatment that usually goes into any therapeutic intervention. In order to devise a preventive approach it is important to take heed of the available information with regard to the complexity of the disease and human health by studies that help to build a foundation for understanding the genetic variation and the factors including contributing to its severity (Hasin et al. 2017). Comprehending human illness and health necessitate deciphering the intricate molecular intricacies and modifications occurring at different levels, encompassing the genome, proteome, and metabolome widely known as the omics study (Yan et al. 2018). It is necessary to have an overall idea of how the different omics approaches are affected by hereditary, environmental, or lifestyle differences.

Multi-omics includes thorough evaluations of several kinds of biological components such as RNA or metabolites and examination of large amounts of transcripts, protein molecules, in addition to the genome (Manolio 2010). Omics studies include the individual approach as well as the integrated approach. Individual omics approach encompasses genetic sequencing, biomarker identification, biomolecules complexity analysis etc. However they often stand inadequate to provide information to monitor illnesses or suggest efficient therapies, hence adopting an integrated omics approach helps to fill in the lacunae created by individual study (Geyer et al. 2017; Wishart 2016). The data from individual omics are combined using integrated approaches either consecutively or concurrently in an integrated omics approach to explore the molecular interactions. By facilitating the examination of information transfer between different omics levels, these methods successfully bridge the genetic and phenotypic divergence (Bersanelli et al. 2016).

Consisting of transcriptomics, proteomics, metabolomics, or other omics, integrating them can help go far beyond the capabilities of genomics alone, providing valuable functional insights thus providing a fresh perspective on the intricate molecular complexities of diseases. Recent advancements in various omics technologies, including proteomics and metabolomics, along with improved computing capabilities, have facilitated the emergence of multi-omics analysis. By combining data from multiple sources, this approach captures the intricate molecular interactions driving both health and disease, harnessing the strengths of certain data types (Ritchie et al. 2015). The availability of data from many omics had a dramatic influence on the biological and medical fields, opening up new opportunities for system-level integration research. Using medical information coupled with multi-omics data has become critical in generating useful insights into cellular functioning. The incorporation of multi-omics data, which includes biomolecules from several levels, holds enormous potential for methodically and fully deciphering biology's intricacies (Yan et al. 2018). The fundamental driving factor behind the omics research has been technological improvements, which have enabled cost-effective and high-throughput study of biological molecules which includes microarrays, based on cDNA hybridization; ChIP-Seq and RNA-Seq etc. (Chakraborty et al. 2018)

Individual omic investigations such as genomes, epigenomics, transcriptomics, proteomics, and metagenomics have been successfully applied in a variety of domains. These researches have aided in gaining knowledge of particular concerns in human disease, agriculture, plant science, microbiology, and the environment (Krassowski et al. 2020). Conducting multiple omics research sometimes entails dealing with datasets including a variety of data modalities derived from distinct assay types, resulting in increased complexity. A multi-omics technique, such as profiling RNA, protein, or metabolites, can create transcripts (including isoforms) from the transcriptomics dataset generated by RNA-seq efforts. As a result, the transcriptome's information load may eclipse more immediately applicable findings obtained, which have a closer connection to the types (Fiehn 2002). The utilization of multi-omics plays a crucial role in identifying predictive or prognostic biomarkers, as well as potential therapeutic targets for repurposing or the development of new drugs. The fundamental goal of applied multi-omics is to improve health diagnostics, illness prediction, and agricultural results by developing a thorough grasp of the genotype-to-phenotype link.

14.2 Different Omics Technologies

See Table 14.1.

14.3 Overview of Omics Technologies

1. *Genomics*: It involves the study of the whole DNA of an organism (genome), and its organization, function, and evolution. It includes techniques such as whole-genome sequencing, genome-wide association studies (GWAS), and comparative genomics. Whole-genome sequencing: This technology allows the determination of an organism's entire DNA sequence. This has transformed our understanding of genetic variations, disease-causing mutations, and evolutionary relationships (Lander et al. 2001). The Human Genome Project represented a groundbreaking research endeavor in genomics, marking the initial complete sequencing of the human genome. GWAS use a comparative analysis of the genomes of individuals with and without a specific phenotype to identify genetic variants linked to complex traits or diseases. This methodology has significantly contributed to the identification of numerous genetic loci associated with various diseases (Visscher et al. 2012).
2. *Transcriptomics*: It involves the examination of the complete set of RNA molecules found within a cell, tissue, or biological organism, also known as transcriptomes. It involves the quantification and characterization of mRNA, non-coding RNA, and splice variants, which provides insights into gene expression levels, alternative splicing, and regulatory mechanisms.

   RNA-Seq: RNA-Seq is a powerful technique that uses next-generation sequencing to determine the RNA sequences and quantify gene expression levels. This technological advancement has transformed transcriptomic research, enabling the identification of new transcripts and investigation of gene regulation networks.

   Microarrays: Although RNA-Seq has largely replaced microarrays, these technologies have been widely used to measure gene expression levels. Microarrays use hybridization to detect and quantify the mRNA expression (Martin and Wang 2011).
3. *Proteomics*: It is the systematic study of proteins within a biological sample. The complete collection of proteins expressed by a biological system at a specific time and under specific conditions needs to be identified, quantified, and characterized. The field of proteomics plays a critical role in understanding protein function, interactions, post-translational modifications, and their roles in various cellular processes and diseases. Key techniques used in proteomics include:
   (a) Mass Spectrometry (MS): Used for protein identification, quantification, and post-translational modification analysis.

Table 14.1 Different omics technologies and its functions

Types of omics	Functions
Genomics	Emphasizes finding genetic variations connected to an illness, the effectiveness of a medication, or the prognosis of the patient. GWAS, genotyping arrays, and NGS (whole-genome and exome sequencing) are examples of technologies
Epigenomic	Describes reversible alterations to DNA (such as DNA methylation or histone acetylation) influenced by environmental and genetic factors. NGS technology is used to modify DNA
Transcriptomics	Examines genome-wide RNA levels qualitatively (transcript presence) and measurably (transcript expression levels). Includes coding and non-coding RNAs
Proteomics	Quantifies peptide abundance, modifications, and interactions. MS-based methods enable high-throughput analyses of thousands of proteins. Detects protein-nucleic acid interactions (e.g., ChIP-Seq) and post-translational modifications
Metabolomics	Assesses a variety of tiny molecules simultaneously (e.g., amino acids, fatty acids) to reflect metabolic function. Perturbations in metabolite levels indicate disease. Technologies: MS-based approaches for relative and targeted small molecule quantification
Microbiomics	Investigates microorganisms in a given community, such as the gut microbiota. Shotgun metagenomics sequencing or profiling using the amplification and sequencing of certain bacterial 16S rRNA gene sections

(b) Liquid Chromatography (LC): Often coupled with mass spectrometry, enabling high-throughput protein analysis.

Technological advancements in mass spectrometry and high-throughput proteomic techniques have revolutionized the field, allowing analysis of complex protein mixtures on a global scale. Proteomics has applications in various areas, including biomarker discovery, drug target identification, and understanding disease mechanisms (Geyer et al. 2017).

4. *Metabolomics*: It is a thorough examination of the metabolites or small compounds found in biological systems. It entails the methodical identification and measurement of all metabolites, including endogenous ones, within a biological system. (e.g., amino acids, lipids, nucleotides) and exogenous compounds (e.g., drugs, toxins) in cells, tissues, or organisms. Metabolomics helps in understanding cellular metabolism, metabolic pathways, and their regulation in response to various biological and environmental factors. This field has applications in disease diagnosis, personalized medicine, and monitoring the effects of drugs and nutritional interventions (Johnson et al. 2016).
5. *Microbiomics*: It is also known as metagenomics, is the study of microbial communities within a particular environment or host organism. It involves the analysis of genetic material (DNA or RNA) directly extracted from a complex microbial community to gain insights into the composition, diversity, and functional potential of microorganisms. Due to the emergence of high-throughput sequencing technologies, metagenomics has offered enhanced insights into microbial ecology, the human microbiome, and their effects on human health and disease. It has led to the discovery of novel microorganisms and functional genes and revealed the complex interactions between the host and its associated microbial communities (Lloyd-Price et al. 2019).
6. *Epigenomics*: It is a research field that focuses on studying the epigenetic modifications occurring in a cell or organism. It examines the patterns of chemical modifications on DNA and its associated proteins, without changing the underlying DNA sequence itself. These modifications play a crucial role in gene expression and regulation, ultimately influencing an organism's phenotype and development. In recent years, epigenomics has gained considerable attention due to its potential impact on various biological processes and diseases. Researchers can now assess epigenetic alterations on a genome-wide scale due to improvements in high-throughput sequencing methods, which has helped them gain a better grasp of the epigenetic landscape. Through integrative analysis of multiple cancer types, the Aran et al. (2019) found distinct epigenetic patterns associated with different cancer subtypes, suggesting potential implications for diagnosis and treatment strategies (Aran et al. 2019). Additionally, the role of epigenomics in understanding human diseases is highlighted in the research carried out by Angermueller et al. (2017). The researchers harnessed epigenomic profiling to investigate the interaction between genetics and environmental factors in complex diseases, highlighting the importance of studying the epigenome for a

comprehensive understanding of disease etiology (Angermueller et al. 2017). These studies, along with numerous others, contribute to the expanding knowledge surrounding epigenomics and its implications in various biological processes and diseases.

14.4 Multi-omics Data

Multi-omics data are a subset of biological data that come from all omic research as well as other levels of biological structure. This data is employed to study the interactions among various biological molecules and pathways, leading to a more in-depth comprehension of intricate biological processes. Single-omics data can provide valuable insights into biological processes, but it is often limited in its ability to explain the complexity of these processes. Multi-omics data can overcome this limitation by providing a more holistic view of biological systems. For example, by combining genomic data with transcriptomic data, researchers can gain insights into how genetic variation influences gene expression. Multi-omics data integration enables researchers to connect molecular changes at different levels of biological regulation, providing a more complete picture of cellular processes. For instance, the correlation between genetic variants and gene expression patterns in the context of disease can be enhanced by integrating genomics and transcriptomics data (Teschendorff et al. 2016).

A significant application of multi-omics data analysis is observed in cancer research. The integration of this data has played a crucial role in identifying potential driver mutations, understanding tumor heterogeneity, and uncovering new therapeutic targets. In this regard, Hoadley et al.'s (2018) study employing multi-omics analysis on more than 10,000 tumors across 33 diverse cancer types revealed distinct molecular subgroups associated with prognosis and response to therapy. (Hoadley et al. 2018). Research on the microbiome has also shown the value of multi-omics data analysis. Our health is significantly impacted by the human microbiome, which is made up of trillions of microbial cells that live inside of our bodies. Integrating metagenomics, metatranscriptomics, and metabolomics data allows for a better comprehension of the practical interactions between microbial populations and hosts. A study by Lloyd-Price et al. (2019) investigated the microbiome's composition and metabolic pathways to uncover potential links between the microbiome and diseases like inflammatory bowel disease, obesity, and diabetes (Lloyd-Price et al. 2019). The development of bioinformatics tools and methodology has tremendously aided computational analysis and interpretation of multi-omics data. Statistical approaches including clustering, network analysis, and machine learning algorithms are commonly used to recognize molecular patterns and predict biological outcomes. Additionally, frameworks like MultiPLIER aid in the integration and interpretation of different omics layers (Taroni et al. 2018).

Databases play a significant role in omics research by providing access to vast amounts of biological data, including nucleotide sequences, protein sequences, and metabolite information. There are several databases available that provide access to multi-omics data, enabling researchers to explore and analyze integrated molecular datasets. These databases play a crucial role in promoting data sharing, facilitating collaborations, and advancing multi-omics research. Here are a few examples of databases that host multi-omics data:

1. *The Cancer Genome Atlas (TCGA)*: TCGA is a comprehensive resource that integrates multiple omics data types, for various cancer types. It provides a wealth of data for researchers to explore tumor molecular profiles and identify potential biomarkers or therapeutic targets (Weinstein et al. 2013).
2. *The Genotype-Tissue Expression (GTEx) Project*: GTEx consists of multi-omics data from multiple human tissues. It looks into how genetic diversity affects gene expression in various tissues, offering information on how genes are regulated and how certain tissues operate (Aguet et al. 2017).
3. *The Human Protein Atlas*: It is a database that integrates multi-omics data, focusing on protein expression and localization in human tissues and cells. It provides information on protein expression patterns, subcellular localization, and tissue-specificity, aiding in the understanding of protein function and disease associations (Uhlén et al. 2015).
4. *The MetaboLights database*: It is a repository for metabolomics data, hosting multi-omics datasets that include metabolite information. It enables researchers to explore metabolite profiles in various biological systems and link them to additional omics data, like genomics or transcriptomics (Haug et al. 2013).

These databases act as valuable assets for researchers who aim to retrieve and analyze multi-omics data. By utilizing these databases, researchers can gain insights into complex biological systems and facilitate the discovery of novel biological markers or therapeutic targets.

14.5 Applications in Disease Research

The characteristics of the illness being studied must be taken into account while designing a multi-omic research. When a simple disease is caused by a single gene mutation, there are often few variables that play deterministic roles in the disease's progression. In these situations, tailored omics studies at certain time periods might offer informative information on the changes in the genome caused by the causal factor, allowing for a more detailed understanding of potential therapy options. It's important to note that the primary cause need not always be hereditary; it might possibly be an infectious pathogen (Hasin et al. 2017).

Complex illnesses, on the other hand, have a more complicated etiology that is not centered on a single component. Instead, these illnesses could develop as a result of diverse causes coming together in different ways, resulting in phenotypically identical states. Furthermore, when there is an absence of a clearly identifiable deterministic process underlying the disease, findings based on a single layer of data often represent associations and should be interpreted as such. This is due to the fact that in biological processes, reactive effects frequently outweigh causal consequences. Therefore, to fully understand complex illnesses, coordinated sets of numerous omics data must be integrated, taking into account both genetic and environmental variables, and gathered from many disease-relevant tissues at different times (Hasin et al. 2017).

One of the most important components of omics studies is the downstream analysis, sample sizes, and statistical power. The power to detect associations and extract meaningful information from large datasets depends on effects, noise-related heterogeneity, sampling size, and other variables. In human studies, confounding factors like diet and lifestyle choices can introduce challenges, making it essential to have sufficient sample sizes to ensure statistical power and account for variation in outcomes. Additionally, proper attention should be given to the demands of data analysis both prior to and throughout data gathering. Each omics field has general analytical pipelines, but tailoring using statistics to analyze a certain dataset is necessary due to the complexity and size of the data (Hasin et al. 2017).

Omics investigations on humans as well as animal models offer insightful information about illnesses. Human studies offer a larger potential for translation, but also have drawbacks such the convergence of multiple cell types in complicated disorders. Models based on animals provide repeatability, environmental variables control, accessibility to pertinent tissues, accurate phenotyping, and the capacity to test hypotheses experimentally with a greater number of duplicates. Animal models have proven highly valuable in examining the impact of environmental stressors and offering detailed mechanistic insights. The biological relevance of animal models may be verified and our understanding of illnesses improved by comparing human and animal omics data. However, animal models also have limitations, including potential differences in genetic background and the inability to fully replicate human biology in complex diseases (Hasin et al. 2017).

14.6 Biomarker Discovery

A biomarker is any detectable element, organ, or bodily function that can affect, foretell, or play a role in the onset of a disease, according to the World Health Organization (WHO) (Ferreira de Lima et al. 2003). At all phases of the illness process, including diagnosis, subtype identification, prognosis, and therapy modification, biomarkers have the potential to be therapeutically beneficial (Leclercq et al. 2019). Utilizing techniques like genome-wide association studies (GWAS) to find common variations in sizable sets of controls and cases, omics technologies have been widely employed to explore genetic risk factors and mutations linked to neoplasia. In neoplasia, next-generation sequencing (NGS) is used to study mutation profiles in the whole genome, transcriptome, exome, or specific genes known to be important in the disease. The ideal approach for tumor mutation profiling by NGS includes concurrent analysis of several target sets, such as DNA sequencing of a gene panel and the tumor transcriptome. Integrating data from different levels of biological information is crucial for studying the progression of complex diseases, as biological information influences and regulates a variety of biological processes (Shi et al. 2017).

The study of the quantitative expression of every gene in a specific biological sample is called transcriptomics, often known as gene expression profiling (Cappola and Margulies 2011). To measure all expressed transcripts, RNA sequencing or mRNA expression microarrays might be used. Comparative analysis of gene expression in healthy and sick tissues is the main use of transcriptomics. Researchers are able to learn more about the pathophysiology of diseases and discover possible biomarkers by comparing gene expression in healthy and sick situations. RNA sequencing, microarrays, and quantitative real-time polymerase chain reaction (qRT-PCR or qPCR) are frequently used techniques for evaluating the transcriptome in clinical applications. These techniques offer precise gene expression analysis, are affordable, and are appropriate for large-scale sample experiments. But they are constrained by having to concentrate on a small number of potential transcripts (Dar et al. 2023).

The exploration of all proteins expressed and the analysis of protein networks within cells, organs, and living things is at the heart of proteomics. It is uncertain how many proteins and peptides there are, although it is believed to be in the hundreds of thousands (Nalbantoglu and Karadag 2019). Proteomic investigations extensively detect, characterize, and identify proteins, making it a viable method for finding biomarkers (Vaiopoulou et al. 2012). Two-dimensional electrophoresis (2DE) and mass spectrometry are frequently used to detect proteins that differ in quantity between controls and patients. Once the proteins have been extracted by 2DE, they may be identified using mass spectrometry techniques like surface-enhanced laser desorption/ionization time-of-flight (SELDI-TOF) or matrix-assisted laser desorption/ionization time-of-flight (MALDI-TOF). For choosing potential biomarkers based on changes in abundance between controls and cases, quantitative proteomics is crucial. Although mass spectrometry does not produce quantitative data by nature, new techniques have been created to do so. Multiple reaction monitoring is a quantitative method that is appropriate for proteomic biomarker discovery investigations since it allows for absolute quantification and is sensitive to detecting low-abundance peptides (Wasinger et al. 2013).

The primary goal of metabolomics is to understand the metabolome, which covers microscopic particles and how they interact with biological systems under certain nutritional, genetic, and environmental conditions. When gene expression, illness vulnerability, protein expression, and environmental factors interact and alter, the metabolome is the result. It is the omics strategy that is most phenotype-relevant and serves as an excellent source of disease-associated markers (Guijas et al. 2018). Integrating multi-omics data has gained significant interest in bioinformatics as a potent and responsive method for harnessing various genomic data. Multi-omics data integration plays a crucial role in various human diseases (Dar et al. 2023).

14.7 Precision Medicine

Precision medicine strives to improve accuracy in prediction, preventing, and treating illnesses by taking into account each person's unique genetic composition, environmental effects, and lifestyle choices. This approach contrasts with traditional medicine's customary "one-size-fits-all" and reactive tactics. Furthermore, by adapting therapies to the specific features of individual patients, this strategy has the potential to improve healthcare efficiency and quality. Precision medicine seeks to personalize medical therapies for individual patients based on their genetic, genomic, and molecular traits. Multi-omics techniques, which integrate data from several omics technologies to give a full and thorough knowledge of a patient's health and disease state, have emerged as valuable tools in precision medicine (Wörheide et al. 2021).

Cancer research and therapy is a prominent use of multi-omics in precision medicine. Researchers can uncover particular genetic abnormalities and molecular pathways driving tumor development and progression by evaluating various layers of molecular data, including genomes, transcriptomics, epigenomics, proteomics, and metabolomics. Pancreatic cancer with a poor prognosis, and a very small number of patients living beyond 5 years, ranging from 5% to 15% impose a dire need of effective immunotherapies, comprehensive comprehension of the tumor environment and the identification of significant biomarkers. Radiomics is emerging as a potential assist in predicting pancreatic cancer prognosis. It has been used to estimate survival trends and classify patients into discrete prognostic categories by applying various models. It can predict biomarkers such as p53 mutations non-invasively, allowing for non-invasive surveillance outperforming traditional approaches. In identifying treatment response, for example, a Bayesian regularization backpropagation neural network got an excellent area under the curve (AUC) of 0.92. Furthermore, in predicting pancreatic cancer outcomes, multiple machine learning models attained AUC values ranging from 0.95 to 0.98. This advancement suggests that radiomics has the potential to dramatically improve our knowledge and therapy of pancreatic cancer (Tabari et al. 2022, 2023).

The study primarily examined cutaneous melanoma, a kind of skin cancer with a high tumor mutation load that makes it challenging to find significantly mutated genes (SMGs). To tackle this challenge, the researchers merged a multi-omic investigation of 470 cases from The Cancer Genome Atlas with an analysis of over 1000 melanoma exomes. In a component of the holoenzyme, co-occurrence was observed between loss-of-heterozygosity and loss-of-function mutations. A protein kinase signature associated with oxidative phosphorylation in melanoma Analysis of bulk tumor transcriptomes identified a route. Furthermore, it has been demonstrated that certain loss-of-function mutations in male patients affect SMGs on the X chromosome, such as the RNA helicase DDX3X. Furthermore, the researchers discovered that tumor mutation load and immune infiltration could serve as predictive indicators for the prognosis of melanoma patients. In summary, multi-omic research shed light on how specific mutations impact the observed sex bias in melanoma and offered valuable insights into its origin (Alkallas et al. 2020). Yet another melanoma study illustrated the value of using multi-omics data to spot possible treatment targets and forecast medication responses (Arozarena and Wellbrock 2019).

In the multi-omic investigation of genome-wide association studies (GWAS), more than 50 genetic variants have been connected to coronary artery disease (CAD) and myocardial infarction (MI), with additional findings predicted in the future and with a tiny portion of heritable risk, the study of gene interactions (epistasis) is progressing quickly and may help to understand disease causes. While the combined impact of several genetic variants may explain complicated inheritance patterns, current GWAS research provides vital insights into the pathophysiology of CAD and MI, establishing the foundation for customized treatment methods. A significant objective of contemporary research is precision medicine, with an emphasis on tailored treatment techniques. Systems medicine techniques, such as integrative analysis of microRNA-mRNA pairings, which have shown enrichments of CAD-associated variations, may make promising contributions to this research. Continued study and collaboration in these areas has the capacity to enhance our comprehension of these complicated diseases and ultimately to more effective treatment choices (Kessler et al. 2016). Multi-omic study for neurological disorders looked at genetic variants in schizophrenia patients and discovered that, while protein-coding de novo mutations (DNMs) are important in other brain illnesses, their influence on schizophrenia risk is minimal. Gene and exome studies show that the mutations were more prevalent in brain-expressed genes linked to neurodevelopmental disorders. There was no one gene that had a significant influence on schizophrenia risk, however, 16 genes had recurrent alterations. However, these mutations explain just a tiny portion of the risk for schizophrenia, and bigger research is needed to discover particular genes connected to the disorder, as numerous genes contribute to schizophrenia risk in the general population (Farhan et al. 2020). Moreover, multi-omics approaches have also been applied to complex diseases beyond cancer. In cardiovascular diseases, the molecular pathways underlying the onset and course of illness have been better understood because to multi-omics investigations. Similarly, in neurodegenerative diseases, multi-omics data integration has helped unravel the genetic and molecular basis of disease pathology.

Finally, multi-omics methods have proven essential in precision medicine, allowing for a thorough knowledge of illness at the molecular level. Researchers and clinicians can find novel biomarkers, therapeutic targets, and individualized treatment methods by combining data from diverse omics technologies.

14.8 Data Integration Challenges and Solutions

Data integration is a vital part of current biomedical research, as enormous amounts of heterogeneous data from numerous sources must be merged to get deeper insights into complicated illnesses and biological processes. To solve these problems and assure successful data integration in biomedical research, several strategies have been offered. The variety in data formats and terminologies used across different databases and repositories is one of the key obstacles in data integration. Standardized data formats, such as the usage of common data models and controlled vocabularies, are critical for enabling smooth integration and interoperability (Wilkinson et al. 2016).

Another significant difficulty is the disparity in data quality and trustworthiness among different sources. To resolve mistakes and inconsistencies and increase overall data quality, data cleaning and curation procedures are required. Furthermore, data integration frequently requires large-scale data processing, which may be computationally and time-consuming. Using distributed computing frameworks and cloud-based resources can help data integration pipelines scale and be more efficient. Biomedical data is frequently collected from various sources with varied levels of privacy and security, causing issues in data sharing and access. Adherence to data privacy requirements, as well as ethical issues, are critical for successful data integration in a collaborative research context.

14.8.1 Types of Omics Data and Databases

See Fig. 14.1.

14.8.2 Approach for Selection of Tools

See Fig. 14.2.

Based on these approaches the commonly used tools are employed that provide the data set of patients.

14.8.3 Data Repositories

See Fig. 14.3.

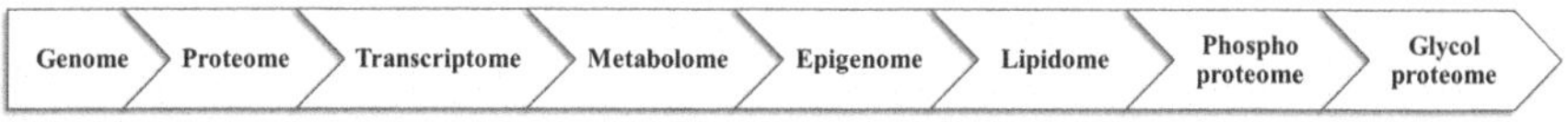

Fig. 14.1 Levels of data generation: The data to be mined is initially gathered at the level of the genome, proteome, transcriptome, metabolome, and epigenome. Data can also be collected in the form of lipidome, phosphoproteome, and glycolproteome levels (Subramanian et al. 2020)

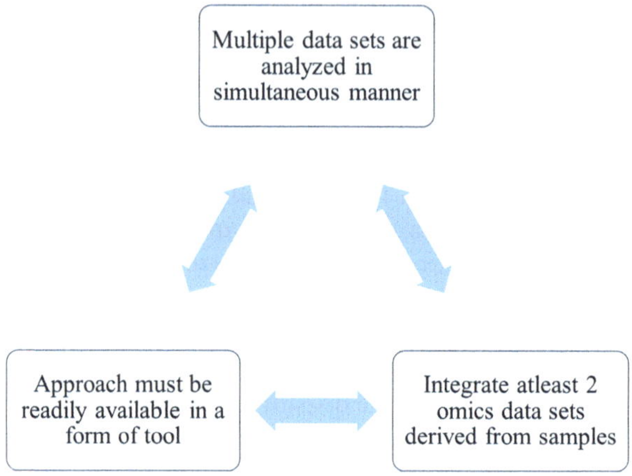

Fig. 14.2 Approaches for data segregation (Subramanian et al. 2020)

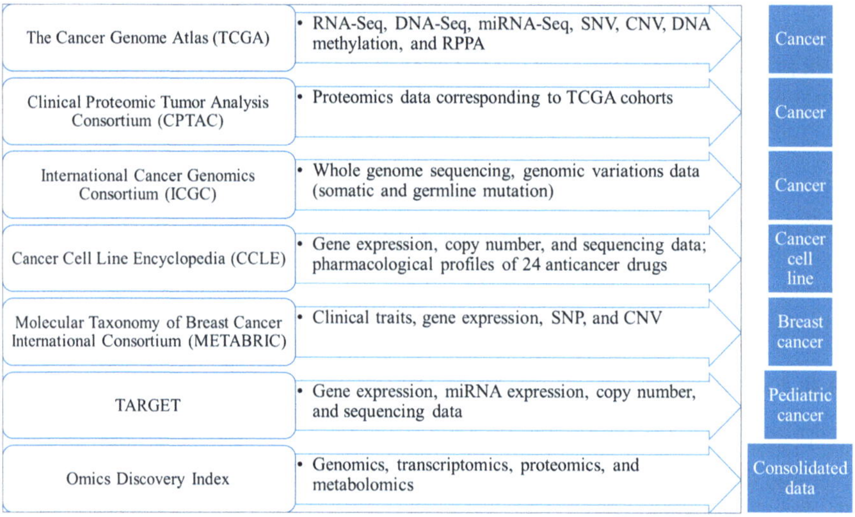

Fig. 14.3 Different data repositories (Subramanian et al. 2020)

The Cancer Genome Atlas (TCGA) is a massive resource that houses multi-omics data sets from over 33 distinct forms of cancer as well as 20,000 individual tumor samples. It tries to examine and understand tumor DNA, RNA, protein, and epigenetic alterations, as well as clinical data. A consortium manages it, ensuring data quality and sharing among scientific communities (Weinstein et al. 2013; Tomczak et al. 2015).

The International Cancer Genomics Consortium (ICGC) organizes large-scale genomic research from many nations and programs. It focuses on genetic profiles from original tumor samples of various malignancies, creating large amounts of data for research communities to utilize to generate new discoveries regarding cancer development and therapy (Bailey et al. 2016; Thompson et al. 2018).

The Cancer Cell Line Encyclopedia (CCLE) is a database with information on 947 human cell lines, 36 different tumor types, and copy number, gene expression, and sequencing. Furthermore, it gives anticancer medication pharmacological profiles across 479 cancer cell lines, helping the discovery of biomarkers and drug response mechanisms (Barretina et al. 2012).

The Molecular Taxonomy of Breast Cancer International Consortium (METABRIC) is a joint Canadian-UK study. It involves clinical characteristics, expression, SNP, and CNV data from breast cancers, aiming to classify breast cancer by utilizing multi-omics molecular markers. As a result, new pharmacological targets have been identified and improved treatment approaches have been developed (Curtis et al. 2012).

TARGET, like TCGA, focuses on juvenile malignancies. It includes clinical data, gene expression, miRNA expression, copy number, and sequencing data from 24 distinct molecular categories of pediatric cancer, enabling a more precise assessment of genomic alterations in various pediatric malignancies (Ma et al. 2018).

Omics Discovery Index (OmicsDI) is an open-source platform that aggregates data sets into a single framework. It enables access to and integration of genomes, transcriptomics, proteomics, and metabolomics data from multiple species (Perez-Riverol et al. 2017).

14.8.4 Multi-omics Data Integration Tools

See Fig. 14.4.

Tools based on the approaches: (a) Disease subtyping, (b) Disease insights, and (c) Biomarker prediction. Each of the given tools have subtypes that correspond to the classification of the biological problem into the three approaches. Among them

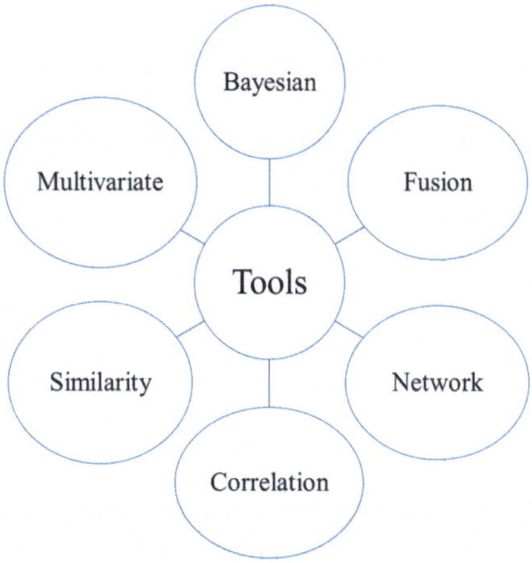

Fig. 14.4 Different tools in data integration (Subramanian et al. 2020)

similarity and fusion focuses more on disease subtyping; correlation focuses on the disease insights; Bayesian, Network and Multivariate deals with all the three approaches.

Bayesian: Two important integrative approaches that entail all the three approaches are PARADIGM and iCluster Plus. Pathway Recognition Algorithm using Data Integration (PARADIGM) from multi-omics data, the activity of biological processes unique to each patient is inferred using genomic models. In a factor graph, a gene is represented by a network of connected parameters that store details regarding its expression as well as the known activity of the gene and its products. This modeling method makes it easier to incorporate diverse forms of omics data as supporting evidence (Vaske et al. 2010). iClusterPlus is an enhanced iteration of iCluster, that uses extended linear regression to create a comprehensive model for both categorical and numerical (continuous and count) variables. This model was created utilizing data from merged genomic, epigenomic, and transcriptome profiling. By employing a set of hidden variables representing "k" driving forces, it effectively predicts the fundamental genomic variables and captures the inherent biological variation (Mo et al. 2013).

Fusion: An important method for disease subtyping is pattern fusion analysis (PFA), which identifies integrated sample patterns from multiple genetic profiles within a condensed feature space. PFA begins by using principal component analysis (PCA) to generate local sample patterns. The resulting patterns are subsequently matched to a shared characteristic space, yielding a global sample pattern that incorporates many different types of data. The quantitative contributions of each kind of data (or individual sample) to the global sample spectrum are reviewed during this process, and biases or systematic noises are steadily minimized to generate a better data fit. This iterative correction process is repeated until convergence is achieved. Following adaptive optimum alignment, the resultant combined sample pattern thoroughly describes the data, more precisely reflecting intrinsic linkages (Shi et al. 2017).

Network: Similarity network fusion (SNF) is a network-based technique that employs network fusion to integrate information from various omics sources. SNF generates multiple networks for each type of information and then combines them into a unified similarity network using a nonlinear fusion approach. In this fusion process, the concept of message-passing is utilized, and the network becomes more similar to the others with each iteration. The advantage of this approach is that weak connections, which represent noise, are gradually eliminated through repetitions, while strong connections persist until convergence is achieved. Although this method is classified as a "Network" approach, SNF also includes "Fusion" and "Similarity" approaches as part of its methodology (Wang et al. 2014).

Correlation: The commonly employed tool is CNAmet, a software program that facilitates the integrated analysis of copy number variation, DNA methylation, and gene expression data. To use CNAmet, researchers need to apply it to the same sample set, ensuring a consistent and coherent analysis. It consists of three main steps: (1) computing weights that link expression levels to copy number and methylation data, (2) calculating a score by combining these weights to generate a single score for each gene, and (3) using corrected P values, assess the statistical relevance of the awarded score. The resultant score assists in identifying genes with amplification, hypomethylation, and upregulation, as well as deletion, hypermethylation, and downregulation (Louhimo and Hautaniemi 2011).

Similarity: PINSPlus (Perturbation Clustering for Data Integration and Disease Subtyping) is an unsupervised clustering technique for detecting subtypes in multi-omics data. The algorithm accomplishes this by estimating the probability with which patients in specific clusters are grouped together under various settings: (1) when the data is disrupted, (2) when several forms of omics data are used, and (3) when various clustering approaches are applied. Patients that show persistent strong links across all of these circumstances are grouped together to form a distinct subtype (Nguyen et al. 2019).

Multivariate: Feature selection multiple kernel learning (FSMKL) is a supervised classification technique that employs multiple kernels to identify disease progression characteristics and capture commonalities in the dataset. Every collected information is converted into a foundational kernel, which is then concatenated to form composite kernels. A wide range of kernels, each with different numbers of features per data type, is utilized. Statistical methods are employed for feature selection, and the algorithm chooses the most relevant kernel and associated features for a specific classification task. The kernel coefficients in the final decision function indicate the importance or weight of the various datasets. One advantage of this approach is that it incorporates existing knowledge, such as pathways from databases like KEGG, in the computation of fundamental kernels (Seoane et al. 2014).

14.8.5 Multi-omics Visualization Tools

An important part of data interpretation is the visualization of data. To visualize data it is important to use the correct tools that ensure the optimal translation of the biological problem. Visualization tools help in integration of data which provides insight on the examination and interpretation of clinical data through multi-omics research (Subramanian et al. 2020). Some of the commonly used visualization tool includes the below mentioned (Fig. 14.5).

cBioPortal: It is a web-based tool for investigating, illustrating, and analyzing cancer data that include multiple genetic information. It now contains information from 233 cancer research investigations that span greater than 30 distinct forms of cancer. The platform offers user-friendly datasets about cancer summaries, data transfer access, network diagramming and analysis, and correlation between data sets and patient-centric inquiries. It is popular among academics because of its purpose of integrating different data kinds to offer particular query-based results (Gao et al. 2013).

UCSC Xena: It is an online application that allows to visualize and analyze the sets of cancer data by TCGA, Cancer Cell Line Encyclopedia (CCLE), and other published studies. It enables cross-study comparison, incorporating diverse

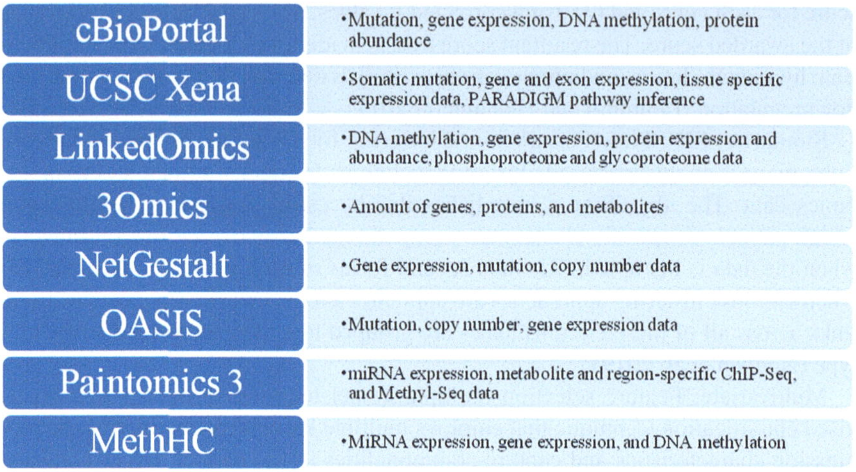

Fig. 14.5 Visualization tools for omics data (Subramanian et al. 2020)

information from omics studies and clinical data, and includes statistical tools for quantification and important analysis (Goldman et al. 2015) .

LinkedOmics: It allows for the comparison and investigation of TCGA data from different types of cancer, as well as data on proteins obtained from CPTAC for chosen samples. It provides three analytic modules for omics data and clinical characteristics association analysis, study through a comparison within and across data sets, and molecular interpretation via pathway and enrichment analysis (Vasaikar et al. 2018).

3Omics: It is a web-based gateway for three forms of omics: transcriptomics, proteomics, and metabolomics. Within a single platform, users may do multiple studies on each data set, such as correlation analysis, co-expression profiling, pathway enrichment, and gene ontology analysis (Kuo et al. 2013).

NetGestalt: It aggregates multi-omics data across biological networks, employing a hierarchical modular design to reduce display complexity. It enables data visualization and navigation on several scales, statistical analysis, route analysis, and cross-data comparisons (Wang et al. 2017).

OASIS: Pfizer created an online analytical tool designed to explore, analyze, and visualize cancer multi-omics data. It includes data from the TCGA, CCLE, GTEx, and other studies on transformation, duplicate quantity, and expression of genes in 55 tumors (Guberman et al. 2011).

Paintomics 3: It is a web-based platform for visualizing multi-omics datasets, capable of analyzing gene expression data, metabolite information, and region-specific data to construct a multi-omics pathway network and assess its biological significance (Hernández-de-Diego et al. 2018).

MethHC: It is a database and analytical gateway for TCGA's 18 human malignancies' DNA methylation, gene expression, and miRNA expression. It integrates the UCSC genome browser, miRStart, and KEGG pathways to enhance interpretation and offers graphical visualization for identifying differentially methylated genes, grouping, and conducting association analysis (Huang et al. 2015).

14.9 Ethical Considerations

The ethical ramifications of the collection, storage, and use of omics data have grown in significance in the era of enhanced genomics and data-driven research. Omics data, which includes genomes, proteomics, transcriptomics, and metabolomics, hold a lot of potential for improvements in biomedicine. To ensure responsible usage and defend individual privacy and rights, ethical concerns are essential. Ethical issues continue to be crucial when omics data are used to promote scientific understanding and medicinal breakthroughs. To preserve public confidence and ethically advance the field, informed consent, privacy protection, responsible data sharing, and equity are essential.

14.9.1 Informed Consent and Data Sharing

It is important to understand what's being requested of you when considering about respect for people. This becomes more difficult in the case of omics research since functional literacy in a broad range of omics words and techniques is necessary for permission. To get informed consent from a research participant, it is vital to clarify the study's aims, anticipated outcomes, beneficiaries of the findings, and any risks and incentives. As a result, the participant's decision to participate can be made willingly. However, obtaining properly informed permission might be difficult unless the researcher goes above and beyond to clearly communicate the advantages and disadvantages. Even then, the choice to participate is influenced by a number of variables, including the patient's particular disease, any accompanying anxiety, treatment expectations and many other factors (Corrigan 2003; Mathaiyan et al. 2013).

14.9.2 Privacy and Confidentiality

When working with delicate omics data, protecting the privacy of study participants is of the highest significance. It is crucial to put suitable safeguards in place before sharing or publishing the data, such as de-identification, anonymization, or pseudonymization. Although protecting privacy is seen as a fundamental value, it is not an absolute and has both advantages and disadvantages. Recently, a number of organizations have been engaged to get their opinions on how to protect the privacy of genetic or medical data. Public opinion polls have offered insightful information on how societal norms surrounding these topics are changing. Based on the study, a significant majority of participants (74%) had not familiarized themselves with the guidelines for data sharing. Nevertheless, a considerable number (86%) believed that the findings would not affect their insurance accessibility, and 64% felt that their employers were unlikely to utilize their genetic test results against them. It is worth mentioning that most respondents (74%) were not actively involved in biomedical research (Haeusermann et al. 2017; Anderlik and Rothstein 2001; Corrigan 2003).

14.10 Future Directions

Phenotyping and reference populations are essential for using multi-omics in disease research. The variability among these categories must be taken into account when comparing omics data between healthy and sick people. Variability can be introduced by variables including population structure, and unidentified confounding factors, which can also have an impact on how the data should be interpreted. The reductionist approach, which seeks to match patient and control groups as closely as possible to exclude environmental influences, is one method for addressing this variability. This strategy has drawbacks since it can only take into account recognized causes of variation and could not fully represent the disease population. In contrast, an integrative omics approach seeks to incorporate multiple sources of variability by analyzing large datasets that accurately capture the background population (Hasin et al. 2017).

With the growing ubiquity of lifestyle monitoring devices and social media, it is becoming easier to compile huge and varied datasets that capture causes of variance in healthy persons. When combined with omics data, these datasets can improve tailored treatment strategies by revealing important information about the environmental variables that influence the onset and course of diseases. Another crucial component of omics investigations is taking the context of sex specificity into account. Sex has a significant role in determining how biological processes work, and many illnesses show sex dimorphism. Therefore, sex should be considered in individualized treatment strategies. The National Institutes of Health have stressed the value of sex-balanced studies in biomedical research in response to this requirement (Hasin et al. 2017).

Significant insights may be gained from reference or populations that qualify examined at various omics levels or in different environmental contexts. For instance, the MuTher project, which focuses on female twins in the UK, has produced important insights into the heritability of gut microbiota, metabolic syndrome pathways, and the genetic regulation of molecular characteristics. Important discoveries about metabolic and cardiovascular features have also been made by the Finnish males who make up the METSIM cohort. The progress of multi-omics research relies on the essential development of technology. The importance of the gut microbiome in maintaining homeostasis has received a lot of attention, and advancements in data collecting and analysis platforms for substances generated from bacteria can offer insightful information about the makeup of microorganisms and how it affects general health. Novel signaling pathways and their connections to human illnesses can be clarified by advances in proteomics and protein modification data processing, notably in phosphoproteomics (Hasin et al. 2017).

Understanding tissue heterogeneity and locating new cell populations have benefited from the introduction of high-throughput single-cell analytic tools. With the expansion of single-cell analysis to a range of omics levels, such as DNA methylome and genome sequencing, it is now possible to investigate the functions of numerous cell types in disease physiology. Analysis of omics data presents a number of analytical difficulties. The correlative nature of the data makes it particularly challenging to distinguish between causal and reactionary changes. It is still unclear how to distinguish between correlation and causation in omics analysis, and prospective and time-course investigations together with in vivo and in vitro molecular research are essential for proving causation. Personalized therapies, prospective monitoring of individual health markers, and integration into clinical practice are all possibilities for the future of omics research. Lifestyle modifications may be used to possibly prevent disease through prospective health monitoring utilizing a variety of omics techniques. Utilizing genome sequencing as a guide, omics technology may be utilized in personalized medicine to discover specific variations that have an impact on clinical outcomes (Hasin et al. 2017).

Further advancements in data collection, processing, and cost-effectiveness are needed to accomplish normal population-scale adoption of omics data. Effective use of publically available data requires standardization of language, technique, and analytic pipelines across research. To tackle the complexity of multi-scale omics data production, analysis, and functional follow-up for diverse illnesses, coordinated activities of several research groups and the creation of substantial consortia are required (Hasin et al. 2017).

14.11 Conclusion

Multi-omics approaches have revolutionized our understanding of biological systems and their implications for disease research. Through the integration of data from diverse omics disciplines, such as genomics, transcriptomics, proteomics, and metabolomics, researchers can obtain valuable insights into the molecular complexities of diseases, pinpoint potential biomarkers, and devise personalized treatment approaches. Multi-omics research holds immense potential for advancing precision medicine and improving patient outcomes. The field of multi-omics research is continuously evolving, and future advancements hold great promise. One area of potential growth is the integration of multi-omics data with clinical data, enabling more comprehensive and personalized healthcare approaches. Additionally, advancements in data analysis techniques, such as machine learning and artificial intelligence, will enhance our ability to extract meaningful insights from complex multi-omics datasets. Collaborative efforts across scientific disciplines and the development of standardized protocols and guidelines will further propel the field of multi-omics research.

References

Aguet F, Brown AA, Castel SE, Davis JR, He Y, Jo B et al (2017) Genetic effects on gene expression across human tissues. Nature 550(7675):204

Alkallas R, Lajoie M, Moldoveanu D et al (2020) Multi-omic analysis reveals significantly mutated genes and DDX3X as a sex-specific tumor suppressor in cutaneous melanoma. Nat Cancer 1:635–652. https://doi.org/10.1038/s43018-020-0077-8

Anderlik MR, Rothstein MA (2001) Privacy and confidentiality of genetic information: what rules for the new science? Annu Rev Genomics Hum Genet 2:401–433

Angermueller C, Lee HJ, Reik W, Stegle O (2017) DeepCpG: accurate prediction of single-cell DNA methylation states using deep learning. Genome Biol 18(1):1–13

Aran D, Looney AP, Liu L, Wu E, Fong V, Hsu A et al (2019) Reference-based analysis of lung single-cell sequencing reveals a transitional profibrotic macrophage. Nat Immunol 20(2):163–172

Arozarena I, Wellbrock C (2019) Phenotype plasticity as enabler of melanoma progression and therapy resistance. Nat Rev Cancer 19(12):377–391. https://doi.org/10.1038/s41568-019-0144-3

Bailey P, Chang DK, Nones K et al (2016) Genomic analyses identify molecular subtypes of pancreatic cancer. Nature 531:47–52

Barretina J, Caponigro G, Stransky N et al (2012) The Cancer Cell Line Encyclopedia enables predictive modelling of anticancer drug sensitivity. Nature 483:603–607

Bersanelli M, Mosca E, Remondini D et al (2016) Methods for the integration of multi-omics data: mathematical aspects. BMC Bioinform 17:15

Cappola TP, Margulies KB (2011) Functional genomics applied to cardiovascular medicine. Circulation 124(1):87–94

Chakraborty S, Hosen MI, Ahmed M, Shekhar HU (2018) Onco-multi-OMICS approach: a new frontier in cancer research. Biomed Res Int 2018:9836256

Corrigan O (2003) Empty ethics: the problem with informed consent. Sociol Health Illn 25(7):768–792

Curtis C, Shah SP, Chin S-F et al (2012) The genomic and transcriptomic architecture of 2,000 breast tumours reveals novel subgroups. Nature 486:346–352

Dar MA, Arafah A, Bhat KA, Khan A, Khan MS, Ali A, Rehman MU et al (2023) Multiomics technologies: role in disease biomarker discoveries and therapeutics. Brief Funct Genomics 22(2):76–96

Farhan SMK, Howrigan DP, Abbott LE et al (2020) Exome sequencing in schizophrenia-affected parent-offspring trios reveals risk conferred by protein-coding de novo mutations. Nat Neurosci 23(2):185–193. https://doi.org/10.1038/s41593-019-0563-1

Ferreira de Lima RLL, Moretti-Ferreira D, Richieri-Costa A, Murray JC (2003) Identity by descent and candidate gene mapping of Richieri-Costa and Pereira syndrome. Am J Med Genet A 122(1):56–58

Fiehn O (2002) Metabolomics—the link between genotypes and phenotypes. Plant Mol Biol 48:155–171

Gao J, Aksoy BA, Dogrusoz U, Dresdner G, Gross B, Sumer SO, Sun Y, Jacobsen A, Sinha R, Larsson E, Cerami E, Sander C, Schultz N (2013) Integrative analysis of complex cancer genomics and clinical profiles using the cBioPortal. Sci Signal 6(269):pl1

Geyer PE, Holdt LM, Teupser D, Mann M (2017) Revisiting biomarker discovery by plasma proteomics. Mol Syst Biol 13:942

Goldman M, Craft B, Swatloski T, Cline M, Morozova O, Diekhans M, Haussler D, Zhu J (2015) The UCSC Cancer Genomics Browser: update 2015. Nucleic Acids Res 43(Database issue):D812–D817. https://doi.org/10.1093/nar/gku1073

Guberman JM, Ai J, Arnaiz O, Baran J, Blake A, Baldock R, Chelala C, Croft D, Cros A, Cutts RJ, Di Génova A, Forbes S, Fujisawa T, Gadaleta E, Goodstein DM, Gundem G, Haggarty B, Haider S, Hall M, Harris T, Haw R, Hu S, Hubbard S, Hsu J, Iyer V, Jones P, Katayama T, Kinsella R, Kong L, Lawson D, Liang Y, Lopez-Bigas N, Luo J, Lush M, Mason J, Moreews F,

Ndegwa N, Oakley D, Perez-Llamas C, Primig M, Rivkin E, Rosanoff S, Shepherd R, Simon R, Skarnes B, Smedley D, Sperling L, Spooner W, Stevenson P, Stone K, Teague J, Wang J, Wang J, Whitty B, Wong DT, Wong-Erasmus M, Yao L, Youens-Clark K, Yung C, Zhang J, Kasprzyk A (2011) BioMart Central Portal: an open database network for the biological community. Database (Oxford) 2011:bar041

Guijas C, Montenegro-Burke JR, Warth B, Spilker ME, Siuzdak G (2018) Metabolomics activity screening for identifying metabolites that modulate phenotype. Nat Biotechnol 36(4):316–320

Haeusermann T, Greshake B, Blasimme A, Irdam D, Richards M, Vayena E (2017) Open sharing of genomic data: who does it and why? PLoS One 12(5):e0177158

Hasin Y, Seldin M, Lusis A (2017) Multi-omics approaches to disease. Genome Biol 18:83

Haug K, Salek RM, Conesa P, Hastings J, De Matos P, Rijnbeek M et al (2013) MetaboLights—an open-access general-purpose repository for metabolomics studies and associated meta-data. Nucleic Acids Res 52:D640

Hernández-de-Diego R, Tarazona S, Martínez-Mira C, Balzano-Nogueira L, Furió-Tarí P, Pappas GJ Jr, Conesa A (2018) PaintOmics 3: a web resource for the pathway analysis and visualization of multi-omics data. Nucleic Acids Res 46(W1):W503

Hoadley KA, Yau C, Hinoue T, Wolf DM, Lazar AJ, Drill E et al (2018) Cell-of-origin patterns dominate the molecular classification of 10,000 tumors from 33 types of cancer. Cell 173(2):291–304.e6

Huang WY, Hsu SD, Huang HY, Sun YM, Chou CH, Weng SL, Huang HD (2015) MethHC: a database of DNA methylation and gene expression in human cancer. Nucleic Acids Res 43(Database issue):D856–D861

Johnson BT, Hennessy EA (2019) Systematic reviews and meta-analyses in the health sciences: Best practice methods for research syntheses. Soc Sci Med 233:237–251

Johnson CH, Ivanisevic J, Siuzdak G (2016) Metabolomics: beyond biomarkers and towards mechanisms. Nat Rev Mol Cell Biol 17(7):451–459

Kessler T, Vilne B, Schunkert H (2016) The impact of genome-wide association studies on the pathophysiology and therapy of cardiovascular disease. EMBO Mol Med 8(7):688–701. https://doi.org/10.15252/emmm.201505452

Krassowski M, Das V, Sahu SK, Misra BB (2020) State of the field in multi-omics research: from computational needs to data mining and sharing. Front Genet 11:610798

Kuo TC, Tian TF, Tseng YJ (2013) 3Omics: a web-based systems biology tool for analysis, integration and visualization of human transcriptomic, proteomic and metabolomic data. BMC Syst Biol 7:64

Lander ES, Linton LM, Birren B, Nusbaum C, Zody MC, Baldwin J et al (2001) Initial sequencing and analysis of the human genome. Nature 409(6822):860–921

Leclercq M, Vittrant B, Martin-Magniette ML, Scott Boyer MP, Perin O, Bergeron A, Droit A (2019) Large-scale automatic feature selection for biomarker discovery in high-dimensional OMICs data. Front Genet 10:452

Lloyd-Price J, Arze C, Ananthakrishnan AN, Schirmer M, Avila-Pacheco J, Poon TW et al (2019) Multi-omics of the gut microbial ecosystem in inflammatory bowel diseases. Nature 569(7758):655–662

Louhimo R, Hautaniemi S (2011) CNAmet: an R package for integrating copy number, methylation and expression data. Bioinformatics 27:887–888

Ma X, Liu Y, Liu Y et al (2018) Pan-cancer genome and transcriptome analyses of 1,699 paediatric leukaemias and solid tumours. Nature 555:371–376

Manolio TA (2010) Genomewide association studies and assessment of the risk of disease. N Engl J Med 363:166–176

Martin JA, Wang Z (2011) Next-generation transcriptome assembly. Nat Rev Genet 12(10):671–682

Mathaiyan J, Chandrasekaran A, Davis S (2013) Ethics of genomic research. Perspect Clin Res 4(1):100–104. PMID: 23533991; PMCID: PMC3601693. https://doi.org/10.4103/2229-3485.106405

Mo Q, Wang S, Seshan VE et al (2013) Pattern discovery and cancer gene identification in integrated cancer genomic data. Proc Natl Acad Sci U S A 110:4245–4250

Nalbantoglu S, Karadag A (2019) Introductory chapter: insight into the OMICS technologies and molecular medicine. In: Molecular medicine. IntechOpen, London

Nguyen H, Shrestha S, Draghici S, Nguyen T (2019) PINSPlus: a tool for tumor subtype discovery in integrated genomic data. Bioinformatics 35:2843–2846

Perez-Riverol Y, Bai M, Leprevost F et al (2017) Discovering and linking public omics data sets using the Omics Discovery Index. Nat Biotechnol 35:406–409

Ritchie MD, Holzinger ER, Li R, Pendergrass SA, Kim D (2015) Methods of integrating data to uncover genotype-phenotype interactions. Nat Rev Genet 16:85–97

Seoane JA, Day INM, Gaunt TR, Campbell C (2014) A pathway-based data integration framework for prediction of disease progression. Bioinformatics 30:838–845

Shi Q, Zhang C, Peng M et al (2017) Pattern fusion analysis by adaptive alignment of multiple heterogeneous omics data. Bioinformatics 3:2706–2714

Subramanian I, Verma S, Kumar S, Jere A, Anamika K (2020) Multi-omics data integration, interpretation, and its application. Bioinform Biol Insights 14:1177932219899051

Tabari A, Chan SM, Omar OMF, Iqbal SI, Gee MS, Daye D (2022) Role of machine learning in precision oncology: applications in gastrointestinal cancers. Cancer 15(1):63. https://doi.org/10.3390/cancers15010063

Tabari A, Chan SM, Omar OMF, Iqbal SI, Gee MS, Daye D (2023) Role of machine learning in precision oncology: applications in gastrointestinal cancers. Cancer 15(1):63. https://doi.org/10.3390/cancers15010063

Taroni JN, Grayson PC, Hu Q, Eddy S, Kretzler M, Merkel PA, et al (2018) MultiPLIER: a transfer learning framework reveals systemic features of rare autoimmune disease. bioRxiv

Teschendorff AE, Zheng SC, Feber A, Yang Z, Beck S, Widschwendter M (2016) The multiomic landscape of transcription factor inactivation in cancer. Genome Med 8(1):1–15

Thompson JA, Christensen BC, Marsit CJ (2018) Pan-cancer analysis reveals differential susceptibility of bidirectional gene promoters to DNA methylation, somatic mutations, and copy number alterations. Int J Mol Sci 19:2296

Tomczak K, Czerwińska P, Wiznerowicz M (2015) The Cancer Genome Atlas (TCGA): an immeasurable source of knowledge. Contemp Oncol 19:A68–A77

Uhlén M, Fagerberg L, Hallström BM, Lindskog C, Oksvold P, Mardinoglu A et al (2015) Proteomics. Tissue-based map of the human proteome. Science 347(6220):1260419

Vaiopoulou A, Gazouli M, Theodoropoulos G, Zografos G (2012) Current advantages in the application of proteomics in inflammatory bowel disease. Dig Dis Sci 57:2755–2764

Vasaikar SV, Straub P, Wang J, Zhang B (2018) Linked Omics: analyzing multi-omics data within and across 32 cancer types. Nucleic Acids Res 46:D956

Vaske CJ, Benz SC, Sanborn JZ et al (2010) Inference of patient-specific pathway activities from multi-dimensional cancer genomics data using PARADIGM. Bioinformatics 26:i237–i245

Visscher PM, Brown MA, McCarthy MI, Yang J (2012) Five years of GWAS discovery. Am J Hum Genet 90(1):7–24

Wang B, Mezlini AM, Demir F et al (2014) Similarity network fusion for aggregating data types on a genomic scale. Nat Methods 1:333–337

Wang J, Ma Z, Carr SA, Mertins P, Zhang H, Zhang Z, Chan DW, Ellis MJ, Townsend RR, Smith RD, McDermott JE, Chen X, Paulovich AG, Boja ES, Mesri M, Kinsinger CR, Rodriguez H, Rodland KD, Liebler DC, Zhang B (2017) Proteome profiling outperforms transcriptome profiling for coexpression based gene function prediction. Mol Cell Proteomics 16(1):121–134

Wang C, Lye X, Kaalia R, Kumar P, Rajapakse JC (2022) Deep learning and multi-omics approach to predict drug responses in cancer. BMC Bioinformatics 22(Suppl 10):632. https://doi.org/10.1186/s12859-022-04964-9. PMID: 36443676; PMCID: PMC9703655

Wasinger VC, Zeng M, Yau Y (2013) Current status and advances in quantitative proteomic mass spectrometry. Int J Proteomics 2013:180605

Weinstein JN, Collisson EA, Mills GB, Shaw KRM, Ozenberger BA, Ellrott K et al (2013) The cancer genome atlas Pan-cancer analysis project. Nat Genet 45(10):1113–1120

Wilkinson MD, Dumontier M, Aalbersberg IJ et al (2016) The FAIR guiding principles for scientific data management and stewardship. Sci Data 3:160018. https://doi.org/10.1038/sdata.2016.18
Wishart DS (2016) Emerging applications of metabolomics in drug discovery and precision medicine. Nat Rev Drug Discov 15:473–484
Wörheide MA, Krumsiek J, Kastenmüller G, Arnold M (2021) Multi-omics integration in biomedical research—a metabolomics-centric review. Anal Chim Acta 1141:144–162
Yan J, Risacher SL, Shen L, Saykin AJ (2018) Network approaches to systems biology analysis of complex disease: integrative methods for multi-omics data. Brief Bioinform 19:1370–1381

Multi-omics in Prospecting of Genes of Biotechnological Importance

15

Priyancka Arora, Aakash Singh, and Ajay Kumar

15.1 Introduction

Multi-omics is an emerging branch related to different omic groups that are used while analyzing data. Multi-omics can also be called integrative omics because it deals with different types of omics including proteomics, genomics, metagenomics, transcriptomics, and metabolomics. In today's era, multi-omics makes research work very easy with the help of new approaches in the field of data analysis. When combined with bioinformatics, which analyzes large amounts of data, the data produced by multi-omic sciences—such as transcriptomics and deep sequencing, proteomics, metabolomics, and comparative genomics—may significantly improve the procedures involved in synthesizing and using compounds (Amer and Baidoo 2021).

Although the multi-omics strategy is not new, it has recently gained popularity among scientists and still has much scope for updation. Several multi-omics combinations have become a fruitful approach to finding novel antibiotics and has a much higher capacity to create specific metabolites in microorganisms (Palazzotto and Weber 2018). Earlier, when multi-omics technology was not applied, scientists were not capable of proposing or predicting the composition, role, three-dimensional model, and action mechanism of fungal secondary metabolites (SMs). When there was no uniform approach for studying metabolites, only genomic or proteomic results were utilized, which led to inconsistent outcomes. The quest for accurate data regarding single-gene evaluation and intricate worldwide studies of the entire (fungal) life forms, such as fungal metabolites and their evolution, interaction study

P. Arora
Department of Biotechnology, Amity University, Sahibzada Ajit Singh Nagar, Mohali, Punjab, India

A. Singh · A. Kumar (✉)
Department of Biotechnology, Faculty of Engineering and Technology, Rama University, Kanpur, India

I. Mani, V. Singh (eds.), *Multi-Omics Analysis of the Human Microbiome*,
https://doi.org/10.1007/978-981-97-1844-3_15

of fungal and its substrates, fungal applications in pharmaceutical products, and their SM study, has followed the development of high productivity omic technology (Shankar and Sharma 2022).

Researchers may be able to better understand and connect the information such as from the primary basis of disease (genetic, epigenetic, or developmental) and its functional implications or significant interactions by using multi-omics integration as opposed to single omics data (Gibson 2015). By definition, omics investigations require comparing a large number of data, specialized statistical evaluation, and a substantial investment of capital, skilled professionals, and time (Hasin et al. 2017). In other studies, multi-omics and systems biology are used to amalgamate knowledge of phenotype to genotype as well as genotype to phenotype models, which may help to enhance the cultivation of crops in environments under stress (Yang et al. 2021). This required a thorough understanding of the linked gene networks and proteins that are part of various cell signaling. Multi-omics also provides details necessary to interpret its potential applicability to the application of medicinal herbs in the current global environment (Pandita et al. 2021).

15.2 Omics Integrated Biotechnological Approaches

The creation of numerous methodologies that allowed for thorough genome-wide association studies and approaches to better comprehend the complexity of human diseases gave rise to the field of "integrated omics" (Wörheide et al. 2021). "Omics" technologies are finding more and more use in the biotechnological process development for numerous major product productions. Utilizing "omics" technologies, researchers have been able to identify and forecast phenotypes in order to more effectively describe and comprehend biological phenomena (Karahalil 2016). Because of this, it is now simpler to streamline biological engineering processes for producing more economically applicable products in terms of both number and quality. Various omics approaches are discussed in the coming sections.

15.2.1 Genomics

Genomic science is the study of the complete or partial genetic or epigenetic sequence information of an organism. It gives information on the genome structure, function, evolution, mapping, mutagenomic, epigenomic, and genome editing (Muthamilarasan et al. 2019). Clarifying gene variance may be aided by genomics, which could improve crop species genetically and boost crop breeding efficiency. With the aid of structural genomics, biologists can generate physical and genetic maps to identify characteristics of interest, like sequence heterogeneity and chromosomal organization. Functional genomics, on the contrary, gives information about how genes regulate the desired attribute. The phenomenon known as epigenomics is the result of epigenetic changes that are independent of nucleotide sequences but change the expression of genomes, a few of such changes are histone modifications, changes in chromatin architecture, and DNA methylations (Mehrmohamadi et al. 2021).

In certain aspects, genomics can be thought of as the foundation of multi-omics analysis because variations in the nucleotide sequence within a particular gene locus may alter the epigenetic control of a gene expression, by affecting the protein expression encoded by the gene and the metabolic composition of the organism. Genetic variations among individuals can influence their susceptibility to sickness and produce variability in the degree of sensitivity to toxic substances in different people (Bradburne et al. 2015). By scanning the genome of interest and locating secondary metabolite biosynthetic gene clusters (BGCs), genome mining has become a highly effective method for estimating the genetic potential of a strain (Blin et al. 2017). With advances in various high-throughput sequencing such as NGS technologies, complete genome sequencing information for numerous agricultural crops is currently accessible. However, linking each gene's structure and function remains a critical issue. By combining single-dimensional sequencing data with three-dimensional protein structure, structural genomics helps the scientific community discover new connections between distantly related proteins (Singh et al. 2021).

15.2.2 Chemoinformatics

Frank Brown defined chemoinformatics as combining information sources to turn data into knowledge and information into data in order to speed up decision-making in the field of drug lead optimization and classification (Brown 1998). Chemoinformatics, according to Greg Paris, is a word used to describe a broad range of activities related to the development, recovery, evaluation, distribution, representation, and application of chemical information (Paris 1999). Besides official definitions, for almost 40 years, chemical informatics, also known as chemoinformatics, has been a field of study. Although drug development is arguably the most prevalent usage for chemoinformatics, there are innumerable other sectors in which it finds utility, indicating its multidisciplinary nature. For instance, the syntheses of organic chemicals and fields like diagnostic and agricultural science have long made use of cheminformatics. Recently, it has been used in many different sectors, including "materials informatics" and "food informatics" (Martinez-Mayorga et al. 2020).

15.2.3 Metabolomics

A metabolome is a complete set of small molecules called metabolites found in cells or biofluids. Numerous small molecules, such as amino acids, fatty acids, carbohydrates, or other byproducts of cellular metabolic activity, are counted simultaneously in metabolomic profiling. It takes into consideration substrates, intermediates, and final products of cellular metabolism. Metabolic flux analysis is a crucial metric for comprehending molecular biological function since it quantifies the net flow of matter through an organism's network of metabolic pathways (Cohen et al. 2022). Deviations from the standard range are often suggestive of sickness or variation. Metabolite levels and relative ratios show metabolic activity. The identification of new genetic locus that control smaller molecules or their corresponding ratios in the plasma and other organs has been enabled by quantitative assessments of metabolite levels (Hasin et al. 2017). Furthermore, metabolomics has received extensive research using metabolomics in conjunction with modeling. MS-based ways of quantifying the proportional and focused tiny-molecule quantities are among related technologies (Dettmer et al. 2007; Madsen et al. 2010).

Metabolomics provides the best tool for the biotechnology sector, agriculture, medicine, and nutraceuticals, therefore, it is used in developing products and other research such as diagnostic biomarkers and drug safety issues (Shankar and Sharma 2022). Two favorable procedures, termed (a) targeted and (b) untargeted procedures, are primarily employed for SM profiling in the metabolome investigation. Targeted procedures investigate a fixed selection of known metabolite molecules, mainly confirming the absolute quantification; untargeted procedures detect all compounds and demonstrate the whole metabolic pathway's capacity, including undiscovered products. As a result, untargeted techniques are typically favored to identify changes in the metabolome, and they are now dominating the scientific community. The best methods for analyzing complex secondary metabolites are gas chromatography-mass spectrometry (GC-MS) and liquid chromatography high-resolution mass spectrometry (LC-HRMS), however, MS/MS-based instruments are employed in untargeted approaches (Klitgaard et al. 2014).

15.2.4 Transcriptomics

Transcriptomics examines the full collection of RNA molecules, or transcripts, in a particular organism, both quantitatively (amount of each transcript expressed) and qualitatively (which transcripts are available, unique splice sites, RNA editing sites).The RNA are the primary functional readers of DNA, thereby serving as a connecting link between DNA and proteins. Its functions as structural and ribozymes are also observed in cells but are seen rarely. Large-scale transcriptome research has emerged in the last 10 years, showing that up to 80% of the genome is transcribed yet only 3% of it codes for proteins (Consortium EP 2012). RNA-Seq analyses revealed thousands of new isoforms and revealed a more complicated protein-coding transcriptome than previously thought. Nonetheless, the advancement of the non-coding RNA sector was an even more noteworthy outcome of these investigations. Numerous long non-coding RNAs that are transcribed in animal cells are now known to be crucial for numerous physiological processes, including neuron formation, endocrine regulation, and brown adipose differentiation (Hasin et al. 2017).

To find out the metabolic changes during transcription, translational, and post-translational modifications, biosynthesis of metabolites, and gene regulation, RNA profiling is helpful. It also helps in screening the gene manipulations done using CRISPR/Cas9 in native genes at the transcriptional level (Palazzotto and Weber 2018). Advanced non-coding RNA approaches have produced greater findings of research than classic protein-coding transcriptome methods because of their lower complexity. A number of RNA-based technologies have been used recently, and the development of transcriptomic research has been greatly aided by three primary strategies: microarray expertise, next-generation sequencing, and serial analysis of gene expression (SAGE) (Shankar and Sharma 2022). The mechanism of various transcriptomics approach and their applications are discussed in Table 15.1.

To find the gene sets with quantitative changes in their expression levels and their pathways, which are mostly in charge of producing the metabolites, the transcriptome investigation is heavily engaged using statistical programs like limma (Ritchie et al. 2015), DESeq2 (Love et al. 2014), gene set enrichment analysis (GSEA) (Zito et al. 2021). For the analysis and interpretation of transcriptomic information, scholars have created systems like GenePattern and UTAP (User-friendly Transcriptome Analysis Pipeline).

The most common of these models, probabilistic graphical models, describes the order of distribution of transcript data, which is essential to RNA-Seqtools (Palazzotto and Weber 2018).

15.2.5 Proteomics

Proteomics is a method for identifying the structure, interactions, functions, and content of peptides. Recently, advances in protein detection and measurement have been applied to large-scale studies of hundreds of proteins in cells or biological fluids (Selevsek et al. 2015).

Bioinformatic breakthroughs now allow for a thorough investigation of fungal biochemistry. Some methods, such as MS, 2-D mass spectrometry, and matrix-assisted laser desorption ionization (MALDI), need proteomic research as part of the procedure used for analysis such as mass spectrometry with time of flight

Table 15.1 Transcriptomics approach

Tools and techniques	Mechanism	Application	Reference
Shotgun sequencing	Identification of RNA using cDNA segments	Determine the coding and non-coding RNA and annotate it	Rivas and Eddy (2001)
Probe-based arrays	mRNA assessment using labeled samples and a connection to the CAGE gene expression analysis tool	Investigate global gene expression by screening the SM BGC cluster with southern blots	Hasin et al. (2017)
Next-generation sequence (NGS)	Identifying sequences by modifying DNA sequencing	Secondary metabolite gene cluster verification, biomarker, medicinal targets	Liu et al. (2021)

(MALDI-TOF MS) for secondary metabolite recognition and evaluation (Téllez-Téllez and Diaz-Godinez 2019). In functional analysis, the protein analysis methods are crucial because they provide knowledge of the protein expressed, that does not correspond to transcriptome data. Due to the enormous amount of available transcriptome and proteome cloud data, proteomic tools have been developed for more than a decade (Jain et al. 2020).

Another technique that is useful for protein analysis is affinity purification, which is based on the isolation of using its interaction with an antibody or a genetic marker. Then, using MS, any related proteins are identified. These affinity techniques have been used to investigate the global interactions between proteins and nucleic acids, occasionally in combination of chemical crosslinking (e.g., ChIP-Seq). Finally, there are several post-translational modifications (PTM), namely, proteolysis, methylation, phosphorylation, acetylation, and O-/N-linked glycosylation that mediate the actions of an enormous proportion of proteins (Hasin et al. 2017).

The learning of relative proteome analysis, their regulation, and alteration in various genetic constitutions or stages of development and stress situations in crops were substantially helped by increasing the proteome scope (Gong et al. 2015). Although PTM and proteins' interactions with one another or ligands reflect the intricate nature of the process of translating genetic information into proteins, understanding them is helpful in understanding how proteins function physiologically, developmentally, and adaptively (Hu et al. 2015). To describe strains, proteomics is a complementary method to transcriptomics and genomes. The rare actinomycete *Planobisporarosea*, which makes the thiopeptide GE2270, was subjected to genomic, transcriptomic, and proteomic analyses. The evaluation and integration of several "omics" data exposed the *P. rosea's* physiology during GE2270 biosynthesis (Tocchetti et al. 2015).

15.2.6 Glycomics

Glycans have long been recognized as being crucial to all living forms and playing important functions in systems of biology. Glycans that are complex and varied seem to be present in everything nature has cells everywhere. 70% or more of human glycosylation occurs on proteins (Cao et al. 2020). In addition to the two main forms, N-/O-linked glycans, other forms of glycosylation are also observed in cells such as the formation of glycol conjugates which may include glycolipids, and unbound carbohydrate chains (Varki et al. 2009).

Glycomics was developed to identify the glycome of a living system to more accurately evaluate the properties of the glycans. An effort has been made to get complete glycome data. However, because of the tremendous diversity, distinctive features, and low large number of glycans, a quantitative analysis of the glycome remains extremely challenging in terms of constituents, organization, and complexity of glycome analysis. Therefore, mass spectroscopy played a major role in identifying the glycan properties. MS is appropriate for glycomic analyses due to several significant characteristics such as excellent sensitivity, small detection limits and broad dynamic range, and the ability to provide glycan structure information (Wells and Hart 2013; Rakus and Mahal 2011).

One essential first step in glycomic analysis is the efficient breakdown of glycans from glycoproteins. The two basic types of release techniques are enzymatic and chemical. Several novel ways for releasing information have lately been created. Glycans more efficiently and quickly. The enzymatic release is commonly employed due to its low conditions of reaction and great specificity, mainly in the release of N-linked glycans. The process of preparing the sample is slowed substantially by the lengthy incubation required for the conventional enzymatic release of glycans. This problem has primarily been solved recently by the application of technologies that promote release efficiency, like the creation of swift processes and microwave-assisted enzymatic hydrolysis, as well as immobilized reproducibility. Most alternative chemical release techniques have the same issue as peeled glycans (Cao et al. 2020). For the total release of glycans, three primary chemical release techniques such as hydrazinolysis, elimination, and oxidation techniques have been established. Methods of hydrazinolysis might cause glycans to isomerize or degrade (Zhang et al. 2018).

Glycan isoform analysis, site-specific glycan identification techniques, MS-based glycan profiling, and the assessment of the spatial arrangement of glycans in beings are a few examples of sophisticated glycomic characterization methods (Cao et al. 2020). Many glycomics informatics tools such as software that helps in searching and glycan libraries were launched in recent years to aid in glycan analysis. Software solutions available for MS-based glycomic analysis may be divided into two categories: tandem-MS (MS/MS)-based translation and LC-MS (MS1)-based annotation. Within the first technique (MS1), glycans are labeled with pre-made combined libraries and databases that include data in terms of molecular masses and retention times. Numerous excellent software repositories are available for glycomic studies (Wu et al. 2010, 2021a, b).

15.2.7 Microbiomics

The fast-developing field of microbiomics investigates all the microorganisms in a given population simultaneously. The human epidermal layer, mucosa, and the stomach lining are infected by bacteria, viruses, and fungi, which together create the microbiota (and their genes, the microbiome). The human microbiome is extremely diverse; for instance, the gut is home to around 100 trillion bacteria comprising of thousand different species. Individuals' microbiota composition varies significantly due to seed at birth and development, dietetics and other climatic variables, medications, and aging (Org et al. 2015a, b). Numerous studies have connected alterations in gut flora to a number of diseases, including autism, diabetes, obesity, cancer, colitis, and heart disease. By increasing the genomic content and sequencing specific hypervariable regions of the bacterial 16S rRNA genes, and then classifying the results into functional taxonomic units, the microbiome can be analyzed. The method of sequencing and analyzing all genes in the sample is called shotgun metagenomics sequencing. It is useful to find out entire bacterial diversity in the environment and can give extra precision for differentiating genetically similar microorganisms. Numerous analytical methods, like QIIME (quantitative insights into microbial ecology), have been created to analyze NGS data from focused 16S or metagenomics studies. These enable precise quantitative identification of taxa that might be linked to illness or other phenotypes of interest (Org et al. 2015a, b).

15.2.8 Phenomics

The term phenomics has been coined to refer to a class of technologies designed to acquire phenotypic data like the omics linked with various molecular technologies. It is the study of an organism's collection of phenotypic traits, which includes qualitative and quantitative characteristics, including clinical, biochemical, and imaging approaches, for phenotype improvement and characterization (Lanktree et al. 2010). As a result, it encompasses a wide range of techniques, the majority of which contain some form of automated sampling or non-invasive ways of obtaining repeated samples from a single person or group (Egea-Cortines and Doonan 2018). With the use of non-invasive imaging techniques including color imaging of biomass, far-infrared imaging of the canopy, lidar (light detection and ranging) to evaluate growth parameters, and magnetic resonance imaging it is possible to quickly examine the phenotype (Furbank and Tester 2011).

15.2.9 Bioinformatics

Bioinformatics is a research tool that utilizes computers and statistics. It helps in exploring biological databases and is also utilized for comparing genotype and proteome data extensively. It is required to identify the variation between two or three genome sequences such as to identify sequences or proteins that differ between diseased and healthy tissues, or between different phenotypes of the same disease. It uses high-throughput techniques for the analysis of substantial numbers of information such as genes or proteins. It can further be utilized to understand metabolic pathways, comprehend possible therapeutic targets, and develop therapeutic medication designs (Allen and Cagle 2009).

15.3 Role of Multi-omics in Biotechnological Advancement

The development of biotechnological procedures for the manufacture of a wide range of important items is increasingly involving omics technology. The development of next-generation sequencing (NGS) has made way for a new wave of omics research, including transcriptomics, proteomics, and genomes. Researchers are now able to recognize and anticipate traits that support the improvement of biotechnological processes toward improved production (in quality and quantity) of products that are pertinent commercially because of the application of omics approaches for describing and understanding biological systems (Abid et al. 2018).

15.3.1 Omics-Guided Agriculture and Food Industries

New species of plants are being created by farmers using recombinant DNA technology in response to commercial needs and environmental difficulties. Using advanced biotechnology, omics technologies are employed in agricultural science technology to improve desired observable qualities which include, color, taste, tolerance to drought, resistance to pesticides, and so on (Amer and Baidoo 2021). Omics has played a major role in strengthening crop quality, yield, and uniformity as well as in promoting the growth of food crops with more nutritious qualities. Furthermore, omics-driven systems biology gives insight into the relationships between "omes" and the processes involved, as well as correlations between genes and characteristics (Roberts et al. 2012).

As agricultural land is cultivated more intensively, soil structure, organic matter, minerals, and erosion become increasingly vulnerable. Therefore, agricultural biotechnology is being used to ensure a sustainable supply of nutrients required for crop plant development. The essential component of this strategy is the use of biofertilizers, which are initiatives that incorporate particular living organisms (microbial inoculants) that can address, mobilize, solubilize, or dissolve nutrients and are utilized by means of seed or soil to improve the absorption of nutrients by plants. The continual increase in food production from less accessible cultivable area is dependent on the balanced use of inorganic minerals, organic matter, and biofertilizer plant nutrition resources to accelerate and maintain soil fertility and productivity (Mohanram and Kumar 2019). Systems biology based on omics gives knowledge of the connections between "omes" and the processes involved, as well as correlations between genes and phenotypes. A latest hybrid genomics and exo-metabolomics study discovered that certain kinds of rhizosphere bacteria have an instinctive preference for specific chemically aromatic organic acids oozed by plants, implying that plant emissions and microbial substrate adoption characteristics engage to produce trends in microbial population assembly (Zhalnina et al. 2018). With the study of associated fields, such as food biotechnology, transcriptomics and metabolomics investigations have shown that *Bacillus pumilus* LZP02 enhances glucose metabolism and phenylpropanoid synthesis to increase rice root development (Liu et al. 2020).

Similarly, other examples involved the creation of climate-smart millets. Intricate regulatory systems have been developed in millets to improve stress tolerance, adjust to shifting climatic conditions, and develop greater disease resistance (Ceasar and Maharajan 2022). The widespread use of NGS methods has permitted the complete understanding of millet gene set (Varshney et al. 2017). Multi-omics data from millets have been documented in several papers. In this regard, a significant amount of data involving mRNA sequencing, genome-wide sequencing, and phenomics has been produced by extensive research on genetic resources related to millet stress (Awan et al. 2022; Zhang et al. 2021).

Another omics technique is ionomics, which proved favorable. Ionomics is the quantitative simultaneous assessment of an organism's elemental composition and how this composition varies in response to genetic mutations, developmental stages, and physiological stimuli. It helps to assess the composition of an organism's mineral nutrients and trace elements (Salt et al. 2008). In one of the methods, genetics with high-throughput elemental analysis has identified many genes governing the ionome. Moreover, genes governing inherent variance in the plant ionome have been found by Utilizing technological developments in genomic mapping methods such as deletion mapping, and DNA microarray-based bulk segregant analysis (BSA) (Huang and Salt 2016). One of the genes controlling the ionome is *frd3*. A citrate transporter that is essential to the Fe-deficient response is encoded by FRD3.A mutation in FRD3 causes an excess of certain metals, such as Mn and Co, to accumulate in leaves. To further support the notion that mutants with a comparable ionomic phenotype are dependably aggregated together based on this phenotype, *frd3* cultivated in various growth trays and soil batches are exclusively clustered together (Huang and Salt 2016).

Furthermore, the use of multi-omics has been made to comprehend the mechanism of stress tolerance, the identification of genes, and the interaction between genetic and environmental factors of certain crops. The capacity to define genetic variation in the gene reservoir for practically any crop species has been greatly enhanced by the use of molecular markers. The regulatory loci controlling a specific trait are identified by QTL mapping and GWAS techniques, which can then be utilized to find potential genes. Several online databases are developed for tomatoes for integrated omics are Tomatoma (Micro Tom database), MiBASE, KaFtom. In addition, scan analyzer 3D is an online phenomic database that is used to analyze physiological parameters in tomatoes in a drought environment. The genome sequencing of certain crops serves as a high-quality reference genome which is beneficial in providing the library of annotated gene models, genomic organization, genes present in the same chromosomes, repeats, and forms the foundation for identifying genetic variants (Chaudhary et al. 2019).

Therefore, various omics tool studies and integrated methodologies inform us about the present and potential directions for effectively managing abiotic stress in various crops, understanding the ionome of the crop, efficient growth and development of the crop, and disease management.

15.3.2 Industrial Application of Integrative Omics

The use of integrative genomics in industry entails applying these methods to enhance industrial processes, including producing biofuels more efficiently or creating novel pharmaceuticals. In industries, optimizing microalgae and comprehending the metabolism of microalgal cells through multi-omics techniques are studied. The microalgae find applications in wastewater treatment, metal toxicity and remediation, biofuel generation, and pharmaceuticals (Mishra et al. 2019). Genomics is required for acquiring the sequencing of several algal genomes such as *Chlamydomonas reinhardtii (green alga)* (Merchant et al. 2007), *Phaeodactylum tricornutum (diatom)* (Scala et al. 2002), *Cyanidioschyzon merolae* (red alga) (Nozaki et al. 2007). The identification and explanation of multiple metabolic pathways, including nitrate assimilation routes, have been made possible by the functional annotation of multiple genes through genomic sequencing of these microalgae (Grossman et al. 2007). Furthermore, proteomics and metabolomics methodologies are used to extract information such as mRNA expression analysis of genes and the effects of different factors in increasing or decreasing the expression levels of different genes and in the survival of the cell (Jamers et al. 2009). The understanding of benefits of the symbiotic relationship between microalgae and bacteria include low-cost oxygenation support, minimal environmental impact, increased nutrient recovery, and the possibility of producing valuable metabolites like lipids, pigments, and proteins for industrial use. This comprehension is best made with the utilization of multi-omics in contrast with conventional methods of cell sorting using Fluorescence-activated cell sorting (FACS), mass spectrometry, and genetic modification. Multi-omics is involved in deciphering the effects of environmental conditions on microalga genes, mRNA expression, protein expression, metabolite production, and its regulation (Mishra et al. 2019). Microalgae are also proven to be used in wastewater treatment, and screening of microalgae using multi-omics helped to identify phenotypic and genotypic traits responsible for bio-remediation. Additionally, the attention is shifting to microalgae as a means of identifying viable alternatives for lipid production, which can then be refined to produce biodiesel and serve as a third-generation biofuel source (Rai et al. 2021).

15.4 "Omics" Technologies' Application in the Creation of COVID-19 Therapeutics

Multi-omics techniques, including transcriptomics, proteomics, metabolomics, metallomics, and genome study, have aided in the understanding of the structural and molecular characteristics of the virus, facilitating the creation of prospective treatments and hastening the process of developing a COVID-19 vaccine (Guo et al. 2023). Omics methodologies contributed to our understanding of infection and illness development. Multi-omics are utilized for finding new drugs, which could be antibodies or small molecules as these can be used to thoroughly research the pathogenesis of diseases, by helping in exploiting targets that can be treated, to anticipate drug resistance, and much more. In this way, drug discovery can proceed at a significantly faster pace, which was otherwise a drawn-out procedure that requires finding novel targets and creating methods to either activate or inhibit them. To start with the process of drug designing for COVID-19, the researchers analyzed surface proteins and transcripts from peripheral blood mononuclear cells (PBMCs) and extracted serial plasma samples from several patients at different stages of COVID-19. They also quantified plasma proteins and metabolites (Su et al. 2020).

The immunological profile of COVID-19 was revealed by utilizing extensive transcriptome data from single-cell analysis. This allowed the scientists to identify changes in different subpopulations of circulating leukocytes and to identify patient characteristics associated with COVID-19 pneumonia, such as the seriousness and stage of the illness (Ren et al. 2021).

Multi-omics turned out to be beneficial since it made it easier to comprehend the molecular mechanisms underlying the development and genesis of the disease as well as the host's reactions to COVID-19 transmission. Based on multiple omics techniques, including transcriptomics, proteomics, metabolomics, and lipidomic profiles, host response analysis provided insights into patients' deficiencies in circulating arginine, hyperactivation of neutrophils, and build-up of tryptophan metabolism metabolites that resulted in dysfunctional T cells in critically ill patients (Wu et al. 2021a, b). Another omics technique, called metallomics, was also applied for screening drugs designed to treat COVID-19. Metallomics is the study of the interaction of molecular mechanisms of life processes with metal ions. Using this approach, bismuth ranitidine citrate, which is used to treat ulcers, was anticipated as an effective drug for treatment against SARS-CoV-2 (Li et al. 2022).

During the pandemic time, vaccine development and immunization have been viewed as the primary and most appropriate treatment due to the high mortality of the infection and the strain on medical organizations. Genetic linkage studies have provided insight into the specifics of the angiotensin-converting enzyme 2 (ACE2) receptors on the cell surface that permit the SARS-CoV-2 virus to infect the host cell (Muus et al. 2021). Several other techniques such as sequencing antibody repertoires, HLA gene polymorphism, and high-throughput epitope characterization, also, guided the innovation of efficient and safe vaccines (Brooks et al. 2022).

Two main omics technology, namely, vaccinomics and adversomics is advantageous as far as the effectiveness of vaccines is concerned. Vaccinomics studies the mechanisms underlying population- and individual-level differences in vaccine effectiveness and side effects (Poland et al. 2011). Adversomics uses systems biology and omics to study the molecular and genetic factors underlying individual variations in vaccination adverse effects. These two techniques are an amalgamation of immunogenetics, and immunogenomics, which also requires the involvement of system biology and immune profile to elucidate the genetic factors involved in the negative consequences of vaccinations (Omersel and Karas Kuželički 2020). Studies done with the help of omics such as system biology and genomics has made it easier to comprehend genetic variables contribute to the negative effects of vaccinations (Whitaker et al. 2015). Moreover, studies examining the relationship between the genome and epigenome have started to shed light on COVID-19 patients who are more likely to experience severe interstitial lung pneumonia following the infection. Additionally, exome and genome sequencing have contributed to discovering a rare variant of TLR7 that may be associated with severe COVID-19 outcomes (Butler-Laporte et al. 2022). Genes linked to T- and NK cell depletion and inhibition may be found by immunocytomics and immunogenomics and linked to unfavorable responses to COVID-19 mRNA vaccinations (Syenina et al. 2022). Transcriptome sequencing taken from the samples from patient's pre- and post-vaccination could help in understanding existing diseases in patients which can contribute to negative effects of vaccines (Syenina et al. 2022). It had assisted in understanding the development of allergic reactions in some patients to certain vaccine components (Greenhawt et al. 2023). Omics technologies may be helpful in identifying people who may be at risk of experiencing side effects years following vaccination by identifying the presence and features of the idiotypes of antibodies generated by the vaccine (Shukla and Misra 2022).

15.5 Conclusion and Future Prospects

The identification of genetic processes, growth, development, and stress tolerance in a variety of crops has been made possible in large part by multi-omics analysis. In addition, biofuel and bioproduct metabolic engineering is also an important application of multi-omics. It has quintessential role in the production of DNA and peptide vaccines and understanding its side effects and suitability to different group of people. The advantages of utilizing different omics techniques are tremendous, but it also comes with certain limitations and it requires further study for proper integration of these techniques. There are technological limits in every methodology involved, as it requires multiple steps, such as sample preparations and the needs for sequencing depth. Moreover, since multi-omics platforms have intrinsic data disparities, integrating them remains a challenge (Koh and Hwang 2019). Multi-omics required automation and use of artificial intelligence, which is systematic. DENDRAL is the robot using AI technology for MS data retrieval (Vavricka et al. 2020). This opens the scope for many traditional laboratories works and data understanding to be automated.

References

Abid F, Zahid MA, Abedin ZU, Nizami SB, Abid MJ, Kazmi SZH, Gul A et al (2018) Omics approaches in marine biotechnology: the treasure of ocean for human betterments. In: Omics technologies and bio-engineering. Academic Press, Boston, MA, pp 47–61

Allen TC, Cagle PT (2009) Bioinformatics and omics. In: Basic concepts of molecular pathology, pp 55–60

Amer B, Baidoo EE (2021) Omics-driven biotechnology for industrial applications. Front Bioeng Biotechnol 9:613307

Awan SA, Khan I, Tariq R, Rizwan M, Wang X, Zhang X, Huang L (2022) Genome-wide expression and physiological profiling of pearl millet genotype reveal the biological pathways and various gene clusters underlying salt resistance. Front Plant Sci 13:849618

Blin K, Wolf T, Chevrette MG, Lu X, Schwalen CJ, Kautsar SA, Suarez Duran HG, de Los Santos ELC, Kim HU, Nave M et al (2017) antiSMASH 4.0-improvements in chemistry prediction and gene cluster boundary identification. Nucleic Acids Res 45:W36–W41

Bradburne C et al (2015) Overview of 'omics' technologies for military occupational health surveillance and medicine. Mil Med 180:34–48

Brooks BD, Beland A, Aguero G, Taylor N, Towne FD (2022) Moving beyond titers. Vaccine 10(5):683

Brown FK (1998) Chapter 35. Chemoinformatics: what is it and how does it impact drug discovery. In: Bristol JA (ed) Annual reports in medicinal chemistry, vol 33. Academic Press, Foster City, CA, pp 375–384

Butler-Laporte G, Povysil G, Kosmicki JA, Cirulli ET, Drivas T, Furini S, Zeberg H et al (2022) Exome-wide association study to identify rare variants influencing COVID-19 outcomes: results from the host genetics initiative. PLoS Genet 18(11):e1010367

Cao WQ, Liu MQ, Kong SY, Wu MX, Huang ZZ, Yang PY (2020) Novel methods in glycomics: a 2019 update. Expert Rev Proteomics 17(1):11–25

Ceasar SA, Maharajan T (2022) The role of millets in attaining united Nation's sustainable developmental goals. Plants People Planet 4:345–349

Chaudhary J, Khatri P, Singla P, Kumawat S, Kumari A, Vikram A, Deshmukh R et al (2019) Advances in omics approaches for abiotic stress tolerance in tomato. Biology 8(4):90

Cohen JD, Tang Q, Hegeman AD (2022) Using targeted metabolomics to elucidate the indole-auxin network in plants. In: Methods in enzymology, vol 676. Academic Press, Boston, MA, pp 239–278

Consortium EP (2012) An integrated encyclopedia of DNA elements in the human genome. Nature 489:57–74

Dettmer K, Aronov PA, Hammock BD (2007) Mass spectrometry-based metabolomics. Mass Spectrom Rev 26:51–78

Egea-Cortines M, Doonan JH (2018) Phenomics. Front Plant Sci 9:678

Furbank RT, Tester M (2011) Phenomics–technologies to relieve the phenotyping bottleneck. Trends Plant Sci 16(12):635–644

Gibson G (2015) A primer of human genetics, 1st edn. Sinauer Associates, Sunderland, MA

Gong F, Hu X, Wang W (2015) Proteomic analysis of crop plants under abiotic stress conditions: where to focus our research? Front Plant Sci 6:418. https://doi.org/10.3389/fpls.2015.00418

Greenhawt M, Shaker M, Golden DB, Abrams EM, Blumenthal KG, Wolfson AR, Dwamena BA et al (2023) Diagnostic accuracy of vaccine and vaccine excipient testing in the setting of allergic reactions to COVID-19 vaccines: a systematic review and meta-analysis. Allergy 78(1):71–83

Grossman AR, Croft M, Gladyshev VN, Merchant SS, Posewitz MC, Prochnik S, Spalding MH (2007) Novel metabolism in Chlamydomonas through the lens of genomics. Curr Opin Plant Biol 10(2):190–198

Guo M, Xiong M, Peng J, Guan T, Su H, Huang Y, Chen C et al (2023) Multi-omics in COVID-19: driving development of therapeutics and vaccines. Natl Sci Rev 10:nwad161

Hasin Y, Seldin M, AldonsLusis. (2017) Multi-omics approaches to disease. Genome Biol 18(1):1–15

Hu X, Yang Y, Gong F, Zhang D, Zhang L, Wu L, Li C, Wang W (2015) Protein sHSP26 improves chloroplast performance under heat stress by interacting with specific chloroplast proteins in maize (*Zea mays*). J Proteomics 115:81–92. https://doi.org/10.1016/j.jprot.2014.12.009

Huang XY, Salt DE (2016) Plant ionomics: from elemental profiling to environmental adaptation. Mol Plant 9(6):787–797

Jain KK, Kumar A, Shankar A, Pandey D, Chaudhary B, Sharma KK (2020) De novo transcriptome assembly and protein profiling of copper-induced lignocellulolytic fungus Ganoderma lucidum MDU-7 reveals genes involved in lignocellulose degradation and terpenoid biosynthetic pathways. Genomics 112(1):184

Jamers A, Blust R, De Coen W (2009) Omics in algae: paving the way for a systems biological understanding of algal stress phenomena? Aquat Toxicol 92(3):114–121

Karahalil B (2016) Overview of systems biology and omics technologies. Curr Med Chem 23(37):4221–4230

Klitgaard A, Iversen A, Andersen M, Larsen T, Frisvad J, Nielsen K (2014) Aggressive dereplication using UHPLC-DAD-QTOF: screening extracts for up to 3000 fungal secondary metabolites. Anal Bioanal Chem 406(7):1933–1943

Koh EJ, Hwang SY (2019) Multi-omics approaches for understanding environmental exposure and human health. Mol Cell Toxicol 15:1–7

Lanktree MB, Johansen CT, Joy TR, Hegele RA (2010) A translational view of the genetics of lipodystrophy and ectopic fat deposition. Prog Mol Biol Transl Sci 94:159–196

Li H, Yuan S, Wei X, Sun H (2022) Metal-based strategies for the fight against COVID-19. Chem Commun 58(54):7466–7482

Liu H, Wang Z, Xu W, Zeng J, Li L, Li S et al (2020) Bacillus pumilus LZP02 promotes rice root growth by improving carbohydrate metabolism and phenylpropanoid biosynthesis. Mol Plant Microbe Interact 33:1222–1231. https://doi.org/10.1094/MPMI-04-20-0106-R

Liu R, Wang Y, Li P, Sun L, Jiang J, Fan X, Liu C, Zhang Y (2021) Genome assembly and transcriptome analysis of the fungus *Coniella diplodiella* during infection on grapevine *Vitis vinifera* L. Front Microbiol 11:3470

Love MI, Huber W, Anders S (2014) Moderated estimation of fold change and dispersion for RNA-seq data with DESeq2. Genome Biol 15(12):550

Madsen R, Lundstedt T, Trygg J (2010) Chemometrics in metabolomics—a review in human disease diagnosis. Anal Chim Acta 659:23–33

Martinez-Mayorga K, Madariaga-Mazon A, Medina-Franco JL, Maggiora G (2020) The impact of chemoinformatics on drug discovery in the pharmaceutical industry. Expert Opin Drug Discovery 15(3):293–306

Merchant SS, Prochnik SE, Vallon O, Harris EH, Karpowicz SJ, Witman GB, Grossman AR et al (2007) The Chlamydomonas genome reveals the evolution of key animal and plant functions. Science 318(5848):245–250

Mehrmohamadi M, Sepehri MH, Nazer N, Norouzi MR (2021) A comparative overview of epigenomic profiling methods. Front Cell Dev Biol 9:714687

Mishra A, Medhi K, Malaviya P, Thakur IS (2019) Omics approaches for microalgal applications: prospects and challenges. Bioresour Technol 291:121890

Mohanram S, Kumar P (2019) Rhizosphere microbiome: revisiting the synergy of plant-microbe interactions. Ann Microbiol 69:307–320. https://doi.org/10.1007/s13213-019-01448-9

Muthamilarasan M, Singh NK, Prasad M (2019) Multi-omics approaches for strategic improvement of stress tolerance in underutilized crop species: a climate change perspective. Adv Genet 103:1–38. https://doi.org/10.1016/bs.adgen.2019.01.001

Muus C, Luecken MD, Eraslan G, Sikkema L, Waghray A, Heimberg G, Ziegler CG et al (2021) Single-cell meta-analysis of SARS-CoV-2 entry genes across tissues and demographics. Nat Med 27(3):546–559

Nozaki H, Takano H, Misumi O, Terasawa K, Matsuzaki M, Maruyama S, Kuroiwa T et al (2007) A 100%-complete sequence reveals unusually simple genomic features in the hot-spring red alga Cyanidioschyzon merolae. BMC Biol 5(1):1–8

Omersel J, Karas Kuželički N (2020) Vaccinomics and adversomics in the era of precision medicine: a review based on HBV, MMR, HPV, and COVID-19 vaccines. J Clin Med 9(11):3561

Org E, Mehrabian M, Lusis AJ (2015a) Unraveling the environmental and genetic interactions in atherosclerosis: central role of the gut microbiota. Atherosclerosis 241:387–399

Org E, Parks BW, Joo JW, Emert B, Schwartzman W, Kang EY et al (2015b) Genetic and environmental control of host-gut microbiota interactions. Genome Res 25:1558–1569

Palazzotto E, Weber T (2018) Omics and multi-omics approaches to study the biosynthesis of secondary metabolites in microorganisms. Curr Opin Microbiol 45:109–116

Pandita D, Pandita A, Wani SH, Abdelmohsen SA, Alyousef HA, Abdelbacki AM, Elansary HO et al (2021) Crosstalk of multi-omics platforms with plants of therapeutic importance. Cells 10(6):1296

Paris G (1999) Meeting of the American Chemical Society. http://www.warr.com/warrzone.htm

Poland GA, Ovsyannikova IG, Kennedy RB, Haralambieva IH, Jacobson RM (2011) Vaccinomics and a new paradigm for the development of preventive vaccines against viral infections. Omics 15(9):625–636

Rai V, Patel SK, Muthuraj M, Gandhi MN, Das D, Srivastava S (2021) Systematic metabolome profiling and multi-omics analysis of the nitrogen-limited non-model oleaginous algae for biorefining. Biofuel Res J 8(1):1330–1341

Rakus JF, Mahal LK (2011) New technologies for glycomic analysis: toward a systematic understanding of the glycome. Annu Rev Anal Chem 4:367–392
Ren X, Wen W, Fan X, Hou W, Su B, Cai P, Zhang Z (2021) COVID-19 immune features revealed by a large-scale single-cell transcriptome atlas. Cell 184(7):1895–1913
Ritchie ME, Phipson B, Wu DI, Hu Y, Law CW, Shi W, Smyth GK (2015) Limma powers differential expression analyses for RNA sequencing and microarray studies. Nucleic Acids Res 43(7):e47
Rivas E, Eddy SR (2001) Noncoding RNA gene detection using comparative sequence analysis. BMC Bioinform 2:1–19
Roberts LD, Souza AL, Gerszten RE, Clish CB (2012) Targeted metabolomics. Curr Protoc Mol Biol 98(1):30–32
Salt DE, Baxter I, Lahner B (2008) Ionomics and the study of the plant ionome. Annu Rev Plant Biol 59:709–733
Scala S, Carels N, Falciatore A, Chiusano ML, Bowler C (2002) Genome properties of the diatom Phaeodactylum tricornutum. Plant Physiol 129(3):993–1002
Selevsek N, Chang CY, Gillet LC, Navarro P, Bernhardt OM, Reiter L et al (2015) Reproducible and consistent quantification of the *Saccharomyces cerevisiae* proteome by SWATH-mass spectrometry. Mol Cell Proteomics 14:739–749
Shankar A, Sharma KK (2022) Fungal secondary metabolites in food and pharmaceuticals in the era of multi-omics. Appl Microbiol Biotechnol 106(9–10):3465–3488
Shukla AK, Misra S (2022) Clinical implications of anti-idiotype antibodies in COVID-19. J Basic Clin Physiol Pharmacol 33(6):727–733
Singh RK, Sood P, Prasad A, Prasad M (2021) Advances in omics technology for improving crop yield and stress resilience. Plant Breed 140(5):719–731
Su Y, Chen D, Yuan D, Lausted C, Choi J, Dai CL, Heath JR et al (2020) Multi-omics resolves a sharp disease-state shift between mild and moderate COVID-19. Cell 183(6):1479–1495
Syenina A, Gan ES, Toh JZ, de Alwis R, Lin LZ, Tham CY, Ooi EE et al (2022) Adverse effects following anti–COVID-19 vaccination with mRNA-based BNT162b2 are alleviated by altering the route of administration and correlate with baseline enrichment of T and NK cell genes. PLoS Biol 20(5):e3001643
Téllez-Téllez M, Diaz-Godinez G (2019) Omic tools to study enzyme production from fungi in the *Pleurotus* genus. Bioresources 14(1):2420–2457
Tocchetti A, Bordoni R, Gallo G, Petiti L, Corti G, Alt S, Cruz JC, Salzano AM, Scaloni A, Puglia AM et al (2015) A genomic, transcriptomic and proteomic look at the GE2270 producer Planobispora rosea, an uncommon actinomycete. PLoS One 10:e0133705
Varki A, Cummings RD, Esko JD et al (2009) Essentials of glycobiology, 2nd edn. Cold Spring Harbor Laboratory Press, Cold Spring Harbor, NY
Varshney RK, Shi C, Thudi M, Mariac C, Wallace J, Qi P, Zhang H et al (2017) Pearl millet genome sequence provides a resource to improve agronomic traits in arid environments. Nat Biotechnol 35:969–976
Vavricka CJ, Hasunuma T, Kondo A (2020) Dynamic metabolomics for engineering biology: accelerating learning cycles for bioproduction. Trends Biotechnol 38(1):68–82
Wells L, Hart GW (2013) Glycomics: building upon proteomics to advance glycosciences. Mol Cell Proteomics 12(4):833–835
Whitaker JA, Ovsyannikova IG, Poland GA (2015) Adversomics: a new paradigm for vaccine safety and design. Expert Rev Vaccines 14(7):935–947
Wörheide MA, Krumsiek J, Kastenmüller G, Arnold M (2021) Multi-omics integration in biomedical research—a metabolomics-centric review. Anal Chim Acta 1141:144–162
Wu S, Tao N, German JB et al (2010) Development of an annotated library of neutral human milk oligosaccharides. J Proteome Res 9(8):4138–4151
Wu B, Sun M, Zhang H, Yang D, Lin C, Khan I, Wang X et al (2021a) Transcriptome analysis revealed the regulation of gibberellin and the establishment of photosynthetic system promote rapid seed germination and early growth of seedling in pearl millet. Biotechnol Biofuels 14:94

Wu P, Chen D, Ding W, Wu P, Hou H, Bai Y, Chen G et al (2021b) The trans-omics landscape of COVID-19. Nat Commun 12(1):4543

Yang Y, Saand MA, Huang L, Abdelaal WB, Zhang J, Wu Y, Wang F (2021) Applications of multi-omics technologies for crop improvement. Front Plant Sci 12:563953

Zhalnina K, Louie KB, Hao Z, Mansoori N, Da Rocha UN, Shi S et al (2018) Dynamic root exudate chemistry and microbial substrate preferences drive patterns in rhizosphere microbial community assembly. Nat Microbiol 3(4):470–480

Zhang Y, Yang P, Yang L et al (2018) Advances in sample preparation strategies for MS-based qualitative and quantitative N-glycomics. TrAC Trends Anal Chem 99:34–46

Zhang A, Ji Y, Sun M, Lin C, Zhou P, Ren J, Luo D et al (2021) Research on the drought tolerance mechanism of *Pennisetum glaucum* (L.) in the root during the seedling stage. BMC Genomics 22:568

Zito A, Lualdi M, Granata P, Cocciadiferro D, Novelli A, Alberio T, Casalone R, Fasano M (2021) Gene set enrichment analysis of interaction networks weighted by node centrality. Front Genet 12:221

Artificial Intelligence and Machine Learning for Analysis of Multi-omics

16

Neeraj Agarwal, Nupur, Prabir Kumar Paul, and Santosh Kumar Mishra

16.1 Introduction

To fully understand human health and ailment, it is essential to interpret molecular complexity and its variations at different levels, including the epigenome, proteome, transcriptome, metabolome, and genome. Since the invention and development of advanced techniques of sequencing technology, biology is now relying more and more on data produced at several levels, collectively referred to as "multi-omics" data. High-throughput experimental methods are generating enormous and complex genomic data sets. By integrating data from these omics disciplines, researchers aim to uncover complex relationships and interactions within biological systems, identify biomarkers for diseases, understand how genetic and environmental factors influence health and disease, and develop personalized medicine approaches (Chung and Kang 2019; Biswas and Chakrabarti 2020; Arjmand et al. 2022). Due to their extreme scale and diversity, the ensuing problems run the risk of becoming computationally intractable. Multi-omics approaches require advanced computational tools, including artificial intelligence and machine learning, to analyze, investigate, and decipher the vast amounts of data generated from each omics discipline (Ahmed 2020; Takahashi et al. 2021). Fortunately, advanced algorithmic techniques result in software that can efficiently answer important biomedical problems. The

N. Agarwal
Department of Biotechnology, Noida Institute of Engineering and Technology, Greater Noida, Uttar Pradesh, India

Nupur · S. K. Mishra (✉)
Department of Life Sciences, School of Basic Sciences and Research, Sharda University, Greater Noida, Uttar Pradesh, India

P. K. Paul
Department of Biotechnology, Manav Rachna International Institute of Research and Studies, Faridabad, Haryana, India

I. Mani, V. Singh (eds.), *Multi-Omics Analysis of the Human Microbiome*,
https://doi.org/10.1007/978-981-97-1844-3_16

availability of multi-omics data has opened up new possibilities for integrated system-level approaches, which has had a transformative impact on the domains of biology and medicine and system biology (Phan et al. 2019; He et al. 2022). Multi-omics aims to acquire a full and all-encompassing comprehension of biological systems through the simultaneous study of multiple biological information layers (Nicora et al. 2020; Reel et al. 2021).

The use of artificial intelligence (AI) and machine learning (ML) in the field of system biology has become increasingly prominent, serving as potent tools for the examination of multi-omics data (Li et al. 2022; Kang et al. 2022). AI and ML methodologies can be harnessed to amalgamate multi-omics datasets, aligning them based on shared attributes like genes, proteins, or metabolites. This amalgamation facilitates the creation of a more comprehensive perspective on biological processes (Misra et al. 2019). ML algorithms are adept at identifying pertinent features or biomarkers within multi-omics data that are linked to specific biological conditions or diseases (Zhou et al. 2023). The process of feature selection aids in reducing data dimensionality and enhancing the comprehensibility of outcomes. ML models can be trained to categorize samples into distinct disease groups or prognosticate disease outcomes using multi-omics data, with particular utility in personalized medicine for patient stratification and early diagnosis (Wani and Raza 2019; Usova et al. 2021; Onoja 2023).

AI possesses the capability to unearth novel biomarkers that have potential applications in diagnostics, prognostics, or therapeutic interventions. Biomarkers serve as specific molecular indicators tied to diseases or biological phenomena. Additionally, AI-driven methodologies are proficient in forecasting potential drug candidates or repurposing existing medications for new indications by scrutinizing multi-omics data to pinpoint target molecules and pathways (Malandraki-Miller and Riley 2021; Tanoli et al. 2021). Furthermore, ML models can prognosticate the functions of genes, proteins, or metabolites by analyzing their associations with other molecules and biological processes, thereby aiding in functional genomics investigations (Ahmed 2022). AI techniques can also enhance the visualization of intricate multi-omics data, simplifying the process for researchers to explore and interpret findings. Moreover, ML can be effectively applied to create and scrutinize biological networks, including gene co-expression networks, or protein–protein interaction networks enabling the recognition of pivotal regulators and interactions within these intricate systems (Boniolo et al. 2021; Pun et al. 2023).

16.2 Omics Data and Its Analysis

Omics data are large, high-dimensional data sets of biological information that are gathered and analyzed using high-throughput methods and techniques from a variety of domains, including genomics, transcriptomics, proteomics, and metabolomics. With the use of various omics tools and approaches, researchers may gather vast amounts of data and gain understanding of the metabolic and physiological functions of the living beings on the earth. Omics data, which includes information on a variety of biological molecules such as genes, transcripts, metabolite, and proteins, is essential for comprehending the intricate relationships and operations of biological systems (Veenstra 2021). Hence, for in-depth understanding and insight into a variety of aspects of health, illness, and other biological processes, multi-omics data can play a crucial role in revealing complex interactions and networks inside the biological entities (Graw et al. 2021). Also, for the advancement of biomedical research, the creation of personalized therapy, the general comprehension of biology and disease mechanism can be analyzed with the help of omics data (Misra et al. 2019; Subramanian et al. 2020).

16.3 Types of Multi-omics Data

Multi-omics data mainly include that data which have been generated from research in the biological systems through genomics, proteomics, transcriptome, metabolomics, and epigenetic. Further spectrum of omics can be extended to several other biological data including phospho-proteome, lipidome, and glycol-proteome etc. (Reel et al. 2021). The summary of multi-omics data is given below (Picard et al. 2021; Subramanian et al. 2020):

- *Genomics*: The study of an organism's whole gene pool is known as genomics. Genomic data comprises details about the genetic makeup, variations, and gene architecture of an organism (Bian et al. 2022).
- *Transcriptomics*: It is the study of gene expression patterns, specifically the transcription of genes into RNA molecules. Transcriptomic data provides insights into which genes are active in a particular cell or tissue and to what extent (Biswas and Chakrabarti 2020).
- *Proteomics*: Proteomics focuses on the study of an organism's complete set of proteins. Proteomic data reveals information about the types, quantities, modifications, and interactions of proteins within a biological system (Zhang and Kuster 2019).
- *Metabolomics*: It is involved with the analysis of small molecules, called metabolites, which play a vital role in metabolic pathways. Metabolomic data provides information about the metabolic state of cells or organisms and can be used to identify biomarkers or metabolic changes in response to stimuli (Jendoubi 2021).
- *Epigenomics*: The study of epigenetic alterations, including histone modifications, DNA methylation, and chromatin structure, is referred to as epigenomics. Researchers can better grasp how gene expression is controlled without affecting the underlying DNA sequence by analyzing and understanding the epigenomic data (Hamamoto et al. 2019).
- *Phenomics*: Phenomics involves the comprehensive study of an organism's observable traits or phenotypes. It can encompass a wide range of characteristics, including physical traits, behaviors, and physiological measurements.

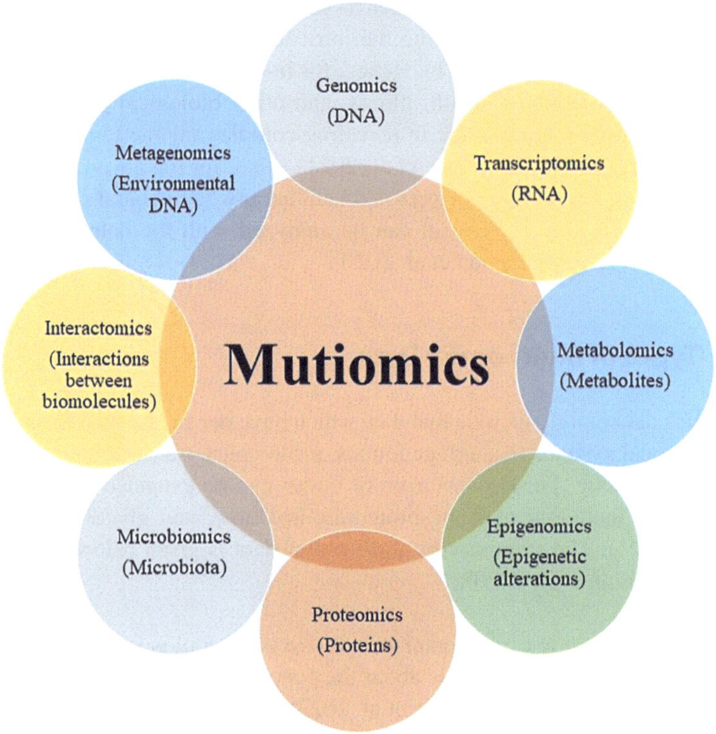

Fig. 16.1 Types of multi-omics data

- *Microbiomics*: Microbiomics is the analysis of the microbial communities (microbiota) present in different environments, such as the human gut or soil. It provides insights into the diversity and functions of microorganisms.
- *Interactomics*: Interactomics aims to understand the interactions between biomolecules, such as interaction between protein–DNA, protein–protein, or protein–ligand interactions. This data helps in building biological networks and understanding cellular processes (Samaras et al. 2020).

Different types of multi-omics data have been summarized in Fig. 16.1.

16.4 Benefits and Challenges of Multi-omics Data

Multi-omics data, which integrates multiple types of biological data, offers various benefits and poses several challenges in the field of biology and biomedical research. Here are the key benefits and challenges associated with multi-omics data (Subramanian et al. 2020; Graw et al. 2021; Picard et al. 2021; Reel et al. 2021):

16.4.1 Benefits of Multi-omics Data

- *Comprehensive Insights*: The holistic perspective of biological systems provided by multi-omics data enables researchers to simultaneously study many molecular information levels. When evaluating each omics data type separately, it may be difficult to detect intricate interactions and linkages that can be found using an all-encompassing method (Vahabi and Michailidis 2022).
- *Biomarker Identification*: Multi-omics data analysis can be used to locate biomarkers linked to ailments, therapeutic reactions, or other biological processes. These biomarkers can be applied to customized therapy, early illness detection, and treatment monitoring.
- *Enhanced Understanding of Diseases*: A deeper knowledge of the molecular pathways behind diseases can be attained by integrating genomes, transcriptomics, proteomics, and other omics data. This information may help researchers identify potential therapeutic targets and treatments.
- *Pharmacological Discovery*: By identifying target molecules and comprehending pharmacological mechanisms, multi-omics data can help in drug discovery (Gupta et al. 2021).
- *Personalized Medicine*: By customizing medical therapies to a patient's unique genetic and molecular profile, multi-omics data might enhance treatment success and minimize side effects.
- *Biological Network Analysis*: Multi-omics data allows the construction of intricate biological networks, revealing how genes, proteins, and other molecules interact within a cell or organism. This can lead to insights into regulatory pathways and system-level behavior.

16.4.2 Challenges of Multi-omics Data

The challenges of multi-omics data have been summarized below (López de Maturana et al. 2019; Pinu et al. 2019; Planell et al. 2021; Reel et al. 2021; Vahabi and Michailidis 2022):

- *Data Integration*: Due to the differences in data formats, sizes, and quality, integrating data from various omics technologies and platforms can be difficult. Proper data harmonization and normalization are essential but complex tasks.
- *Dimensionality*: Multi-omics datasets are often high-dimensional, with many variables or features. Analyzing and visualizing such data can be computationally intensive and may require dimensionality reduction techniques.
- *Biological Complexity*: Biological systems are highly complex and dynamic. Understanding the functional relationships between genes, proteins, metabolites, and epigenetic modifications can be intricate, and interpretation can be challenging.

- *Computational Resources*: Analyzing multi-omics data demands substantial computational power and storage capacity. Researchers need access to advanced computing resources and bioinformatics expertise.
- *Data Interpretation*: Identifying biologically meaningful patterns and distinguishing between true associations and noise can be difficult in multi-omics datasets. Careful statistical and bioinformatic analysis is essential.
- *Experimental Validation*: Findings from multi-omics data analysis often require experimental validation to confirm their biological significance. This can be time-consuming and resource-intensive.
- *Ethical and Privacy Concerns*: Handling and sharing multi-omics data, especially when it involves human subjects, raises ethical and privacy concerns. Adherence to data protection standards and appropriate anonymization of data are essential.

16.5 Steps in Multi-omics Data Analysis

The following steps are followed in multi-omics data analysis (Subramanian et al. 2020; Pinu et al. 2019; Planell et al. 2021; Reel et al. 2021; Picard et al. 2021)

- *Data Preprocessing*: Clean and normalize the data, deal with missing values, and correct for batch effects.
- *Data Integration*: Combine multiple omics datasets to create a unified dataset.
- *Feature Selection*: Identify relevant features or variables for analysis.
- *Exploratory Data Analysis (EDA)*: Visualize and explore the data to gain insights.
- *Statistical Analysis*: Apply statistical methods to identify significant associations and patterns.
- *Machine Learning*: Use machine learning algorithms for classification, clustering, or prediction.
- *Network Analysis*: Construct biological networks to understand interactions.
- *Pathway Analysis*: Identify enriched biological pathways.
- *Validation*: Validate findings through experiments or cross-validation.

16.6 Tools and Software for Multi-omics Data Analysis

Various bioinformatics tools and software packages are available for multi-omics data analysis, such as R/Bioconductor, Python libraries (e.g., scikit-learn, pandas), and specialized software for omics data integration (e.g., OmicsIntegrator, Multi-omics Data Integration).

16.7 Understanding Multi-omics and Artificial Intelligence

The simultaneous analysis of several omics data sets, including transcriptomics, proteomics, metabolomics, and genomics, is known as multi-omics. This method enables a thorough comprehension of biological systems on several levels. The scope of multi-omics extends beyond individual omics technologies, as it aims to integrate data from different sources to provide a more holistic view of biological processes. By harnessing the power of artificial intelligence (AI), multi-omics data can be analyzed and interpreted in a more efficient and accurate manner. AI methods like deep learning and machine learning make it possible to extract insightful information from complicated, large-scale multi-omics datasets. The potential for this multi-omics and AI integration to transform healthcare through individualized treatment, enhanced disease diagnostics, and expedited drug development (Ahmed 2020; Arjmand et al. 2022; He et al. 2022; Bansal et al. 2023). However, there are issues that must be resolved, including as standardizing and integrating data, taking privacy and ethics into account, and validating the outcomes. Despite these obstacles, multi-omics and artificial intelligence have a bright future ahead of them. Technological developments and the use of these strategies in clinical settings could result in substantial gains for public health and healthcare systems. (Pinu et al. 2019; Planell et al. 2021; Ahmed 2022).

The study of developing intelligent machines that can carry out tasks without any requirement for human intervention is known as AI. AI has grown in significance in the multi-omics environment for the analysis and interpretation of massive biological data. New insights and discoveries can be made by using AI techniques like machine learning and deep learning to find connections and patterns within multi-omics datasets. Researchers can better understand complex processes in biology and advance individualized treatment plans, medication development, and diagnostics by combining AI and multi-omics (Biswas and Chakrabarti 2020). However, the use of AI in multi-omics also presents challenges, such as the need for data integration and standardization, ethical considerations, and the validation of results. As technology continues to advance, the integration of multi-omics and AI in clinical practice holds great promise for transforming healthcare and improving public health outcomes (Reddy et al. 2020; Reska et al. 2021; Ahmed 2022).

16.7.1 Importance of Integrating Multi-omics and AI

Integrating multi-omics and AI is essential in advancing scientific research and healthcare. By combining the vast amounts of data from different omics technologies, for example, proteomics and genomics, etc., with the analytical power of artificial intelligence, new insights and discoveries can be made. This integration enables a deeper understanding of complex biological processes, disease mechanisms, and personalized treatments (Tong et al. 2023). AI algorithms can process and analyze these multi-omics datasets, identifying patterns, biomarkers, and therapeutic targets that would be difficult to uncover using traditional methods alone

(Biswas and Chakrabarti 2020; Zhou et al. 2023). Healthcare might be completely transformed by the integration of AI with multi-omics since it would enable precision treatment, enhance disease diagnosis and risk assessment, and hasten the creation of new drugs (Reddy et al. 2020). However, challenges such as data integration and standardization, ethical considerations, and interpretation/validation of results need to be addressed to fully harness the power of this integration (Gupta et al. 2021). Looking to the future, advancements in technology and data analytics, along with the integration of multi-omics and AI into clinical practice, have the potential to significantly impact public health and healthcare systems (Hamamoto et al. 2019; Pinu et al. 2019; Planell et al. 2021; Ahmed 2022; Bansal et al. 2023).

16.7.2 Applications of Multi-omics and AI in Healthcare

The objective of AI applications in the health sector is to examine the connections between patient outcomes and clinical data. AI-dependent systems can be used in various fields, including patient monitoring, medication discovery, precision medicine, diagnosis, and the creation of efficient treatment protocols (Wang and Preininger 2019; Schwalbe and Wahl 2020; Babu and Snyder 2023).

16.7.2.1 Precision Medicine and Personalized Treatment

Precision medicine is a rapidly changing and evolving field that targeting to customize medical treatments to individual patients. It can offer individualized treatment recommendations based on a patient's genetic, environmental, and lifestyle characteristics by leveraging multi-omics and artificial intelligence. This method minimizes needless procedures and side effects while enabling a more focused and efficient treatment plan. AI algorithms can be used to evaluate multi-omics data in order to find biomarkers and potential treatment targets (Ahmed 2020). By optimizing disease diagnosis and risk assessment, improving patient outcomes, and strengthening drug development procedures, this integration of multi-omics and AI has the potential to completely transform the healthcare industry (Pinu et al. 2019; Planell et al. 2021). However, challenges such as data integration and standardization, ethical considerations, and result interpretation need to be addressed for the successful implementation of precision medicine using multi-omics and AI technologies (Haleem et al. 2019; Schwalbe and Wahl 2020; Xu et al. 2021).

16.7.2.2 Disease Diagnosis and Risk Assessment

By integrating data from various omics technologies with advanced AI algorithms, healthcare professionals can identify biomarkers and patterns that aid in the diagnosis of diseases (Reddy et al. 2020). This approach enables a more comprehensive understanding of disease mechanisms and improves diagnosis accuracy. AI and multi-omics are also utilized to determine a person's likelihood of contracting specific diseases, which enables early intervention and individualized preventive measures. AI and multi-omics together have the power to completely change how diseases are diagnosed and risks are evaluated, resulting in more focused and efficient medical interventions (Johnson et al. 2021).

16.7.2.3 Drug Discovery and Development

One essential use of multi-omics and AI is drug discovery and development. Researchers can expedite the process of identifying and optimizing possible drug candidates by utilizing AI algorithms and multi-omics data analysis. This integration enables a deeper comprehension of the many biological systems implicated in disease, which leads to the identification of new therapeutic targets and more individualized methods of therapy (Chung and Kang 2019). AI can augment the traditional drug discovery pipeline by aiding in target identification, predicting drug efficacy and toxicity, optimizing drug design, and enabling virtual screening of large compound libraries. Additionally, multi-omics data, offers a holistic view of a patient's molecular profile, helping to identify disease mechanisms and predict patient response to specific drugs. Overall, the combination of multi-omics and AI has enormous potential to transform the process of finding and developing new drugs, which will ultimately result in more precise and potent treatments for a range of illnesses (Pinu et al. 2019; Planell et al. 2021; Ahmed 2020; Bohr and Memarzadeh 2020; Gupta et al. 2021; Iqbal et al. 2021).

16.7.3 Challenges and Limitations of Multi-Omics and AI

There are multiple challenges innate to all types of biological data irrespective of analytical methodology. But integration of multi-omics data creates the problems in data scaling, normalization, standardization of each individual dataset because of heterogenous nature.

16.7.3.1 Data Integration and Standardization

In the fields of multi-omics and AI, data standardization and integration are crucial. Integrating and standardizing a variety of data formats and types is crucial when working with numerous omics data sets. This procedure makes sure that the information can be evaluated and contrasted with other datasets or research in an efficient manner. Furthermore, standardization makes it possible to combine multi-omics data with other clinical or phenotypic data, facilitating thorough analysis and interpretation (Pinu et al. 2019; Planell et al. 2021). Despite its importance, data integration and standardization present significant challenges due to the complexity and heterogeneity of omics data. Addressing these challenges is crucial for the successful application of AI in multi-omics research and its potential impact on healthcare (Reddy et al. 2020; Subramanian et al. 2020; Miao et al. 2021; Picard et al. 2021).

16.7.3.2 Ethical and Privacy Considerations

Ethical and privacy considerations play a crucial role in the field of multi-omics and AI. The integration of these technologies raises concerns about the responsible use and protection of personal data (Pinu et al. 2019; Planell et al. 2021). Given the sensitive nature of genomic and health information, it is important to ensure that proper safeguards are in place to maintain patient privacy and confidentiality. Additionally, the potential for unintended consequences, such as unintended discrimination or biases, must be carefully addressed. As multi-omics and AI continue to advance, it is essential to establish robust ethical frameworks and regulations that uphold patient autonomy, informed consent, and transparency. The responsible development and deployment of these technologies will be essential for their acceptance and long-term success (López de Maturana et al. 2019; Canzler et al. 2020; Rhem 2021).

16.7.3.3 Interpretation and Validation of Results

Interpretation and validation of results is a crucial step in multi-omics and artificial intelligence research. This process involves analyzing the data generated from multi-omics experiments and applying machine learning algorithms to extract meaningful patterns and insights. By interpreting the results, researchers can identify key biomarkers, pathways, and molecular networks associated with various diseases and conditions. Validation is then performed to ensure the reliability and accuracy of these findings. This may involve cross-referencing with existing biomedical knowledge, conducting experimental validations, or comparing results with independent datasets. The interpretation and validation of results provide the foundation for further research, clinical translation, and the development of targeted therapies and interventions (Kaur et al. 2021; Reel et al. 2021).

16.8 Integration of Multi-omics and Artificial Intelligence

By integrating technologies that rely on multiple parameters, researchers can uncover not only the details of a single regulatory layer but also the strong connections between various layers. This allows researchers to create multi-dimensional molecular landscapes and draw significant biological conclusions (Pinu et al. 2019; Planell et al. 2021).

16.8.1 Techniques for Data Integration in Multi-omics Studies

Techniques for data integration in multi-omics studies: It delves into the different approaches and algorithms employed to combine and analyze data from different omics disciplines. It covers methods such as statistical analysis, machine learning, network-based approaches, and data fusion techniques (Kang et al. 2022). By effectively integrating multi-omics data, researchers can uncover novel insights and potential biomarkers that may have been missed when analyzing each omics layer in isolation (Thomas et al. 2021; Locke and Hoyt 2023; Terranova et al. 2023).

16.8.2 Benefits of Artificial Intelligence in Multi-omics Integration

When it comes to integrating multi-omics data, artificial intelligence (AI) offers various advantages. First of all, large-scale datasets may be efficiently analyzed and interpreted by AI algorithms, making it possible to find intricate patterns and links in the data. This makes it possible for researchers to find novel biomarkers for a variety of diseases as well as hidden insights. Second, AI-based methods make it easier to integrate various omics data types by offering thorough frameworks and instruments for data fusion and integration. A more comprehensive understanding of biological processes and disease mechanisms is made possible by this integration. AI algorithms can also help anticipate treatment results and drug response, which can lead to the creation of personalized medicine techniques. (Thomas et al. 2021). Moreover, AI methods can also expedite the process of finding and developing new drugs by allowing virtual compound screening and molecular structure optimization based on omics data. All things considered, there is a great deal of promise for improving healthcare and the biological sciences when AI is incorporated into multi-omics research (Reddy et al. 2020; Thomas et al. 2021; Locke and Hoyt 2023).

16.9 Applications of Multi-omics and Artificial Intelligence

In the past few years AI has been employed to solve some complex and most challenging issues of bioinformatics which includes protein structure prediction, homology, sequence alignment and phylogenetic analysis, genomic sequence analysis, and gene finding, etc.

16.9.1 Disease Biomarker Discovery Using Multi-omics and AI

A key application that combines the use of numerous omics technologies and artificial intelligence (AI) algorithms to find biomarkers for different diseases is disease biomarker discovery utilizing multi-omics and AI. With this method, prospective disease markers can be found by thoroughly analyzing many biological data types, including proteomic, metabolomic, transcriptomic, and genomic data. Through the integration of diverse omics data sets, artificial intelligence systems can discern patterns and correlations that would not be readily discerned through the solo analysis of each data type. By offering more precise and individualized methods, this strategy has the potential to completely transform illness diagnosis, prognosis, and therapy. The possibility to identify new therapeutic targets and create creative treatment plans for complex diseases is another benefit of combining multi-omics with AI. Overall, disease biomarker discovery using multi-omics and AI holds great promise for advancing precision medicine and improving patient outcomes (Schauperl and Denny 2022).

16.9.2 Precision Medicine and Personalized Treatment Approaches

The combination of artificial intelligence and multi-omics has transformed personalized medicine and precision medicine. Through the application of AI algorithms to analyze various omics data types, researchers can detect patient-specific molecular markers and trends. This enables a more tailored approach to treatment, allowing healthcare professionals to select the most effective therapies based on a patient's unique genetic and molecular profile Precision medicine presents significant potential for enhancing patient outcomes by enabling a more customized and focused approach to treatment (Reddy et al. 2020). AI techniques play vital role in analyzing complex multi-omics datasets, identifying correlations, and predicting treatment responses, ultimately guiding clinicians in making informed decisions for personalized treatment strategies. AI and multi-omics together have the potential to completely transform healthcare by offering tailored treatment plans and precision medicine (Xie et al. 2021).

16.9.3 Drug Discovery and Development Using Multi-omics and AI

Drug discovery and development is an important field that can greatly benefit from the integration of multi-omics and AI. Multi-omics approaches provide a comprehensive view of biological systems by analyzing various omics data like genomics. AI algorithms can then be applied to analyze these complex datasets and identify potential drug targets and therapeutic strategies. By combining the power of multi-omics and AI, researchers can gain valuable insights into the molecular mechanisms of diseases and identify novel drug candidates. This strategy has the potential to greatly speed up the medication research and discovery process, resulting in more individualized and efficient therapies for a range of illnesses. Enhancing healthcare outcomes and advancing the biological sciences are potential benefits of integrating AI and multi-omics in drug discovery and development (Wang et al. 2020; Gupta et al. 2021).

16.10 Future Directions and Implications of Multi-omics and Artificial Intelligence

It is expected that in future, harnessing multi-omics data especially clinical data will have the excellent potential to enhance the data-driven treatment as well as enhance diagnostic accuracy. This will be too beneficial for patients suffering a range of critical diseases (Zhao et al. 2022).

- Emerging technologies in multi-omics and artificial intelligence are shaping the future of biomedical research. These technologies hold immense potential for revolutionizing healthcare (Reddy et al. 2020) and advancing precision medicine. One major area of interest is the use of AI approaches to integrate multi-omics data, which allows for a thorough knowledge of biological systems at different levels (Zhao et al. 2022).

- Ethical considerations in multi-omics and AI research are of utmost importance in order to ensure responsible and unbiased practices. These considerations revolve around issues such as data privacy, informed consent, and potential bias in AI algorithms. Researchers must ensure that proper consent is obtained from individuals whose data is being used for multi-omics studies, and that privacy measures are in place to protect sensitive information. Transparency and accountability in the development and use of AI technologies are crucial to uphold ethical standards in multi-omics research (Rhem 2021).
- The potential impact of integrating multi-omics and artificial intelligence on healthcare and biomedical sciences is significant (Reddy et al. 2020). Researchers can learn more about intricate biological processes and disorders by merging AI techniques with numerous omics data sets, including proteomics, metabolomics, transcriptomics, and genomes. Through this integration, new biomarkers for illness diagnosis, prognosis, and therapy can be found, enabling more accurate and individualized medical methods (Chung and Kang 2019; Bohr and Memarzadeh 2020; Bansal et al. 2023).
- AI can also speed up the research process and help with drug discovery and development by making it easier to identify possible therapeutic targets. All things considered, the combination of multi-omics and AI holds the potential to completely transform healthcare by facilitating the creation of novel therapeutics, individualized treatment plans, and more accurate diagnostic instruments. (Chung and Kang 2019; Memarzadeh 2020; Gupta et al. 2021; Bansal et al. 2023).

16.11 Conclusion

In conclusion, multi-omics data offers a wealth of information and holds great promise for advancing our understanding of biology and medicine. However, it comes with significant analytical and computational challenges that require specialized expertise and resources to overcome. Addressing these challenges can lead to valuable insights that have the potential to revolutionize healthcare and biological research.

References

Ahmed Z (2020) Practicing precision medicine with intelligently integrative clinical and multi-omics data analysis. Hum Genomics 14(1):35

Ahmed Z (2022) Multi-omics strategies for personalized and predictive medicine: past, current, and future translational opportunities. Emerg Top Life Sci 6(2):215–225

Arjmand B et al (2022) Machine learning: a new prospect in multi-omics data analysis of cancer. Front Genet 13:824451

Babu M, Snyder M (2023) Multi-omics profiling for health. Mol Cell Proteomics 22(6):100561
Bansal H, Luthra H, Raghuram SR (2023) A review on machine learning aided multi-omics data integration techniques for healthcare. In: Data analytics and computational intelligence: novel models, algorithms and applications. Springer, Cham, pp 211–239
Bian L, Zhang H, Ge Y, Čepl J, Stejskal J, El-Kassaby YA (2022) Closing the gap between phenotyping and genotyping: review of advanced, image-based phenotyping technologies in forestry. Ann For Sci 79(1):1–21
Biswas N, Chakrabarti S (2020) Artificial intelligence (AI)-based systems biology approaches in multi-omics data analysis of cancer. Front Oncol 10:588221
Bohr A, Memarzadeh K (2020) The rise of artificial intelligence in healthcare applications. In: Artificial intelligence in healthcare. Academic Press, Boston, MA, pp 25–60
Boniolo F, Dorigatti E, Ohnmacht AJ, Saur D, Schubert B, Menden MP (2021) Artificial intelligence in early drug discovery enabling precision medicine. Expert Opin Drug Discovery 16(9):991–1007
Canzler S, Schor J, Busch W, Schubert K, Rolle-Kampczyk UE, Seitz H, Hackermüller J et al (2020) Prospects and challenges of multi-omics data integration in toxicology. Arch Toxicol 94:371–388
Chung RH, Kang CY (2019) A multi-omics data simulator for complex disease studies and its application to evaluate multi-omics data analysis methods for disease classification. GigaScience 8(5):giz045
Graw S, Chappell K, Washam CL, Gies A, Bird J, Robeson MS, Byrum SD (2021) Multi-omics data integration considerations and study design for biological systems and disease. Mol Omics 17(2):170–185
Gupta R, Srivastava D, Sahu M, Tiwari S, Ambasta RK, Kumar P (2021) Artificial intelligence to deep learning: machine intelligence approach for drug discovery. Mol Divers 25:1315–1360
Haleem A, Javaid M, Khan IH (2019) Current status and applications of artificial intelligence (AI) in medical field: an overview. Curr Med Res Pract 9(6):231–237
Hamamoto R, Komatsu M, Takasawa K, Asada K, Kaneko S (2019) Epigenetics analysis and integrated analysis of multiomics data, including epigenetic data, using artificial intelligence in the era of precision medicine. Biomol Ther 10(1):62
He X, Liu X, Zuo F, Shi H, Jing J (2022) Artificial intelligence-based multi-omics analysis fuels cancer precision medicine. Semin Cancer Biol 88:187
Iqbal MJ, Javed Z, Sadia H, Qureshi IA, Irshad A, Ahmed R, Sharifi-Rad J et al (2021) Clinical applications of artificial intelligence and machine learning in cancer diagnosis: looking into the future. Cancer Cell Int 21(1):1–11
Jendoubi T (2021) Approaches to integrating metabolomics and multi-omics data: a primer. Metabolites 11(3):184
Johnson KB, Wei WQ, Weeraratne D, Frisse ME, Misulis K, Rhee K, Snowdon JL et al (2021) Precision medicine, AI, and the future of personalized health care. Clin Transl Sci 14(1):86–93
Kang M, Ko E, Mersha TB (2022) A roadmap for multi-omics data integration using deep learning. Brief Bioinform 23(1):bbab454
Kaur P, Singh A, Chana I (2021) Computational techniques and tools for omics data analysis: state-of-the-art, challenges, and future directions. Arch Comput Meth Eng 28:4595–4631
Li R, Li L, Xu Y, Yang J (2022) Machine learning meets omics: applications and perspectives. Brief Bioinform 23(1):bbab460
Locke D, Hoyt CC (2023) Companion diagnostic requirements for spatial biology using multiplex immunofluorescence and multispectral imaging. Front Mol Biosci 10:1051491
López de Maturana E, Alonso L, Alarcón P, Martín-Antoniano IA, Pineda S, Piorno L, Malats N et al (2019) Challenges in the integration of omics and non-omics data. Genes 10(3):238
Malandraki-Miller S, Riley PR (2021) Use of artificial intelligence to enhance phenotypic drug discovery. Drug Discov Today 26(4):887–901
Miao Z, Humphreys BD, McMahon AP, Kim J (2021) Multi-omics integration in the age of million single-cell data. Nat Rev Nephrol 17(11):710–724

Misra BB, Langefeld C, Olivier M, Cox LA (2019) Integrated omics: tools, advances and future approaches. J Mol Endocrinol 62(1):R21–R45

Nicora G, Vitali F, Dagliati A, Geifman N, Bellazzi R (2020) Integrated multi-omics analyses in oncology: a review of machine learning methods and tools. Front Oncol 10:1030

Onoja A (2023) An integrated interpretable machine learning framework for high-dimensional multi-omics datasets. Scuola Normale Superiore, Pisa

Phan NN, Chattopadhyay A, Chuang EY (2019) Role of artificial intelligence in integrated analysis of multi-omics and imaging data in cancer research. Transl Cancer Res 8(8):E7

Picard M, Scott-Boyer MP, Bodein A, Périn O, Droit A (2021) Integration strategies of multi-omics data for machine learning analysis. Comput Struct Biotechnol J 19:3735–3746

Pinu FR, Beale DJ, Paten AM, Kouremenos K, Swarup S, Schirra HJ, Wishart D (2019) Systems biology and multi-omics integration: viewpoints from the metabolomics research community. Metabolites 9(4):76

Planell N, Lagani V, Sebastian-Leon P, van der Kloet F, Ewing E, Karathanasis N, Gomez-Cabrero D et al (2021) STATegra: multi-omics data integration—a conceptual scheme with a bioinformatics pipeline. Front Genet 12:620453

Pun FW, Ozerov IV, Zhavoronkov A (2023) AI-powered therapeutic target discovery. Trends Pharmacol Sci 44:561

Reddy S, Allan S, Coghlan S, Cooper P (2020) A governance model for the application of AI in health care. J Am Med Inform Assoc 27(3):491–497

Reel PS, Reel S, Pearson E, Trucco E, Jefferson E (2021) Using machine learning approaches for multi-omics data analysis: a review. Biotechnol Adv 49:107739

Reska D, Czajkowski M, Jurczuk K, Boldak C, Kwedlo W, Bauer W, Kretowski M et al (2021) Integration of solutions and services for multi-omics data analysis towards personalized medicine. Biocybernet Biomed Eng 41(4):1646–1663

Rhem AJ (2021) AI ethics and its impact on knowledge management. AI Ethics 1(1):33–37

Samaras P, Schmidt T, Frejno M, Gessulat S, Reinecke M, Jarzab A, Wilhelm M (2020) ProteomicsDB: a multi-omics and multi-organism resource for life science research. Nucleic Acids Res 48(D1):D1153–D1163

Schauperl M, Denny RA (2022) AI-based protein structure prediction in drug discovery: impacts and challenges. J Chem Inf Model 62(13):3142–3156

Schwalbe N, Wahl B (2020) Artificial intelligence and the future of global health. Lancet 395(10236):1579–1586

Subramanian I, Verma S, Kumar S, Jere A, Anamika K (2020) Multi-omics data integration, interpretation, and its application. Bioinform Biol Insights 14:1177932219899051

Takahashi S, Takahashi M, Tanaka S, Takayanagi S, Takami H, Yamazawa E, Hamamoto R et al (2021) A new era of neuro-oncology research pioneered by multi-omics analysis and machine learning. Biomol Ther 11(4):565

Tanoli Z, Vähä-Koskela M, Aittokallio T (2021) Artificial intelligence, machine learning, and drug repurposing in cancer. Expert Opin Drug Discovery 16(9):977–989

Terranova N, Renard D, Shahin MH, Menon S, Cao Y, Hop CE, Lu J et al (2023) Artificial intelligence for quantitative modeling in drug discovery and development: an Innovation & Quality (IQ) consortium perspective on use cases and best practices. Clin Pharmacol Ther

Thomas D, Rathinavel AK, Radhakrishnan P (2021) Altered glycosylation in cancer: a promising target for biomarkers and therapeutics. Biochimica et Biophysica Acta (BBA)-Rev Can 1875(1):188464

Tong L, Shi W, Isgut M, Zhong Y, Lais P, Gloster L, Wang MD (2023) Integrating multi-omics data with EHR for precision medicine using advanced artificial intelligence. IEEE Rev Biomed Eng 17:80

Usova EI, Alieva AS, Yakovlev AN, Alieva MS, Prokhorikhin AA, Konradi AO, Baragetti A (2021) Integrative analysis of multi-omics and genetic approaches—a new level in atherosclerotic cardiovascular risk prediction. Biomolecules 11(11):1597

Vahabi N, Michailidis G (2022) Unsupervised multi-omics data integration methods: a comprehensive review. Front Genet 13:854752

Veenstra TD (2021) Omics in systems biology: current progress and future outlook. Proteomics 21(3–4):2000235

Wang F, Preininger A (2019) AI in health: state of the art, challenges, and future directions. Yearb Med Inform 28(01):016–026

Wang D, Geng L, Zhao YJ, Yang Y, Huang Y, Zhang Y, Shen HB (2020) Artificial intelligence-based multi-objective optimization protocol for protein structure refinement. Bioinformatics 36(2):437–448

Wani N, Raza K (2019) Integrative approaches to reconstruct regulatory networks from multi-omics data: a review of state-of-the-art methods. Comput Biol Chem 83:107120

Xie G, Wang L, Chen T, Zhou K, Zhang Z, Li J, Jia W (2021) A metabolite array technology for precision medicine. Anal Chem 93(14):5709–5717

Xu L, Sanders L, Li K, Chow JC (2021) Chatbot for health care and oncology applications using artificial intelligence and machine learning: systematic review. JMIR Cancer 7(4):e27850

Zhang B, Kuster B (2019) Proteomics is not an Island: multi-omics integration is the key to understanding biological systems. Mol Cell Proteomics 18(8):S1–S4

Zhao J, Feng Q, Wei WQ (2022) Integration of omics and phenotypic data for precision medicine. Sys Med:19–35

Zhou XJ, Zhong XH, Duan LX (2023) Integration of artificial intelligence and multi-omics in kidney diseases. Fundam Res 3(1):126–148

The manufacturer's authorised representative in the EU is Springer Nature Customer Service Centre GmbH, Europaplatz 3, 69115 Heidelberg, Germany. If you have any concerns regarding our products, please contact ProductSafety@springernature.com

Printed and bound by CPI Group (UK) Ltd, Croydon, CR0 4YY
01/10/2026
02225850-0016